BIOLOGICAL ORGANIZATION: MACROMOLECULAR INTERACTIONS AT HIGH RESOLUTION

BIOLOGICAL ORGANIZATION: MACROMOLECULAR INTERACTIONS AT HIGH RESOLUTION

Edited by

ROGER M. BURNETT
HENRY J. VOGEL

College of Physicians and Surgeons
Columbia University
New York, New York

1987

ACADEMIC PRESS, INC.
Harcourt Brace Jovanovich, Publishers

Orlando San Diego New York Austin
Boston London Sydney Tokyo Toronto

COPYRIGHT © 1987 BY ACADEMIC PRESS, INC.
ALL RIGHTS RESERVED.
NO PART OF THIS PUBLICATION MAY BE REPRODUCED OR
TRANSMITTED IN ANY FORM OR BY ANY MEANS, ELECTRONIC
OR MECHANICAL, INCLUDING PHOTOCOPY, RECORDING, OR
ANY INFORMATION STORAGE AND RETRIEVAL SYSTEM, WITHOUT
PERMISSION IN WRITING FROM THE PUBLISHER.

ACADEMIC PRESS, INC.
Orlando, Florida 32887

United Kingdom Edition published by
ACADEMIC PRESS INC. (LONDON) LTD.
24–28 Oval Road, London NW1 7DX

Library of Congress Cataloging in Publication Data

Biological organization.

　　Includes index.
　　1. Molecular biology.　2. Macromolecules.
I. Burnett, Roger M.　II. Vogel, Henry J.
(Henry James)
QH506.B545　1987　　　574.8′8　　　86-28800
ISBN 0—12—145948—9 (alk. paper)

PRINTED IN THE UNITED STATES OF AMERICA

87 88 89 90　　9 8 7 6 5 4 3 2 1

Contents

Preface   xiii

1   The Left-Handed Double Helix of DNA
ALEXANDER RICH

Text	1
References	7

PART I   DNA–PROTEIN INTERACTIONS

2   The Structure and Function of *Eco*RI Endonuclease
JOHN M. ROSENBERG, JUDITH A. MCCLARIN, CHRISTIN A. FREDERICK, BI-CHENG WANG, HERBERT W. BOYER, JOHN GRABLE, AND PATRICIA GREENE

Introduction	12
Methods	14
Description and Discussion of the Structure	22
Conclusions	40
References	41

3   The Klenow Fragment Structure Suggests Mechanisms for Fidelity and Processivity of DNA Polymerase I
T. A. STEITZ

Introduction	45
Structure of the Klenow Fragment	47
Function of the Small Domain	48
Function of the Large Domain	49
Co-Crystallization of Klenow Fragment with DNA	50

Relation to Other Polymerases	51
Processivity	51
Fidelity of DNA Synthesis	52
References	55

4 Structural Studies of DNA–Protein Recognition

R. G. BRENNAN, H. M. HOLDEN, Y. TAKEDA, AND B. W. MATTHEWS

Introduction	58
Three DNA-Binding Proteins	58
Models for DNA Binding	58
A Common Helix-Turn-Helix DNA-Binding Motif	61
Studies in Progress	63
References	64

PART II VIRUSES

5 Refinement of the Structure of Tobacco Mosaic Virus

GERALD STUBBS

Introduction	69
Structure Determination	72
Structural Details	73
TMV Assembly	79
References	82

6 The Structure of a Human Common Cold Virus (Rhinovirus 14) and Its Functional Relations to Other Picornaviruses

MICHAEL G. ROSSMANN, EDWARD ARNOLD, JOHN W. ERICKSON, ELIZABETH A. FRANKENBERGER, JAMES P. GRIFFITH, HANS-JÜRGEN HECHT, JOHN E. JOHNSON, GREG KAMER, MING LUO, ANNE G. MOSSER, ROLAND R. RUECKERT, BARBARA SHERRY, AND GERRIT VRIEND

Introduction	86
Structure Determination	89
Structure	94

Immunogenic Sites on HRV14 100
Antigenic Sites in Poliovirus and FMDV 102
The Canyon as Receptor Binding Site 104
Neutralization and Serotypes 106
Assembly 107
References 108

7 Adenovirus Hexon: A Novel Use of the Viral Beta-Barrel

MICHAEL M. ROBERTS AND ROGER M. BURNETT

Text 113
References 124

8 The Structure of an Antineuraminidase Monoclonal Fab Fragment and Its Interaction with the Antigen

P. M. COLMAN AND R. G. WEBSTER

Introduction 125
Fab Structure 126
Binding Properties of S10/1 to Monoclonal Variants 129
Binding Properties of S10/1 to Neuraminidase Field Stains 130
Electron Microscopy of Fab–Neuraminidase Complexes 131
References 132

9 Influenza Hemagglutinin

J. J. SKEHEL AND D. C. WILEY

Text 135
References 138

10 The Budding of Enveloped Viruses: A Paradigm for Membrane Sorting?

KAI SIMONS AND STEPHEN FULLER

The Problem 139
The Enveloped Virus Solution 140
Virus Budding as a Paradigm for Intracellular Membrane Sorting 147
References 149

PART III ANTIBODY AND PROTEIN–PROTEIN INTERACTIONS

11 Intramolecular Localization of Antigenic Determinants by Molecular Immunoelectron Microscopy

J. LAMY

Introduction	153
Structure of Arthropod Hemocyanins: The Subject of This Study	154
Intramolecular Localization of Subunits with Polyclonal Antibodies	164
Intramolecular Localization of Epitopes with Monoclonal Antibodies	180
Perspectives and Conclusions	188
References	189

12 Studies of the Tertiary and Quaternary Structure of Antibody Constant Domains

EDUARDO A. PADLAN, GERSON H. COHEN, AND DAVID R. DAVIES

Introduction	193
An Analysis of the Interface Residues between C_H1 and C_L	194
A Model of IgE Fc	203
References	213

13 Three-Dimensional Structure of a Lysozyme–Fab, Antigen–Antibody Complex

A. G. AMIT, G. BOULOT, G. BRICOGNE, R. A. MARIUZZA, S. E. V. PHILLIPS, AND R. J. POLJAK

Introduction	215
Materials and Methods	216
Crystallographic Studies	218
Discussion	219
References	224

14 The Order–Disorder Paradox in Antigen–Antibody Union: Anti-Peptide Antibodies as a Probe for Structured Regions of Small Peptides

H. JANE DYSON, MARK RANCE,
RICHARD A. HOUGHTEN, PETER E. WRIGHT,
AND RICHARD A. LERNER

Introduction	227
NMR Evidence for a Type II β-Turn	228
NMR Evidence for a Helical Structure	231
Discussion and Conclusions	233
References	233

15 Characteristics of Protein Interfaces

WAYNE A. HENDRICKSON, JANET L. SMITH,
AND WILLIAM E. ROYER, JR.

Introduction	235
Categories of Protein Interactions	235
Complementarity	236
Determinants of Affinity	238
Interfaces in Human Hemoglobin	239
Subunit Entropy	240
Hydration	241
Mobility	242
References	244

16 Contributions of Individual Amino Acids to Protein Stability Determined by X-Ray Crystallographic Analysis of Selected and Directed Mutants of T4 Phage Lysozyme

TOM ALBER, TERRY M. GRAY, LARRY H. WEAVER,
JEFFREY BELL, JOAN A. WOZNIAK, KEITH WILSON,
SUN DAOPIN, AND BRIAN W. MATTHEWS

Introduction	246
Structure and Dynamics of Wild-Type Lysozyme	246
Temperature-Sensitive Lysozyme Mutants	247
Site-Directed Mutagenesis	248
Conclusions	253
References	257

PART IV LIGHT-SENSITIVE PROTEINS

17 The Rhodopsins of *Halobacterium halobium*
WALTHER STOECKENIUS

Introduction	261
Halorhodopsin	262
Sensory Rhodopsin	265
Conclusions	267
References	268

18 Structure and Function in Photosynthetic Reaction Centers
STEVEN G. BOXER

Introduction	271
Magnetic Field Effects in Photosynthetic Reaction Centers	273
Reaction Center Energetics	278
References	281

19 The Structural Basis of Photosynthetic Light Reactions in Bacteria
ROBERT HUBER

Text	283
References	286

PART V MEMBRANE PROTEINS AND SIGNALING

20 X-Ray Analysis of Deaminooxytocin: Conformational Flexibility and Receptor Binding
J. E. PITTS, S. P. WOOD, I. J. TICKLE,
A. M. TREHARNE, Y. MASCARENHAS, J. Y. LI,
J. HUSAIN, S. COOPER, T. L. BLUNDELL, V. J. HRUBY,
H. R. WYSSBROD, A. BUKU, AND A. J. FISCHMAN

Introduction	290
Structure Determination	291
Crystal Structure Description	294
Solution Structure	298
Receptor Binding and Biological Response	301
References	304

21 The Acetylcholine Receptor: What the Three-Dimensional Structure Tells Us about Ion Conductance

ROBERT MICHAEL STROUD
AND JANET FINER-MOORE

Introduction	307
The Ionic Channel Lies between Five Quasi-Equivalent Subunits	309
Topography of Sequences Confirms Existence of an Amphipathic Membrane Crossing	310
A Model of the Ionic Channel That Can Be Tested	313
How Is the Ionic Channel Gated?	316
Location of the Binding Site for Ligands	317
References	318

22 Membrane Protein Folding Motifs: An Examination of Empirical Predictions of Secondary Structure

D. L. MIELKE, M. CASCIO, AND B. A. WALLACE

Introduction	319
Materials and Methods	321
Results	322
Discussion	329
References	331

PART VI CONCLUSION

23 Structure and Function of the LDL Receptor

JOSEPH L. GOLDSTEIN AND MICHAEL S. BROWN

Introduction	337
Itinerary of the LDL Receptor	338
Structure of the LDL Receptor	340
Exon Organization and Protein Domains of the LDL Receptor	345
References	346

Index   349

Preface

We currently are at a threshold in biological organization studies that is as portentous as that reached some twenty-five years ago when the detailed structure of myoglobin was discovered. The powerful methods used have since been applied to over one hundred different molecules. Until recently, however, the structures examined have been, with few exceptions, those of isolated water-soluble enzymes or other proteins. Recently, dramatic advances have been made with membrane-bound proteins as exemplified by the remarkable structure determination of a photosynthetic reaction center, which promises to reveal intimate details of the basic biological process of capturing solar energy. In conjunction with progress in gene cloning, which has led to the production of any protein in quantities adequate for study, these advances are leading to explorations of biological systems instead of isolated molecules. Thus, the main thrust of this volume is to feature important current information on interactions of macromolecules themselves (rather than, say, enzyme–substrate interactions). Viruses, as paradigms of small biological systems, are covered as are the pivotal areas of DNA–protein and of antibody interactions. The treatment of the comparatively new field of membrane structure at high resolution includes the latest results on the photosynthetic reaction center, placed in perspective by contributions on light sensitivity of proteins. Finally, chapters on signal receptors highlight the importance of mechanisms for the control of the other systems presented.

It is a pleasure to acknowledge the advice and help of Dr. Joseph L. Goldstein, Dr. Wayne A. Hendrickson, Dr. Barry Honig, Dr. Barbara W. Low, Dr. Lee Makowski, Dr. Dinshaw Patel, Dr. Alexander Rich, Dr. Benno P. Schoenborn, and Dr. Bonnie A. Wallace in the development of this volume.

We are grateful for the fine support of the College of Physicians and Surgeons (P & S) of Columbia University without which this volume would not have reached fruition. This volume was developed from a P & S Biomedical Sciences Symposium held at Arden House, on the Harriman Campus of Columbia University, from May 31 through June 2, 1985.

Roger M. Burnett
Henry J. Vogel

1
The Left-Handed Double Helix of DNA[1]

ALEXANDER RICH
Department of Biology
Massachusetts Institute of Technology
Cambridge, Massachusetts 02139

The molecular structure of the DNA double helix was discovered in 1953 (1), and for the next 26 years it was thought that this molecule existed largely in one form or conformation. However, in 1979, an X-ray diffraction study of a single crystal of a DNA fragment (with the sequence CGCGCG) (2) revealed a form of the double helix that was dramatically different from the familiar right-handed double helix (B-DNA). The left-handed form had the same two backbone chains, a string of sugar and phosphate residues pointing in opposite or antiparallel directions, and the chains were connected by base pairs that were held together by hydrogen bonds in the manner first described by Watson and Crick in 1953 (1). However, in the left-handed form the molecule had undergone a dramatic conformational change (Fig. 1). The base pairs had in essence "flipped over" so that they had a different relationship to the backbones than they have in the original right-handed B-DNA (see Fig. 2). The bases flipped over by two different mechanisms. In this sequence with alternations of guanine and cytosine residues, the cytidine residues, both sugar and base, rotated. The rotation of the sugar introduced the zigzag configuration (hence Z-DNA). However, the guanine residues rotated about their glycosyl bonds, so they assumed the syn conformation (Fig. 3), in contrast to

---

[1] Based on the Opening Address delivered at the symposium "Biological Organization: Macromolecular Interactions at High Resolution," held at Arden House on the Harriman Campus of Columbia University, May 31 through June 2, 1985.

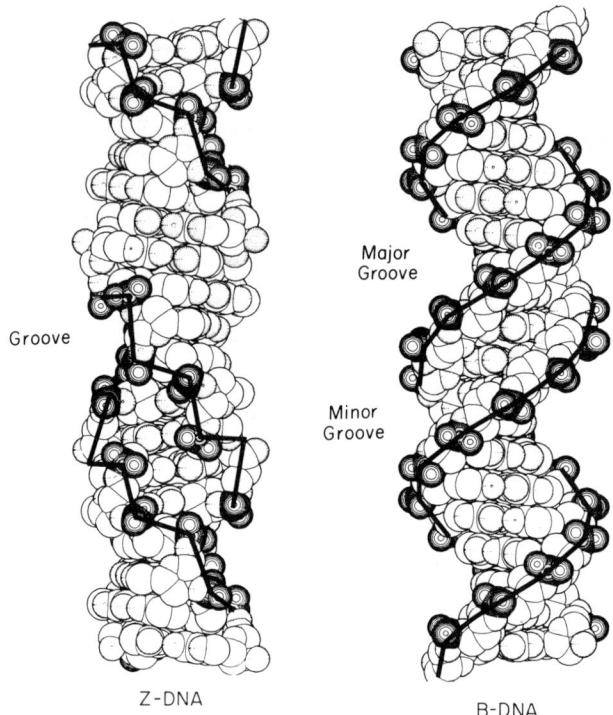

Fig. 1. van der Waals models of Z-DNA and B-DNA. The irregularity of the Z-DNA backbone is illustrated by the heavy lines that go from phosphate to phosphate residue along the chain. The Z-DNA is shown as it appears in the hexamer crystal (2). The groove in Z-DNA is quite deep, extending nearly to the axis of the double helix. In contrast, B-DNA has a smooth line connecting the phosphate groups and two grooves, neither of which extends to the helix axis of the molecule.

the anti conformation in B-DNA. This conformational discovery was somewhat startling since most researchers at the time felt that we knew a great deal about the double helix.

In Z-DNA the bases are set off to the side slightly, compared to their position in B-DNA (Fig. 4). Because of this, the chemical reactivity of the molecule differs in these two conformations. Although these two conformations are made of the same components, they have distinctly different properties.

Since 1979 work has progressed steadily in the direction of uncovering the chemical characteristics and biological roles of this left-handed form of the molecule. The left-handed form is favored in DNA segments with specialized sequences in which there are alternations

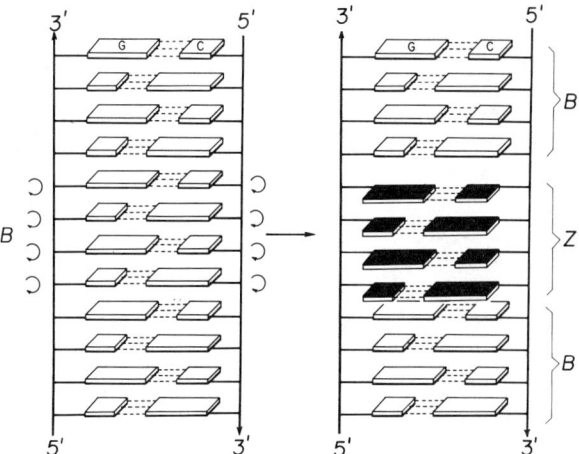

Fig. 2. A diagram illustrating the change in topological relationship if a four-base pair segment of B-DNA were converted into Z-DNA. The conversion is accomplished by rotation or flipping over the base pairs as indicated by the curved arrows.

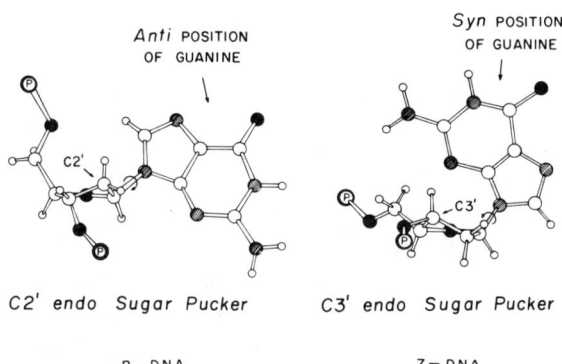

Fig. 3. Conformation of deoxyguanosine in B-DNA and in Z-DNA. The sugar is oriented so that the plane defined by C1'-O1'-C4' is horizontal. Atoms lying above this plane are in the endo conformation. The C3' is endo in Z-DNA whereas in B-DNA the C2' is endo. In addition, Z-DNA has guanine in the syn position, in contrast to the anti position in B-DNA. A curved arrow around the glycosyl carbon–nitrogen linkage indicates the site of rotation.

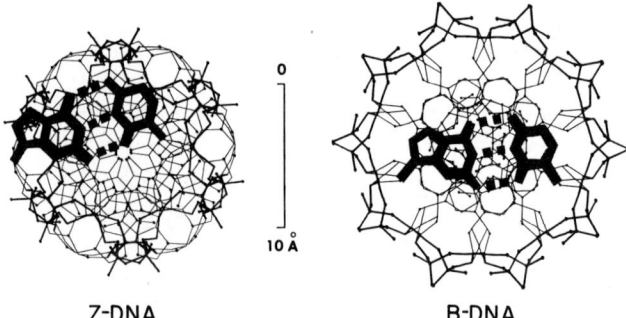

Fig. 4. End views of regular, idealized helical skeletal diagrams of Z-DNA and B-DNA. Heavier lines are used for the phosphate–sugar backbone. A guanine–cytosine base pair is shown by shading and the difference in the positions of the base pairs is quite striking: they are near the center of B-DNA but at the periphery of Z-DNA.

of purines and pyrimidine residues because of the nature of the flipping over process. Purines can assume either syn or anti conformations, but anti conformations are favored in pyrimidines (3).

There is an equilibrium between right-handed B-DNA and left-handed Z-DNA in which the two interconvert. However, the conversion is not symmetric since the left-handed form is at a higher energy state (4). It can maintain that state only when something acts to hold it in that conformation. For example, it is possible to take synthetic DNA that has alternations of guanine and cytosine residues which favor Z-DNA formation [poly(dG-dC)] and put it completely in the left-handed form (5) in which it can be stabilized through chemical modification of the molecule (6). This modified and stable form of Z-DNA can then be used for a variety of chemical and biological studies.

One of the first things discovered about Z-DNA was that it is highly immunogenic and is able to stimulate the production of antibodies (7) even though the right-handed B-DNA is not immunogenic. The antibodies produced by this process are highly specific for Z-DNA and can be obtained in both normal polyclonal and monoclonal varieties (8). These have been useful tools in carrying out a number of studies. It was also discovered that antibodies against left-handed Z-DNA occur naturally in certain autoimmune diseases. For example, during acute attacks in systemic lupus erythematosus, patients are found to have considerable titers of antibodies aginst left-handed Z-DNA in their blood (9) as well as antibodies against a variety of additional cellular components.

Two factors are known to stabilize left-handed Z-DNA in biological systems. One is associated with the fact that long DNA molecules are generally not found in a simple relaxed form but usually are somewhat twisted upon themselves, that is, supercoiled. This supercoiling occurs because DNA is generally constrained in either circles or large loops in chromosomes and the number of turns that the double helical DNA molecule has is not equal to the number it would adopt if the DNA were cut and simply allowed to relax. A class of enzymes, topoisomerases, act to keep DNA in an underwound state. In this underwound state (called negative supercoiling), the DNA begins to store energy in supercoiling which is proportional to the square of the number of negative supercoils (10). For example, if a segment of such a strained DNA forms a left-handed fragment instead of the normal right-handed B-DNA, then the supercoiling is decreased and the supercoiling energy is used to stabilize the left-handed Z-DNA fragment (11,12) (Fig. 5). Thus the normal tension that DNA is under in biological systems is one of the forces for stabilizing Z-DNA.

In addition, proteins have been discovered in biological systems that will bind to left-handed Z-DNA specifically and not to right-handed B-DNA. These Z-DNA binding proteins are currently being studied and are found to have a number of different biological properties.

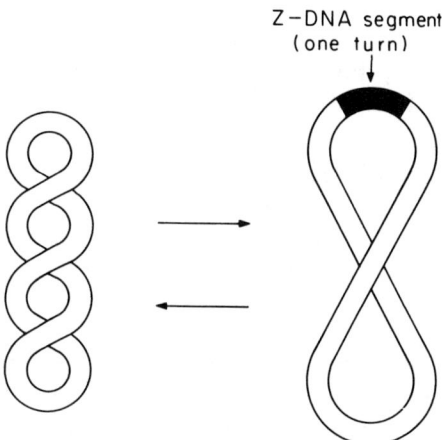

Fig. 5. Schematic diagram in which the double helix is represented as a tubular structure. The diagram on the left represents a negatively supercoiled plasmid with three negative supercoil turns. In the diagram at the right, a 12-bp segment of B-DNA has converted to Z-DNA, and this results in a loss of two superhelical turns. The energy of supercoiling is used to stabilize Z-DNA.

Z-DNA binding proteins can be isolated using the method of affinity chromatography. In this method synthetic DNA polymers that have been stabilized in the Z-DNA form through chemical modification (6) are linked to a matrix that can be formed into columns. The columns are porous to solutions and large molecules. If a mixture of proteins is passed over the column, the proteins that bind to Z-DNA will remain attached to the column while the other proteins will pass through. In this manner, proteins can be isolated that bind to Z-DNA but not to B-DNA. These have been isolated from cells of a variety of organisms, including humans, simians, wheat germ, yeast, and the common fruit fly (13–15). In general, they are found in all organisms in chromatin, the complex of DNA and protein that makes up the material of the chromosomes in cell nuclei. Knowing the proteins are in chromatin, we next have to ask where they are located and what they do.

Recent experiments have been carried out on the characterization of the Z-DNA binding proteins that are found in cells infected with the tumor virus SV40. This virus, which transforms monkey kidney cells into a malignant form, contains a small circular piece of DNA of slightly over 5200 base pairs. The DNA has one region that controls or regulates both the replication of DNA and its transcription. The control or regulatory region is only a few hundred base pairs long, whereas the rest of the DNA is used in coding for RNA transcripts that direct the synthesis of proteins. When the DNA of SV40 is negatively supercoiled, small segments of it form left-handed Z-DNA (16). These segments were found in the control or regulatory region of the viral DNA. This is striking since the control region has only about 8% of the whole genome.

In infected cells the SV40 virus is found in the form of a minichromosome, with its circular 5200-base pair DNA coiled around nucleosomes and organized into a structure that is active in transcription and replication. The minichromosome has many of the same properties as transcriptionally active chromatin in cells. A study has been carried out on the proteins in the minichromosome and it was found that they contain some that bind specifically to Z-DNA (15). These proteins have been isolated and have been found to attach to SV40 DNA in the control region of the virus. These are in a position such that they might play some role in governing expression of the gene. Uncovering this mechanism would significantly enhance our understanding of one aspect of the regulation of gene expression.

Recent experiments have shown that a purified enzyme that carries out genetic recombination in a lower eukaryote (*Ustilago*) is a Z-DNA binding protein (17). Genetic recombination is the process by which two pieces of DNA exchange segments at points where their nucleotide sequences are identical. It is an important process in the generation of genetic diversity. In these experiments, plasmids were found to undergo recombination if they had Z-DNA-forming segments, but only if the plasmids were supercoiled so that the segments were in the Z-conformation (18). This suggests that the enzyme binds to the Z-segment and strand separation ensues there. This must occur with both strands. In this case, Z-DNA formation may facilitate strand separation. This discovery suggests that Z-DNA formation may be a key intermediate in the exchange of DNA sequences, and it enlarges our understanding of another potential role of left-handed Z-DNA in biological systems.

REFERENCES

1. Watson, J. D., and Crick, F. H. C. (1953). *Nature (London)* **171**, 737.
2. Wang, A. H.-J., Quigley, G. J., Kolpak, F. J., Crawford, J. L., van Boom, J. H., van der Marel, G., and Rich, A. (1979). *Nature (London)* **282**, 680–686.
3. Haschemeyer, A. E. V., and Rich, A. (1967). *J. Mol. Biol.* **27**, 369–384.
4. Rich, A., Nordheim, A., and Wang, A. H.-J. (1984). *Annu. Rev. Biochem.* **53**, 791–846.
5. Pohl, F. M., and Jovin, T. M. (1972). *J. Mol. Biol.* **67**, 375–396.
6. Moller, A., Nordheim, A., Kozlowski, S. A., Patel, D. J., and Rich, A. (1984). *Biochemistry* **23**, 54–62.
7. Lafer, E. M., Moller, A., Nordheim, A., Stollar, B. D., and Rich, A. (1981). *Proc. Natl. Acad. Sci. U.S.A.* **78**, 3546–3550.
8. Moller, A., Gabriels, J. E., Lafer, E. M., Nordheim, A., Stollar, B. D., and Rich, A. *J. Biol. Chem.* **257**, 12081–12085.
9. Lafer, E. M., Valle, R. P. C., Moller, A., Nordheim, A., Schur, P., Stollar, B. D., and Rich, A. (1983). *J. Clin. Invest.* **71**, 314–321.
10. Peck, L. J., and Wang, J. C. (1983). *Proc Natl. Acad. Sci. U.S.A.* **80**, 6206–6210.
11. Singleton, C. K., Klysik, J., Stirdivant, S. M., and Well, R. D. (1982). *Nature (London)* **299**, 312–316.
12. Peck, L. J., Nordheim, A., Wang, J. C., and Rich, A. (1982). *Proc. Natl. Acad. Sci. U.S.A.* **79**, 4560–4564.
13. Nordheim, A., Tesser, P., Azorin, F., Kwon, Y. H., Moller, A., and Rich, A. (1982). *Proc. Natl. Acad. Sci. U.S.A.* **79**, 7729–7733.
14. Lafer, E., Sousa, R., Rosen, B., Hsu, A., and Rich, A. (1985). *Biochemistry* **65**, 5070–5076.
15. Azorin, F., and Rich, A. (1985). *Cell (Cambridge, Mass.)* **41**, 365–374.
16. Nordheim, A., and Rich, A. (1983). *Nature (London)* **303**, 674–679.
17. Kmiec, E. B., Angebidies, K. J., and Holloman, W. K. (1985). *Cell (Cambridge, Mass.)* **40**, 139–145.
18. Kmiec, E. B., and Holloman, W. K. (1986). *Cell (Cambridge, Mass.)* **44**, 545–554.

PART I

DNA–PROTEIN INTERACTIONS

2
The Structure and Function of *Eco*RI Endonuclease

JOHN M. ROSENBERG, JUDITH A. MCCLARIN,
CHRISTIN A. FREDERICK,[1] BI-CHENG WANG,[2]
HERBERT W. BOYER,[3] JOHN GRABLE,
AND PATRICIA GREENE[3]

Department of Biological Sciences
University of Pittsburgh
Pittsburgh, Pennsylvania 15260

The 3-Å structure of a cocrystalline recognition complex between *Eco*RI endonuclease and the cognate oligonucleotide TCGCGAATTCGCG has been solved by the ISIR method using a platinum isomorphous derivative. Each subunit of the endonuclease is organized into an α/β domain based on a five-stranded β-sheet and an extension, called the "arm," which wraps around the DNA. The primary β-sheet consists of antiparallel and parallel motifs which contain the sites of DNA strand scission and sequence-specific recognition, respectively. The hydrolytic active site is located in a cleft which binds the DNA backbone in the vicinity of the scissile bond.

The DNA conformation departs significantly from those which have been observed in crystals containing pure DNA, suggesting that the altered conformation has been stabilized by the enzyme. The confor-

---

[1] Present address: Department of Biology, MIT, 77 Massachusetts Avenue, Cambridge, Massachusetts 02139.
[2] Present address: Biocrystallography Laboratory, Box 12055, VA Medical Center, Pittsburgh, Pennsylvania 15260.
[3] Present address: Department of Biochemistry and Biophysics, University of California, San Francisco, California 94143.

mations seen only upon protein binding are termed neoconformations to distinguish them from those which are intrinsically stable in the absence of protein.

Sequence specificity is determined by "modular" interactions based on the crossover α-helices, i.e., those which connect the β-strands of the parallel segment of the principal β-sheet. They point into the major groove of the DNA and amino acid side chains at the amino ends of these helices form bidentate interactions with the bases. The inner recognition module consists of two symmetry-related α-helices which recognize the inner tetranucleotide (AATT), wheras the two symmetry-equivalent outer recognition modules are single α-helices which recognize the GC base pairs.

INTRODUCTION

The recognition by a protein of a specific sequence of bases along a strand of double helical DNA lies at the heart of many fundamental biological processes, including the regulation of gene expression by repressors and activators, site-specific genetic recombination, and host-dependent restriction and modification of DNA. The detailed molecular mechanisms of some of these interactions are beginning to be understood due to efforts in many laboratories using genetic, biochemical, and structural methods. One of the most intriguing questions in molecular biology today is whether the details of several of these recognition mechanisms will form a a small number of simple patterns which would lead to a general understanding of DNA recognition processes. To answer this question, structural data are required from representative DNA–protein cocrystalline complexes.

The structures of three sequence-specific DNA-binding proteins, the *cro* and CI repressors from bacteriophage λ and the *Escherichia coli* catabolic gene activator protein (CAP), have been solved in the absence of DNA (1–5). These three proteins share a common "two-helical motif" at the putative DNA binding site which has led to model building of the recognition complexes (3,6,7). The 7-Å structure of a cocrystalline complex between a tetradecanucleotide and bacteriophage 434 repressor supports the general features of these models (8–10). The common features in these structures suggest that they are examples of one class of DNA recognition protein based on the two-helical motif.

Similarly, structural investigations are in progress on other systems, inculding the Klenow fragment of *E. coli* DNA polymerase I (11), nucleosome core particles (12,13), the histone octamer (14), and the "histonelike" DNA-binding protein II of *Bacillus stearothermophilus* (15). These analyses will facilitate the elucidation of the specific structural features of both protein and DNA in these specific protein–DNA interactions as well as general structural principles of DNA–protein interactions.

The essence of the recognition process is the energy differences between states of the macromolecular system. For a simple DNA-binding protein, such as a repressor, this would be the thermodynamic free energy which partitions the system into three classes of states: (1) cognate binding between the protein and the target (operator) site on the DNA, (2) nonspecific binding of the protein at noncognate sites, and (3) states in which the protein and DNA are physically separate. (Most DNA-binding proteins exhibit fairly strong nonspecific binding with the result that the cognate and nonspecifically DNA-bound states predominate under physiological conditions.) The *in vivo* determinant of functionality in these systems is the fractional occupancy of the target sites on the DNA.

The situation is slightly more complicated for an enzyme which catalyzes a chemical reaction at a specific site on DNA. An enzyme functions by lowering the activation barrier separating the reactant and product states. Hence, a sequence-specific enzyme must selectively lower the activation energy only at cognate DNA sequences. The functional measure of specificity here is the ratio of activation energies at the cognate site(s) to the activation energies at all the noncognate sites (which comprise the bulk of the DNA). *Eco*RI endonuclease possesses both types of specificity (binding in the absence of Mg^{2+} and catalytic in its presence). Although the two specificities appear to be very similar, as indicated above, it should be noted that the mechanistic details of the two cases may not be precisely identical. Regarding either the binding or catalytic specificity, the basic question is: What is the source of this energy difference?

The energy differences could, in principle, come from two types of interaction between the protein and the cognate bases: direct and indirect. Hydrogen bonds between the protein and the cognate bases are clear examples of direct determinants of specificity which undoubtedly are central to most, if not all, DNA recognition systems. However, other types of direct physical interaction are also probably significant, including van der Waals interactions between nonpolar moieties on the protein and bases [e.g., an interaction with a thymine methyl group; see the results reported by Youdarian *et al.* for a strongly suggestive example of this type of interaction (16)]. Intercalation of aromatic amino acid side chains has been proposed as a mechanism of specificity (17) as has cobinding of divalent cations between the protein and DNA (17–19). All these interactions have as a common element the direct interaction between bases and moieties on the recognition protein.

In indirect recognition, the protein is presumed to interact only with the backbone of the DNA and sequence specificity could be generated in one of two ways. First, Dickerson *et al.* have noted that the precise conformation of the DNA backbone varies in a sequence-dependent way (20–22), leading to the suggestion that proteins could have evolved to recognize these differences. Second, we have noted significant distortions of the DNA within the DNA–*Eco*RI endonuclease complex (23–25) (see also below). We suggested that different sequences of DNA might require differential inputs of energy in order to adopt the altered conformations. In the former, the protein would fit (or not fit) intrinsic sequence-dependent differences in the DNA backbone, whereas in the latter both protein and DNA would be driven toward a common (sequence-independent) structure via a sequence-dependent input of energy. In reality these both represent poles of what could be a continuous spectrum.

The highly specific recognition of the double-stranded sequence d(GAATTC) by *Eco*RI endonuclease offers compelling advantages as a model system for investigating DNA recognition. It is a small protein of molecular weight 32,000 (276 amino acids) of known sequence (26,27) which hydrolyzes the phosphodiester bond between the guanylic and adenylic acid residues resulting in a 5' phosphate. This reaction proceeds with inversion of configuration at the reactive phosphorus (28). The protein forms highly stable catalytically active dimers in solution and will form tetramers at higher protein concentrations (29,30). Although *Eco*RI endonuclease requires Mg^{2+} for phosphodiester bond hydrolysis, DNA binding and recognition specificity is retained in the absence of this ion (31–34).

We recently reported (23) our observations based on a preliminary electron density map (see below) of a cocrystalline complex between *Eco*RI endonuclease and a cognate oligonucleotide. Here we report further analysis of this structure.

METHODS

Crystallization conditions and methods of data collection were reported previously (23,35). It should be noted that hydrolysis of the

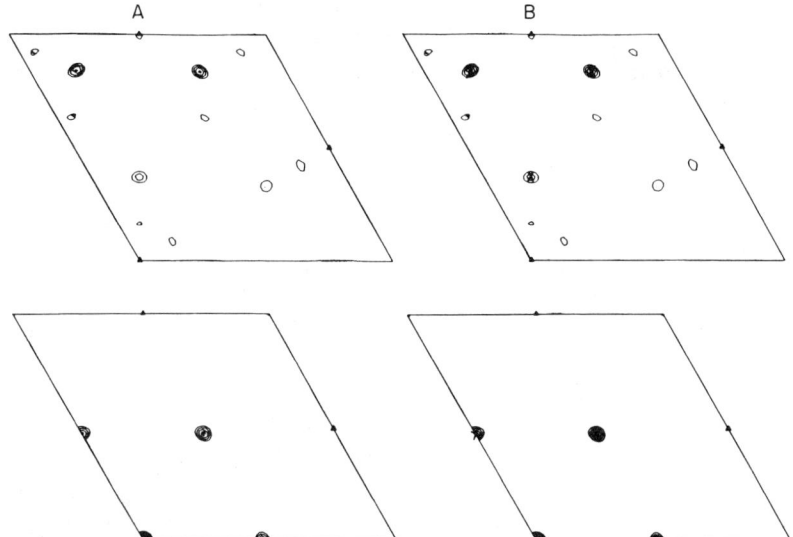

Fig. 1. (A) The $w = 0$ and $w = 0.24$ sections of the platinum difference Patterson function, which contains the Pt-Pt vectors. (B) The indicated positions mark the expected positions for the Pt-Pt vectors based on the coordinates of the major platinum binding site.

DNA was prevented by omitting the required cofactor, Mg^{2+}, from the crystallization medium and substituting EDTA. Platinum and mercury heavy atom derivatives were prepared as reported previously (23). The platinum Difference Patterson Function is shown in Fig. 1. The Cullis R-factors for the Pt and Hg derivatives to 3.5 Å are 49.2 and 52.2%, respectively.

An MIR[4] electron density map was calculated from these two derivatives to 5 Å resolution. The MIR phases were used to calculate a Pt–native Difference Fourier map, which revealed the presence of a single minor heavy atom site. Wang's ISIR method (36) was independently applied to the platinum and mercury data. The general features, such as the solvent regions and the molecular outline, were similar in all three electron density maps, however, the MIR and Hg–ISIR maps contained significant amounts of noise whereas the Pt–ISIR map was clear. We suspect that the noise in both cases is caused by a problem with the mercury derivative.

---

[4] We use the following abbreviations: MIR, multiple isomorphous replacement; SIR, single isomorphous replacement; ISIR, iterative single isomorphous replacement.

Statistics for the platinum derivative exhibited evidence of a slight nonisomorphism at high resolution, as can be seen in Fig. 2, which is a plot of intensity differences as a function of resolution. The plot shows a minimum at 3.5 Å, with a slight increase at higher resolution.[5] We felt that the platinum phase information was dubious beyond 3.5 Å and therefore did not utilize it further.

It should be noted in this context that the accuracy of the data and the absence of nonisomorphism are crucial to the success of the ISIR procedure. This can be seen by considering that the ISIR procedure resolves the phase ambiguity initially present in the SIR phases. If, however, both the probable phases are seriously in error for a significant fraction of the data, then the ISIR procedure most likely will converge to a false minimum. This is in contradistinction to the MIR case, in which a preponderance of valid phase information can tend to overpower inaccuracies. For ISIR, the initial phase information should be carefully selected to ensure that it is accurate.

The positions and occupancies of the major and minor platinum sites were refined as follows. We calculated for each reflection within 5 Å resolution the absolute value of the difference between the native and derivative structure factors. We used all the differences for the centric data as well as the acentric reflections with larger differences, specifically the largest 40% of the acentric data. These data were used as input to a conventional full matrix–least squares refinement calculation with the platinum positions and occupancies treated as variables (program QKREF, W. Furey, unpublished). Statistics from this refinement are shown in Table I.

The ISIR procedure was then used to resolve the phase ambiguity in the platinum SIR data to 3.5 Å; it was used again to extend the data to 3.2 Å and then 3.0 Å resolution, as reported earlier (23). The average figure of merit at the beginning of the process was 0.33 for those 4033 reflections which had both the native and the derivative information, and at the end of the process it was 0.79 for all 5880 observed reflections including those 1847 reflections for which the derivative information had been rejected (see Table II). An electron density map based on these phases was the basis of our earlier report (23). Although this electron density map was very clear in most places and

---

[5] The source of this slight nonisomorphism was obvious once the structure was solved. The Pt was bonded to the sulfur atoms of two methionyl residues, which fortuitously were in close proximity on the surface of the protein. However, the (mean) positions of the two sulfur atoms in the native structure were not precisely those required by the geometry of the bridging reaction, which therefore appears to have required a small structural adjustment in a short segment of the polypeptide chain.

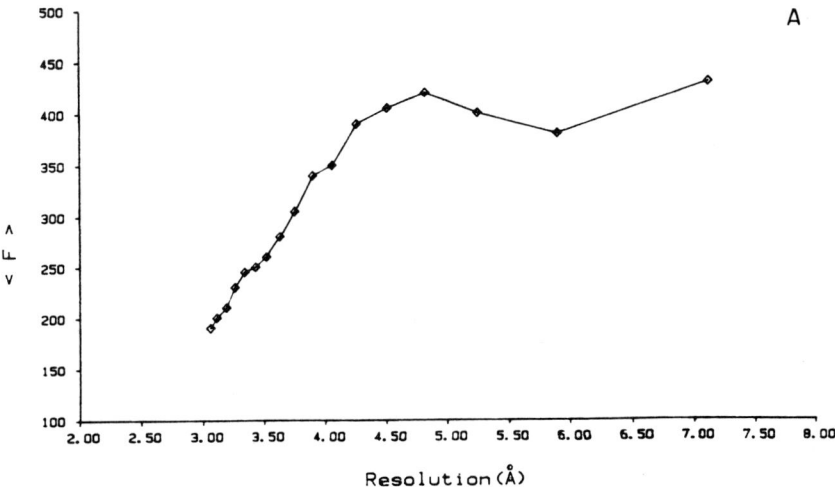

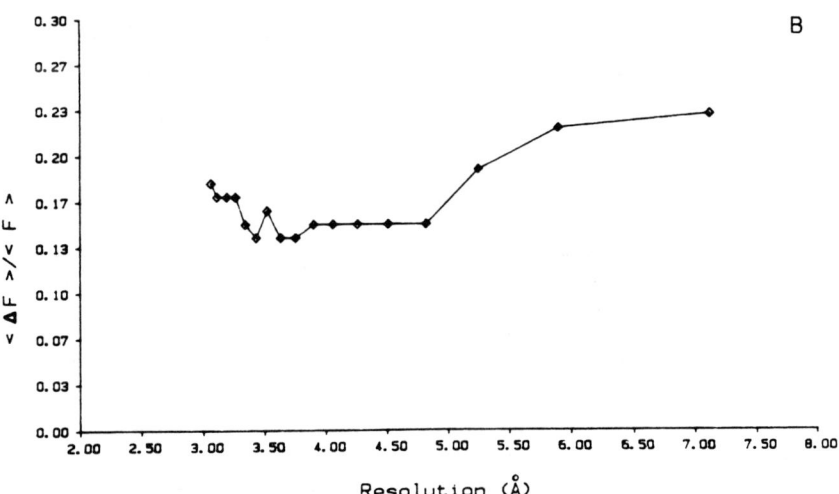

Fig. 2. (A) A plot showing the mean value of the observed structure factors as a function of resolution. (B) A plot showing, as a function of resolution, the ratio of the mean of the absolute values of the differences between native and Pt-derivative structure factors to the mean of the structure factors.

TABLE I
Lattice Parameters and Heavy Atom Refinement Statistics

Lattice parameters[a]
Unit cell: $a = b = 118.4$ Å; $c = 49.7$ Å; $\gamma = 120°$
Space group: P321
Asymmetric unit: one protein subunit and one DNA strand
Solvent content: 58%; $V_M = 2.8$

Heavy atom refinement statistics[b,c,d]

	Initial Pt sites			
Occ.	X	Y	Z	B
1.000	0.1258	0.5707	0.1109	25.0
0.383	0.4045	0.1329	0.3203	26.0
5.0-Å data: 1395 reflections; $R_C = 0.42$; $\langle m \rangle = 0.41$				
3.0-Å data: 5880 reflections; $R_C = 0.58$; $\langle m \rangle = 0.24$				
QKREF refinement: 5 cycles on all centric + 40% largest acentric data within 5 Å				
Initial $R = 0.33$; final $R = 0.30$				

	Final Pt sites			
Occ.	X	Y	Z	B
0.996	0.1247	0.5695	0.1124	30.0
0.399	0.4066	0.1304	0.3138	35.0
5.0-Å data: 1395 reflections; $R_C = 0.41$; $\langle m \rangle = 0.42$				
3.5-Å data: 4073 reflections; $R_C = 0.49$; $\langle m \rangle = 0.33$				

[a] V_M = Matthews' coefficient in Å³/Dal.
[b] Occ. = Occupancy of heavy atom site.
[c] $R_C = \Sigma ||F_{PH} \pm F_P| - F_{H(calc)}|/\Sigma |F_{PH} - F_P|$, where F_P and F_{PH} are the native and Pt-derivative structure factors, respectively, and $F_{H(calc)}$ is the calculated contribution of the heavy atoms.
[d] $\langle m \rangle$ = The mean figure of merit.

allowed us to trace the entire DNA double helix and much of the polypeptide backbone, there were a few regions where the electron density was not easily interpretable.

A fraction of the data were missing from the original data sets and we suspected that the absence of this information was interfering with the ISIR procedure. Three factors led to the absence of data. First, a few reflections at very low resolution were obscured by the beam stop of our Arndt–Wonacott camera. Second, a few reflections were saturated even on the third film of our film packs and were deleted from the data sets by the computer programs we used to process our film

TABLE II
ISIR Refinement Statistics

Filter[a]	Cycles[b]	Res.[c]	N_P[d]	N_{UP}[e]	Shift[f]	$\langle m \rangle$[g]	R[h]
Phase refinement							
1	4	3.5	4033	229	47.0	0.75	0.25
2	4	3.5	4033	229	47.2	0.75	0.23
3	8	3.5	4033	229	50.4	0.78	0.21

Filter	Cycles	Res.	N_P	N_{UP}	Shift	$\langle m \rangle$	R
Phase extension							
4	6	3.2	4033	1228	59.4	0.78	0.22
5	6	3.0	4033	1847	63.3	0.79	0.21

Filter	Cycles	Res.	N_P	N_{UP}	N_G[i]	Shift	$\langle m \rangle$	R
Amplitude extension								
6	4	5.0	4033	1847	293		0.76	0.21
7	4	4.0	4033	1847	576		0.78	0.20
8	4	3.5	4033	1847	992		0.84	0.16
9	4	3.0	4033	1847	2394		0.87	0.15
10	6	3.0	4033	1847	2394	63.6	0.87	0.16
11	6	3.0	4033	1847	2394	63.8	0.87	0.16

[a] Filter = The (sequential) number of the calculated solvent mask.

[b] Cycles = The number of cycles of solvent flattening and Fourier inversion using the current solvent mask.

[c] Res. = The resolution (in Å) for the calculation; for phase refinement and extension this is resolution limit for all the data in the calculation, whereas for amplitude extension it is the resolution limit for the generation of estimates for the unobserved reflections (all the observed data to 3.0 Å were used for this calculation).

[d] N_P = The number of "paired" reflections for which both native and derivative data were available.

[e] N_{UP} = The number of "unpaired" reflections for which only native data were available.

[f] Shift = The mean phase shift from the SIR "best" phase.

[g] $\langle m \rangle$ = The mean figure of merit for all reflections based on the current ISIR phase probability distribution.

[h] $R = \Sigma \mid F_{obs} - F_{calc}\mid/\Sigma \; F_{obs}$, where F_{obs} are the observed (native) structure factors and F_{calc} are the structure factors obtained from Fourier inversion.

[i] N_G = The number of estimates generated for unobserved data.

data (37,38). Third, these programs also deleted a significant proportion of our weakly observed data because they were deemed statistically unreliable.

Efforts were then made to estimate the missing amplitudes and phases and incorporate these estimates in the electron density calculations via an option in the ISIR procedure. This was initiated for reflections within a 5-Å-resolution limit and iterated for four cycles (all the observed data to 3.0 Å were used during this process). At each iteration, the structure factor amplitudes and phases were estimated for the missing reflections by Fourier inversion of the modified electron density map. These estimates were used in electron density calculations during the next cycle. At the end of the fourth cycle, a new solvent mask was calculated based on the 5880 originally observed reflections and the 293 estimates generated by this process. The process was repeated in a similar manner to estimate the missing reflections to 4.0 Å, then to 3.5 Å, and finally to 3.0 Å (see Table II). This process produced 2394 estimated structure factor amplitudes and phases.

At this stage, an electron density map was calculated using all the observed and estimated reflections (8274 in total). This map showed considerable improvement over the original map based on the 5880 observed reflections only, however, it still showed small ripples around some of the threefold axes, although their magnitudes had been diminished considerably from the first map. These ripples around the threefold axes were finally leveled by recalculating a solvent mask using a 10-Å radius in the masking function instead of the 5.1-Å radius used earlier (filters 10 and 11 of Table II). After twelve cycles of iterations and two recalculations of the solvent mask (filters), the final figure of merit and map inversion R-factor remained at 0.87 and 15%, respectively. The electron density based on the third set of phases exhibited an improvement in clarity as shown in Fig. 3 and was used for the final chain tracing and fitting of the chemical sequence of the enzyme as described below.

We compared the electron density maps which preceded and followed both of the "extension" steps (phase extension from 3.5 to 3.0 Å using observed amplitudes as well as estimation of missing intensities) and concluded in both cases that the extensions reduced noise and improved the clarity of the maps while maintaining the basic features which were present in the initial 3.5-Å map. These features included the DNA (the phosphate positions were striking and immediately obvious features of all the maps), as well as several prominent α-helices and strands of β-sheet. (See Fig. 3 for examples of these

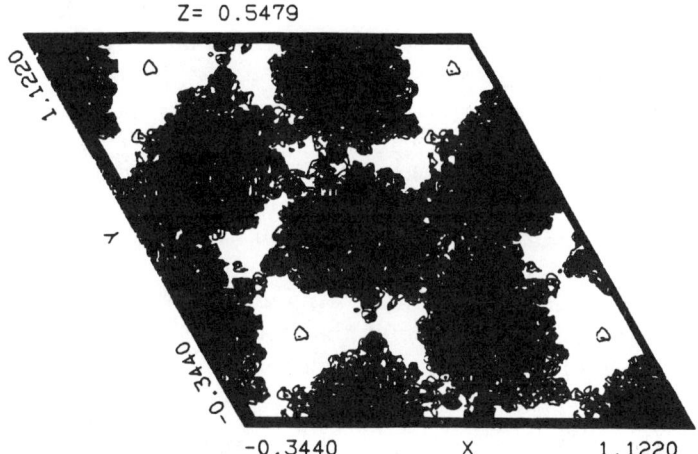

Fig. 3. A projection down the c axis of the Pt–ISIR electron density map of the DNA–*Eco*RI endonuclease complex.

maps.) The improvements in detail were most noticeable in the problematic regions. These include the β-hairpin which forms part of the "arm" (see discussion) and the region surrounding one of the crystallographic axes, which is very densely packed with protein. Some of the loops connecting secondary structure elements and some of the side chains were also clarified. These improvements enabled us to distinguish possibilities which had been ambiguous before the extension.

The final electron density map was displayed on Plexiglas sheets. The DNA and protein secondary structure elements were very clear. Over two-thirds of the amino acid side chains were distinctly visible along with main chain density for all but four amino acid residues. The missing residues were in the immediate vicinity of the major heavy atom site, and it appears likely that their movement is associated with the small nonisomorphism noted previously. Almost all of the large hydrophobic side chains, tryptophans, phenylalanines, and tyrosines were clearly recognizable. Many basic residues, especially arginines, which were located at the DNA–protein interface, were also easily identifiable. Most of the poorly visualized side chains were located at the protein–solvent interface. Both the DNA–protein and protein subunit–subunit interfaces were well ordered and provided useful constraints when we assigned the known amino acid sequence to the electron density map. These amino acid assignments were made via inspection of the electron density map, aided by model building, distance measurements, and the known stereochemistry of proteins. This process led to a unique tracing of the polypeptide chain through the protein–DNA complex.

Coordinates for an α-carbon atom and for either a β-carbon atom or a terminal side chain atom for larger amino acids were taken from the ISIR map on Plexiglas sheets and used to generate atomic coordinates for the entire molecule with the program FRODO (39–41). Electron density fitting continued with FRODO on an Evans and Sutherland PS340 computer graphics system. The coordinates were regularized to approximately ideal geometry alternately with improving the fit to the electron density. At this point, the model has been fit to all of the electron density features noted above. Refinement of this model is in progress.

DESCRIPTION AND DISCUSSION OF THE STRUCTURE

GENERAL FEATURES OF THE COMPLEX

The cocrystalline asymmetric unit contains one protein subunit of 276 amino acid residues of known sequence (26,27) and one strand of the oligonucleotide TCGCGAATTCGCG (*Eco*RI site underlined). The complex is a twofold symmetric dimer in which the protein–protein intersubunit diad, the principal diad of the symmetric DNA double helix, and the crystallographic twofold axis all coincide (35). The molecular boundary, as seen in the Pt–ISIR electron density map, clearly encloses this complex in a well-defined globular structure 50 Å across (see Fig. 3).

The DNA–protein complexes are packed within the crystalline lattice so that the DNA forms a continuous rod parallel to the *c*-axis. The unpaired 5′ thymine residues at each end of the double helix appear to be stacked upon each other leading to a continuous series of stacked bases across a crystallographic twofold axis (the space group is P321). Therefore, although the oligonucleotide is 13 nucleotides long, there are 14 stacked nucleotides per unit cell. While this is consistent with a prediction by Harrison that oligonucleotides 14 base pairs long should be particularly useful in the growth of DNA–protein cocrystals (8), it should be noted that Harrison's prediction was based on assumptions regarding the structure of the DNA, which are violated in this case by the neokinking reported below (i.e., Harrison assumed B-DNA with 10.5 base pairs per turn). The oligonucleotide is actually somewhat larger than the DNA-binding face of the protein. However, its length almost exactly matches the net width of the protein, which tapers slightly at the DNA interface leaving a solvent gap separating the ends of the oligonucleotide from the neighboring protein.

There are three major areas of protein–protein interaction which, together with the DNA–DNA interaction, form the crystalline lattice. First is the subunit–subunit interface within the dimeric complex which contains the determinants of dimer formation. [EcoRI endonuclease forms highly stable dimers in solution in both the presence and absence of DNA (29,30).] Second is the region around the threefold symmetry axis, where three dimers are tightly packed (see Fig. 3). Third is a smaller region of limited protein–protein interactions along the direction of the c axis. These involve contacts between loops at the molecular surface of the protein. The DNA–DNA interaction comprises a significant fraction of the net intermolecular interactions along the c axis. This confirms the concept that stability in DNA–protein cocrystals requires compatibility between the DNA–DNA, protein–protein, and protein–DNA contacts in the direction of the average DNA helix axis and suggests that the length and terminal sequence of the cognate oligonucleotide should be treated as a critical variable in future attempts to form sequence-specific DNA–protein cocrystals.

The conformation of the DNA in this complex departs from that of classical B-DNA. There are three abrupt dislocations in the helix, termed neokinks (see below), which divide the DNA into four blocks of three base pairs each, as shown in Fig. 4. In comparison to the neokinks, the helical parameters within each block are relatively more regular. The recognition sequence is contained in the central two blocks of the structure and the flanking sequence, CGC/GCG, is contained in the terminal blocks.

STRUCTURAL ORGANIZATION OF THE PROTEIN SUBUNIT

EcoRI endonuclease is an α/β protein consisting of a five-stranded β-sheet surrounded on both sides by α-helices (see Fig. 5). Four of the

```
T C G C ᵛG*A A T T C ᵛG C G
G C G C T T A A*G C G C T
```

Fig. 4. The sequence of the synthetic oligonucleotide with the recognition sequence underlined. An asterisk denotes the location of phosphodiester bond hydrolysis resulting in a 5′ phosphate. The hydrolysis reaction requires Mg^{2+} as a cofactor. A dotted line denotes the center of the type I neokink which is coincident with the crystallographic and molecular twofold symmetry axis. The type I neokink unwinds the DNA by 25°. ^ denotes the center of the asymmetric type II neokink, which separates the terminal blocks from the central blocks of nucleotide pairs.

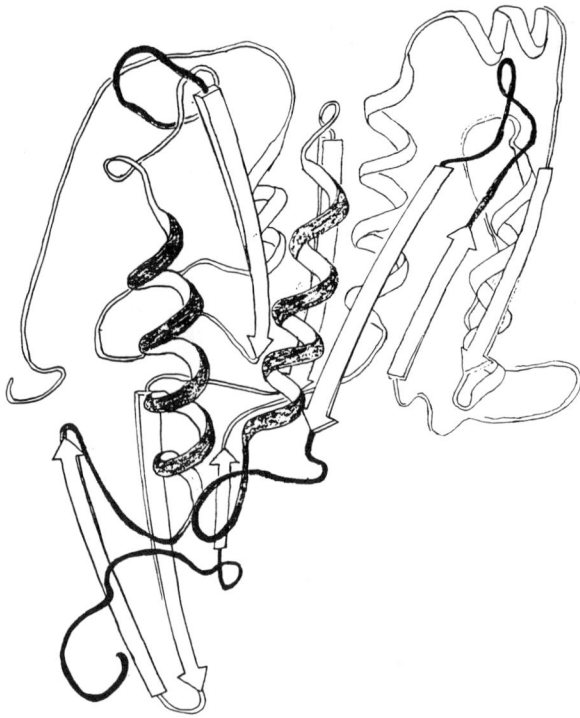

Fig. 5. Schematic backbone drawing of one subunit of (dimeric) *Eco*RI endonuclease and both strands of the DNA in the complex. The arrows represent β-strands, the coils represent α-helices, and the ribbons represent the DNA backbone. The helices in the foreground of the diagram connect the third β-strand to the fourth and the fourth to the fifth. They also interface with the other subunit. The amino terminus of the polypeptide chain is in the arm near the DNA.

five strands are parallel, however, the location of the single antiparallel strand divides the sheet into parallel and antiparallel three-stranded segments (see Fig. 6). Each of these segments forms a sizable structural unit constructed on a simple three-dimensional pattern in which the physically adjacent elements of secondary structure are essentially contiguous within the primary sequence, i.e., a motif. It is interesting that the parallel motif is the locale for the direct contacts between the protein and DNA bases as well as subunit–subunit interaction, whereas the antiparallel motif contains the site of DNA strand scission. We also note that the parallel motif is topologically very similar to one-half of the well-known nucleotide binding domain (42).

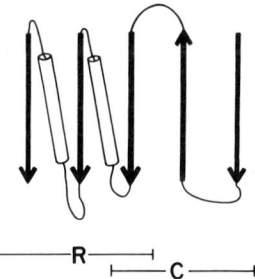

Fig. 6. Topology diagram of the major α/β domain of *Eco*RI endonuclease. The β-sheet is divided into two overlapping topological segments, the parallel and antiparallel motifs, which correspond to the functional division of the β-sheet into a motif primarily responsible for recognition of the specific DNA sequence, R, and a motif primarily involved in catalytic activity, C.

(The nucleotide binding domain is a six-stranded parallel β-sheet constructed out of two three-stranded motifs, which are very similar to each other.)

Following the course of the polypeptide chain, we find the amino terminus of the polypeptide chain located in an extension of the principal α/β domain of the protein, referred to as the "arm," which wraps around the DNA. The polypeptide chain then forms a long α-helix on the surface of the molecule which is followed by a loop into the first strand of the β-sheet. This β-sheet is formed sequentially starting from the outside of the antiparallel motif (see Fig. 6). The next loop, which connects the first and second β-strands, contains another α-helix situated on the surface of the molecule. The loop between the second and third (antiparallel) β-strands projects somewhat into the solvent. The third β-strand is a common element of both the antiparallel and parallel motifs. The parallel motif is formed next, sequentially from the middle of the β-sheet to the fifth strand at the edge of the sheet. The α-helices found at the subunit interface are the crossover helices (43) of the parallel motif, i.e., those connecting the third β-strand to the fourth and the fourth to the fifth. After exiting the fifth β-strand, the polypeptide chain loops around the surface of the complex, placing the carboxy terminus in the proximity of the DNA backbone.

As can be seen in Fig. 5, all the α-helices in the protein are aligned so that their amino terminal ends are pointing toward the DNA. This orients the α-helix dipoles (44) so that they interact favorably with the electrostatic field generated by the negatively charged phosphates on the DNA backbone. The two crossover helices of the parallel motifs are actually oriented so that their amino terminal ends project into the major groove of the DNA. The amino acid side chains which interact with the DNA bases are located at the ends of these helices.

The β-sheet exhibits the conventional twist (43–49) with the individual β-strands approximately perpendicular to the DNA helical axis.

EcoRI ENDONUCLEASE POSSESSES AN ARM WHICH WRAPS AROUND THE DNA

The EcoRI endonuclease "arm" is an extension of the α/β domain (see Fig. 7) which wraps around the DNA, partially encircling it, and thereby clamping it into place on the surface of the enzyme. Because of the twofold symmetry of the complex there are two arms, each of which interacts with the DNA directly across the double-stranded helix from the scissile bond. Jen-Jacobson has demonstrated that these nonspecific contacts between the arm and the DNA are required for DNA cleavage by selective proteolysis in which portions of the arm are selectively removed (49a). Many of the resulting "deletion derivatives" retain sequence-specific DNA binding but lack strand scission capability.

The arm has a structural "identity" of its own. It is composed of the amino terminus of the protein and a β-hairpin sequentially located between the fourth and fifth strands of the primary β-sheet. (A β-hairpin is a structure consisting of two antiparallel β-strands connected by a short turn.) Part of the amino terminal portion of the arm adds a third β-strand so that the structural foundation of the arm is a three-stranded antiparallel β-sheet. Thus, there are two β-sheets in each EcoRI endonuclease subunit, the primary five-stranded sheet described above and the subsidiary three-stranded sheet described

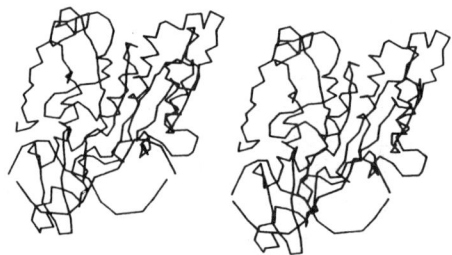

Fig. 7. Stereo drawing of the α-carbon coordinates of one subunit of EcoRI endonuclease and the phosphorus coordinates of the double-stranded DNA. This drawing was generated with the program FRODO.

here. The first 14 amino acid residues of the polypeptide chain form an irregular structure which is sandwiched in between the subsidiary β-sheet and the DNA; many of the nonspecific DNA–protein contacts mediated by the arm are located here. Additional DNA backbone contacts are located in the short segment of polypeptide chain which connects the third subsidiary β-strand with the α-helix which follows it in the primary sequence (this α-helix is the "outer recognition module" described below).

Even though the arm has the structural features described above, it is not a domain, as defined by Richardson (43). It does not appear to have a fully developed hydrophobic core and it is composed of two passes of the polypeptide chain, rather than a single chain segment. Indeed, it is doubtful that it could assemble or maintain its tertiary structure in the absence of the principal domain. It is therefore an extension on the principal domain, but one which has an important functional role.

STRUCTURAL FEATURES OF THE DNA

We have reported (23) that the DNA is kinked in the recognition complex by which we mean that it departs from the B-conformation according to certain criteria (see below). The centers of these distortions occur every three base pairs, as summarized in Fig. 4. We also noted that these kinks appear to be stabilized by the binding of the protein. Our previous report was primarily based on the location of the phosphate peaks, which are very prominent features of the initial electron density map. The electron density corresponding to the deoxyribose and base moieties showed significant improvement in the final ISIR electron density map and it is now clear that these groups are also displaced. Each kink distorts approximately two base pairs on either side of the "kink center."

The most striking departure from B-DNA is centered on the crystallographic and molecular twofold axis (between adenine 6 and thymine 7, where the tridecanucleotide is numbered to maintain consistency with the Dickerson dodecamer):

T pC pG pC pG pA pA pT pT pC pG pC pG
    1   2   3   4   5   6   7   8   9   10   11   12

We refer to this feature as the "type I neokink." It represents a net rotation of the upper half of both strands of the DNA relative to the entire lower half of the double helix so as to unwind the DNA. This can be seen in the relative positions of phosphorus atoms 6 and 7 which show virtually no relative rotation about the average helix axis. The unwinding is approximately 25°. This would clearly propagate through the DNA as a long-range effect on the net winding of the double helix. Kim and co-workers have measured the unwinding of DNA in solution upon *Eco*RI endonuclease binding in the absence of Mg^{2+} and obtained an identical value (50). The effect of a type I neokink on B-DNA can be seen in Fig. 8, in which a single type I neokink was placed between segments of DNA which have the helical parameters associated with "standard" B-DNA, i.e., 10.3 residues per turn and 3.2 Å per residue.

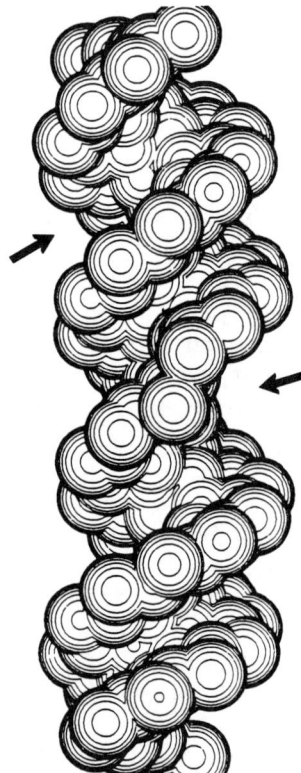

Fig. 8. A drawing of double-stranded B-DNA with a type I neokink inserted. The helical parameters used to generate the DNA double helix were the parameters determined for the central block, GAA, of the oligonucleotide bound to *Eco*RI endonuclease in the cocrystals. The arrows indicate the location of the major groove widened by the presence of the type I neokink and a standard width B-DNA major groove.

The principal effect of the unwinding is that the major groove becomes wider. The phosphate–phosphate distances across the major groove are increased by approximately 3.5 Å. Interestingly, the bases do not significantly increase their separation although the base–base contacts are changed. Thus, the type I neokink represents an effective mechanism for increasing the separation of the backbones of DNA strands without increasing the separation of the bases. This is possible because a helix is a screw, hence twisting it advances the "thread." As we shall see, the increased backbone separation is essential because otherwise the recognition α-helices would not fit in between and therefore could not approach close enough to interact with the bases. This consequence of the type I neokink suggests that it may be a general mechanism for facilitating access by proteins to the major groove of DNA. If so, similar DNA structures should be seen in some other recognition complexes.

There are also significant displacements of the A:T base pairs on either side of the kink center. These base displacements are probably critical to the recognition mechanism because they align adjacent adenine residues within the recognition site (residues five and six). These two purines are both involved in "bridging" interactions with amino acid side chains Glu 144 and Arg 145 (see below). These recognition interactions could not occur without this realignment because the N-6 moieties bridged by Glu 144 and the N-7 moieties bridged by Arg 145 would be too far apart in B-DNA. It is not clear at this time whether the purine realignment and the net unwinding are independent, i.e., whether one could exist without the other, or whether they are causally linked. However, since both are manifestations of a localized reduction in the twist of DNA, we strongly suspect that they are linked.

This realignment of the base pairs reveals another aspect of the type I neokink that may be of general significance, namely, that it creates sites for multiple hydrogen bonds which are absent in B-DNA. Indeed, the notion that *Eco*RI endonuclease creates the detailed features on the surface of the DNA which it then recognizes is a provocative and unexpected feature of this structure.

We have based our usage of the term "kink" on some of the ideas originally introduced by Crick and Klug (51). The current concept invokes two criteria: (1) an abrupt, highly localized disruption of double helical structure and (2) structural effects which propagate through the DNA over long distances. Since DNA is a relatively stiff rod, the simplest way to introduce long-range structural effects is to either bend or twist the double helix. These long-range effects are important both because they could be of functional significance and because they can be detected *in vitro*. We therefore employ the term "kink" to refer to an abrupt disruption of the double helix which includes a sharp bend and/or a highly localized under- or overwinding (twisting) of the DNA.

It has been suggested that separate terms be used to describe bending and twisting. However, we suspect that more "kinks" will be observed experimentally and that many such structures, probably the majority, will combine both bending and twisting at the same locus. Indeed, close inspection of the type I neokink suggests that it could introduce a hinge into DNA, i.e., the kink reported here might represent one member of a family of related structures with similar unwinding but different bending angles. While this possibility requires further study for its confirmation, it does suggest that a single term for bending and/or twisting would be more useful in the long run than a multiplicity of terms. A single term also serves to focus attention on the critical features of "kinking," namely, localized changes which generate long-range structural effects.

The prefix "neo-" adds the additional concept that the departure from B-DNA is induced by an external agent (the protein) and is not seen when DNA is studied in isolation. The oligonucleotide used in these cocrystals is virtually identical to that studied by Dickerson and colleagues (20–22). The structures they report do not contain dramatic kinks, such as the type I neokink. This suggests that the protein provides energy to drive the DNA into conformations which would otherwise be unfavorable. (This should not be taken to exclude the possibility that thermally transient fluctuations in DNA structure would include neokinks in the absence of protein. Indeed, fluctuations of this sort may well be important intermediates in the formation of the DNA–protein complex.)

The other localized departure from D-DNA is tentatively designated the type II neokink. The two-fold symmetry of the recognition complex generates a duplicate of this feature (see Fig. 4). The distortions are centered at phosphate moieties of guanine 4 and guanine 10. The backbone associated with nucleotides on either side of these phosphates is in an unusual conformation. For example, the distance between the phosphorus atoms associated with residues 4 (G) and 5 (A) is 7.3 Å, which is longer than expected for B-DNA. It is interesting to note that this extended segment spans the scissile bond. Similarly, the phosphorus–phosphorus distance between residues 9 (C) and 10 (G) is 7.4 Å. The base pair immediately adjacent to the *Eco*RI hexanucleotide, i.e., that involving cytosine 3 and the symmetry-related equivalent of guanine 10, is clearly anomalous. Its propeller twist is exaggerated and the pyrimidine is at an unusual angle.

We applied the methods for calculating helical properties and bending angles which we used previously (23,52) and obtained similar results: the bend angle of the type II neokink is between 20 and 40°. However, the interpretation of this result is clouded by the fact that these calculations include nucleotides which we now know are not exactly in a helical conformation (which is an assumption of the method). Furthermore, highly refined coordinates (which are not yet available) are required to properly choose which to include in or exclude from these calculations. Consequently, values for the bend angle of the type II neokink should be considered provisional. Unwinding is more readily assessable (e.g., by examining phosphorus positions in projection down the average helix axis) and the type II neokink does not introduce a major change in the net winding of the DNA. Thus the long-range effects of the type II neokink are also provisional although we strongly suspect that they do exist. This accounts for the tentativeness with which we ascribe the properties of a kink to this feature.

Other proteins distort DNA when they form complexes with it. Richmond et al. observed that the DNA in their 7-Å nucleosome structure contained "sharp bends" and/or possible kinks (53) which are not likely to be present in naked DNA; these could be neobends or neokinks, depending on the abruptness of the transition. Similarly, Anderson et al. reported that their 7-Å electron density map suggested that the 434 repressor introduced small perturbations into the structure of its operator (9). As structural information becomes available on additional DNA–protein complexes, we suspect that additional neoconformations will be observed.

DNA–PROTEIN INTERACTIONS

The major groove of the recognition hexanucleotide (GAATTC) appears to be filled with protein, forming a large, complementary interface. All the base–amino acid interactions appear to be located in the major groove of the DNA, whereas the edges of the bases which are exposed in the minor groove are open to the solvent. Thus, the direct interaction of complementary surfaces in the major groove is a major determinant of the recognition specificity. However, there is the additional possibility that the energy required to drive DNA into the neoconformations noted above depends on the base sequence. This would provide an additional, indirect mechanism of sequence recognition which may also contribute to *Eco*RI endonuclease specificity.

A crucial component of the DNA–protein interface is formed by the amino terminal ends of the crossover α-helices in the parallel β-motif. Since the DNA–protein complex possesses twofold symmetry, α-helices from both subunits participate in the formation of a four-helix bundle which inserts into the major groove of the DNA. One of the roles of the type I neokink is to make room for this bundle, which would not fit into the major groove of conventional B-DNA.

The α-helix which connects the third and fourth β-strands makes an angle of approximately 60° with the average DNA helix axis as shown in Fig. 9. The polypeptide chain turns sharply at the end of the α-helix so that the amino acid residues at the amino terminus end of the helix, those in the bend, and those in the adjacent stretch of chain are in close proximity to the DNA. This α-helix is also adjacent to the molecular twofold axis and the amino terminus of the helix is physically

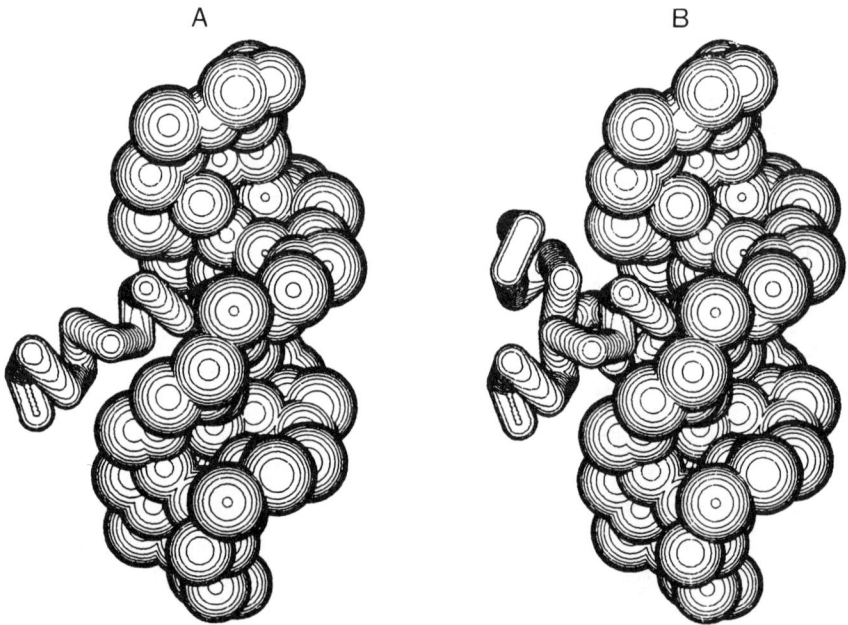

Fig. 9. (A) The double-stranded DNA in the *Eco*RI endonuclease–DNA complex and the α-helix which connects the third and fourth β-strands. (B) The same view as in (A) with the symmetry-related α-helix from the other subunit included. The two α-helices form the inner module, which recognizes the inner tetranucleotide, AATT.

adjacent to the amino terminus of the symmetry-related helix from the other subunit. These two α-helices form a symmetric module which is responsible for the direct interactions between the endonuclease and the bases of the inner tetranucleotide (AATT). This structural unit will be referred to as the inner recognition module. Two symmetry-related pairs of amino acid side chains make contact with two symmetry-related pairs of sequential adenine residues. Interestingly, each pair of adjacent adenines interacts with one amino acid from each subunit (see Fig. 10). In our current interpretation of the electron density map, these residues are glutamic acid 144 and arginine 145. The position of the arginine side chain in the current model is consistent with a hydrogen-bonding interaction with the N-7 moieties of adjacent adenines, whereas the glutamic acid is probably hydrogen bonding to the exocyclic N-6 groups of both adenines.

Separate α-helical modules are responsible for the direct contacts between the protein and the two outermost G-C pairs of the canonical hexanucleotide, GAATTC. These modules are called the outer recognition modules. They are identical by virtue of the twofold symmetry and each independent module consists of the crossover α-helix which connects the fourth and fifth strands of the principal β-sheet. This helix has many interactions with both α-helices of the inner recogni-

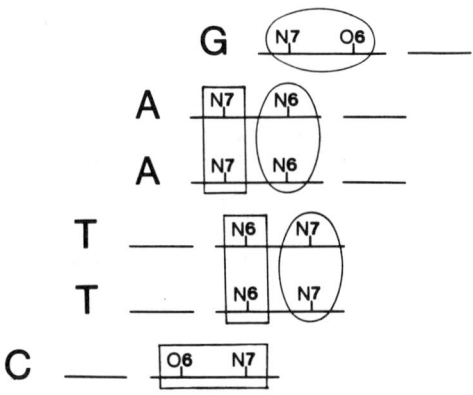

Fig. 10. A schematic drawing depicting the interactions between base pairs in the EcoRI recognition site and amino acid side chains of EcoRI endonuclease. Rectangles denote interactions from one subunit and ovals denote interactions from the symmetry-related subunit. The proposed interactions are arginine 200 hydrogen bonding to the N-7 and O-6 of the GC base pairs, arginine 145 hydrogen bonding to the two N-7 moieties of adjacent adenines, and glutamic acid 144 hydrogen bonding to the N-6 groups of adjacent adenines.

tion module and is thereby positioned so that it projects its amino terminus into the major groove of the DNA. At this stage of our analysis, it appears likely that arginine 200 interacts with the guanine in the manner predicted by Seeman, Rosenberg, and Rich (54).

In addition to the recognition of and precise requirement for the canonical site, GAATTC, *Eco*RI endonuclease also exhibits a dependence on the sequence of the nucleotides flanking this hexanucleotide (32,55,56). These workers noted that the flanking sequence environment can affect the overall rate of hydrolysis by at least an order of magnitude. Furthermore, Modrich and co-workers showed that *Eco*RI endonuclease can dissociate from DNA after making only a single strand nick and that the frequency of such nicking depended on the flanking sequences (57). It is unlikely that these effects are due to direct contacts between the enzyme and bases outside of the canonical hexanucleotide because we have not noted any such interactions in the electron density maps. However, there are extensive contacts between the enzyme and the DNA backbone which extend well beyond the hexanucleotide. This suggests that the conformational free energy of the type II neokink depends on the sequence of oligonucleotides immediately flanking the *Eco*RI site.

There appear to be interactions between the protein and the backbone of the DNA from the second through the ninth phosphates on each strand. Counting from the 5' end of the oligonucleotide, TpCpGpCpGpApApTpTpCpGpCpG, the third, fourth, and seventh phosphates are buried in the protein, i.e., they appear to be inaccessible to solvent. The remaining phosphates in the indicated region interact with the enzyme even though they are partially exposed to solvent. The third, fourth, and fifth phosphates are bound in a large cleft in the protein which forms the active site for DNA hydrolysis. This cleft is partially open to solvent in the vicinity of the fifth phosphate, where the scissile bond is located. It is through this channel that magnesium probably enters the active site.[6]

The cleft surface contains many basic amino acid residues which interact electrostatically with the phosphates. These interactions contribute to the overall stability of the complex. There are, of course, two identical clefts in the surface of the twofold symmetric enzyme. They are too far apart to fit regular B-DNA and we suspect that this separa-

---

[6] We have recently demonstrated that Mg^{2+} can be diffused into the crystals and the hydrolytic reaction carried out *in situ* (57a). The enzyme–product cocrystals still diffract X-rays and their structure determination is in progress.

tion promotes the formation of the type I neokink via long-range electrostatic attraction between the basic clefts and phosphates on the incoming DNA molecule.

The contacts noted above in the X-ray structure of the tridecamer–protein complex probably contain all of the major DNA–protein contacts between the endonuclease and larger natural DNA substrates. The association constant measured for the dodecamer, CGCGAATTCGCG, is within experimental error of that measured for plasmid DNA (27,30,34,58). The lower association constant for an octanucleotide substrate as compared with dodecameric or larger substrates (30,59) suggests strongly that interactions between the enzyme and the flanking regions of the DNA backbone make significant contributions to the net stability of the complex.

INFERRED AND OBSERVED CONTACT POINTS

There is good general agreement between the results of ethylation interference experiments (60) and the phosphate contacts described previously. The largest effects observed by Lu *et al.* correspond to the third, fourth, and seventh phosphates, which as noted above are buried in the protein and protected from solvent. The next largest effect is observed for the reactive phosphate at the fifth position. Small effects are noted for the sixth phosphate, which is probably forming interactions to the protein even though it is partially exposed to the solvent. (We suspect that a stronger ethylation interference would have been observed at lower protein concentrations at which the equilibrium is more sensitive to smaller reductions in the protein–DNA association constant.)

Methylation protection and interference experiments implicated both the major groove and the minor groove as points of DNA–protein contact. It is clear that the predicted contacts in the major groove of the recognition sequence are in good agreement with the X-ray structural data. The implications of the minor groove data must be reevaluated in light of the structural results. The N-3 positions on the central adenine, in the minor groove, were protected from dimethylsulfate by endonuclease and prior methylation at the N-3 blocked subsequent binding of the enzyme. Since there is no density observed in the minor groove of the DNA, and there are no sections of polypeptide chain left unaccounted for in the chain tracing of the protein, the observed effects at the N-3 are probably related to the conformational changes induced in the DNA by protein binding.

STRUCTURAL SUGGESTIONS FOR CONFORMATIONAL MOBILITY

The formation of the *Eco*RI endonuclease–DNA complex requires conformational changes in both substrate DNA and the enzyme in order to achieve specific binding. The DNA adopts neokinks in the specific DNA–protein complex. The protein must also alter it's conformation during the binding event because the arms encircle the DNA to such an extent that it appears unlikely that substrate DNA could enter the active site in the absence of some movement. There are four general possibilities: (1) The arms may move as relatively rigid units retaining their internal structure while they move with respect to the rest of the molecule. (2) The arms may have two stable structures, one in the presence and one in the absence of DNA. One detailed example of this relates to the fact that the amino terminus 14 residues of the arm (which are sandwiched between the rest of the arm and the DNA) appear to be rather loosely associated with the β-hairpin, suggesting that these residues fold against the DNA when it is present and refold in a tighter association with the β-hairpin when DNA is absent. (3) Part of the arms (probably that consisting of the amino terminus 14 residues) may undergo an order–disorder transition in which they are disordered in the absence of DNA and condense on it during complexation. (4) The dimeric endonuclease could undergo a quaternary conformational change in which the subunits move with respect to each other. These possibilities are not all mutually exclusive and the actual changes probably involve a combination of several of these factors. We have also grown crystals of the protein in the absence of DNA and that structure is in progress.

*Eco*RI* SPECIFICITY

*Eco*RI endonuclease loses its normally tight specificity under a variety of buffer conditions and recognizes many nucleotide sequences similar to the canonical site, GAATTC (61–65). These buffer conditions include elevated pH (8–9.5), Mn^{2+}, low ionic strength, and by the addition of organic compounds such as glycerol or ethylene glycol. An analysis of the patterns of reduced specificity led to a correct prediction of the hydrogen bonding sites on the DNA (66). That argument sheds light on recognition issues and therefore a modified form of it is reproduced below.

When Polisky *et al.* first observed the *Eco*RI* phenomenon, they noted that some RI* sequences are hydrolyzed much more rapidly than others (61). In particular, these authors noted that the canonical

EcoRI site is always hydrolyzed much more rapidly than any RI* site. These results were extended by Goodman et al., who noted that there is a clear hierarchical order in the hydrolysis rates at the leftmost base (G in the canonical sequence). Data collected by Gardner et al. (67) and Rosenberg and Greene (66) showed that cleavage rates at all EcoRI* sites can be represented by hierarchies. The hierarchy at the leftmost base is G $\gg$ A > T $\gg$ C (68), that is, GAATTC is hydrolyzed much more rapidly than AAATTC, etc. The hydrolysis rates for A and T are much closer than any others in the hierarchy. The hierarchies at the next two positions (both adenine in the canonical hexanucleotide) are identical: A $\gg$ [G,C] $\gg$ T (66) (GAATTC $\gg$ GAGTTC = GACTTC $\gg$ GATTTC, etc). The hierarchies at the last three positions are complementary to those at the first three because of the intrinsic symmetry of the EcoRI site. T $\gg$ [G,C] $\gg$ A at both positions in which thymine is found in the canonical sequence and C $\gg$ T > A $\gg$ G at the right-hand end of the hexanucleotide.

These hierarchies can be correlated with hydrogen bonds. The theoretical foundation for this analysis is the work of Seeman et al., who showed that if a recognition protein utilized hydrogen bonds for specificity, two hydrogen bonds per base pair would be required to uniquely discriminate between all the bases (54). A single hydrogen bond to a base would result in degeneracies in that two different bases would be recognized. The particular pair of degenerate bases depends on the particular hydrogen bond, as summarized in Table III. These ideas have been applied to restriction enzymes which contain degeneracies in their recognition sequences (33,66,69).

The basic assumptions of the RI* hydrogen-bonding model are as follows: (1) There are two hydrogen bonds between EcoRI endonuclease and each purine in the hexanucleotide, as indicated by the X-ray structure. (2) Under RI* conditions, one or more of these base–protein hydrogen bonds are randomly replaced by hydrogen bonds to water molecules, reducing the recognition specificity. (3) The rate of hydrolysis is correlated with the total number of hydrogen bonds between the enzyme and the bases, i.e., each RI* sequence can be positioned within its appropriate hierarchy by simply counting base–protein hydrogen bonds.

Two protein–guanine hydrogen bonds (see Fig. 10) correlate with the observed hierarchy at the left end of the hexanucleotide: they join arginine 200 to N-7 and O-6 of the guanine. These allow complete discrimination between guanine and other bases. However, if the hydrogen bond to O-6 is "lost" (replaced by hydrogen bonds to water), the remaining bond to N-7 is unable to discriminate between adenine

TABLE III
Single Hydrogen Bond Degeneracies

Bases Equal	Symbol[a]	Compliment[b]	Groove	Hydrogen bonds
A and G	Pu	Py	Major	N-7 (purine)
T and C	Py	Pu	Major	N-7 (purine')[c]
A and C	Ac	Gt	Major	N-6(A) and N-4(C) or O-4(T') and O-6(G')
G and T	Gt	Ac	Major	O-4(T) and O-6(G) or N-6(A') and N-4(C')
A and T	At	At	Minor	N-3(A) and O-2(T)[d]
G and C	Gc	Gc	Minor	N-2(G)

[a] The two are indicated with the purine in capitals and the pyrimidine in lower-case (e.g., Ac represents the simultaneous recognition of adenine and cytosine), except for the purine (Pu) and pyrimidine (Py) degeneracies.

[b] Symbol of degeneracy on complementary strand of DNA.

[c] Purine', C', etc., refers to the purine, C, etc., on the complementary strand. [A degeneracy on one strand of DNA automatically implies another degeneracy on the complementary strand. Hence, degeneracies occur in complementary pairs. This complementarity restores twofold recognition symmetry, which appears to be violated by many of those restriction enzymes which normally recognize degenerate sequences (33).]

[d] An additional contact to exclude N-2(G) is necessary.

and guanine (because both possess identical N-7 atoms), i.e., the Pu degeneracy occurs. If the bond to N-7 is lost while that to O-6 is retained, the enzyme is unable to discriminate between guanine and thymine, i.e., the Gt degeneracy occurs. (The O-4 of thymine occupies a three-dimensional position in the DNA very close to that occupied by O-6 of guanine, so either could accept this hydrogen bond.) Both hydrogen bonds would have to be lost before a cytosine could be accepted at the left-hand end of the hexanucleotide.

The position of an RI* sequence within the hierarchy is simply determined by the number of hydrogen bonds between the protein and the bases: two with guanine (plus ten with the other five base pairs, for a total of twelve hydrogen bonds), one with either adenine or thymine (plus the additional ten for a total of eleven hydrogen bonds), and none between the protein and cytosine (for a total of ten). Thus, our model clearly predicts the observed hierarchy G $\gg$ [A,T] $\gg$ C. Secondary interactions would be expected to order those bases with an equal number of hydrogen bonds (A and T in this case, although there is a relatively small difference in the rate for adenine versus that for thymine).

The hierarchies at the second and third base positions (which are both adenine in the canonical RI site) follow similarly. Recall that the structure showed hydrogen bonds between the protein and the N-7 and N-6 moieties of both adenine residues. Loss of a hydrogen bond at N-6 leads to Pu (failure to discriminate between purines) as before. Similarly, loss of a hydrogen bond at N-7 results in the Ac degeneracy. (Because N-4 of cytosine and N-6 of adenine occupy essentially the same positions.) Thus A, G, C, and T would have two, one, one, and zero protein–base hydrogen bonds, respectively. We thereby obtain the hierarchy $A \gg [G,C] \gg T$, which is what is actually observed.

From a mechanistic viewpoint, the problem is not to "explain" the EcoRI* activity, rather it is to understand its absence under physiological conditions. The hierarchical spectrum of EcoRI* sites is just what one should expect from a recognition mechanism based solely on hydrogen bonds (that form more or less independently of each other). Loss of a single hydrogen bond would be expected to reduce the interaction energy by 1 to 4 kcal. The resulting reduction in association constant or catalytic rate constant would be one or two orders of magnitude, just what is observed in the EcoRI* hierarchies. However, under physiological conditions, there is no detectable activity at RI* sites.

The mystery is compounded by the observation that the "cleavage" specificity of EcoRI endonuclease is higher than the "binding" specificity. Halford and Johnson have shown that the relative cleavage rate of EcoRI endonuclease is more than seven orders of magnitude faster at specific versus nonspecific sites (32). This is probably a very conservative estimate of the ability of this enzyme to discriminate between DNA sequences. Even so, it is substantially higher than the discrimination by binding as noted by Jack et al., who reported that the relative binding affinities of the protein for specific and nonspecific DNA differ by 10^5. [These authors report specific and nonspecific K_D values of 10^{-11} and 10^{-6} M^{-1}, respectively (34).] These data suggest strongly that the enzyme will bind tightly to nonspecific DNA sequences, but it will not cleave them even in the presence of Mg^{2+}. Furthermore, it should be noted that EcoRI endonuclease is under very strong selective pressure to be extraordinarily precise in its cleavage—a single erroneous cleavage of host DNA would be a lethal event. That is, even though the enzyme does have a higher binding affinity for the canonical EcoRI site, the differential specificity cannot be explained via simple binding constant differentials. What is the basis of this incremental specificity and how does the change to EcoRI* buffer conditions allow cleavage at additional sites?

Our current working hypothesis is that there are conformational changes in the endonuclease (and DNA) such that functional recognition and cleavage sites are formed in an obligate temporal order. In the structure described above, we noted that although there were many DNA–protein interactions at the recognition interface, the cleavage site was not fully assembled. Furthermore, we hypothesize that there is physical coupling between the individual components of the DNA–protein interaction. As a result, the conformational change between the initial (inactive) and final (active) forms assumes cooperative properties. By this we mean that the enzyme retains the inactive conformation under physiological conditions until virtually all the sequence-specific DNA–protein interactions have formed. This allows relatively subtle effects to dramatically alter the relative population of these two states. Furthermore, it is not unreasonable to argue that EcoRI* buffer conditions relax the "cooperativity" of the transition and/or alter its point of "onset," i.e., the number of sequence-specific DNA–protein interactions required for an appreciable fraction of the population to assume the active form.

CONCLUSIONS

EcoRI endonuclease specifically binds the canonical sequence, GAATTC, through DNA–protein interactions in the major groove of the DNA. The minor groove of the canonical sequence is not directly involved in sequence specificity. Upon binding to the endonuclease, the DNA adopts new (neo) conformational states not previously seen in protein-free DNA. We have defined a neoconformation as a structural distortion which is imposed on DNA by a binding protein and which is not seen in the absence of protein. Two neokinks have been observed in this structure, the type I neokink and the type II neokink. The type I neokink unwinds the DNA by 25° and renders the major groove accessible to the protein.

EcoRI endonuclease is a dimer with identical, symmetry-related subunits. Each subunit is an α/β domain with an extension called the "arm." The α/β domain is organized into topological motifs which have identifiable functional roles. The three-stranded parallel β-motif is associated with sequence recognition and the subunit interface. The three-stranded antiparallel motif of the β-sheet is associated with phosphodiester bond cleavage. The two segments overlap to form a five-stranded β-sheet. The "arms" which embrace the DNA and clamp it into place are based on a subsidiary three-stranded antiparallel sheet.

DNA sequence recognition is broken down into modular elements based on the crossover α-helices in the α/β domains. The inner recognition module which recognizes the sequence AATT consists of two symmetry-related α-helices, one from each subunit. This inner recognition module appears to hydrogen bond with the adenines in the four adjacent AT base pairs. The outer base pairs at either end of the canonical sequence are recognized by the two outer recognition modules. Each outer module consists of one α-helix which appears to hydrogen bond to the guanines.

REFERENCES

1. Anderson, W. F., Ohlendorf, D. H., Takeda, Y., and Matthews, B. W. (1981). *Nature (London)* **290**, 754–758.
2. Anderson, W. F., Takeda, Y., Ohlendorf, D. H., and Matthews, B. W. (1982). *J. Mol. Biol.* **159**, 745–751.
3. Pabo, C. O., and Lewis, M. (1982). *Nature (London)* **298**, 443–447.
4. McKay, D. B., and Steitz, T. A. (1981). *Nature (London)* **290**, 744–749.
5. Steitz, T. A., Ohlendorf, D. H., McKay, D. B., Anderson, W. F., and Matthews, B. W. (1982). *Proc. Natl. Acad. Sci. U.S.A.* **79**, 3097–3100.
6. Ohlendorf, D. H., Anderson, W. F., Fisher, R. G., Takeda, Y., and Matthews, B. W. (1982). *Nature (London)* **298**, 718–723.
7. Sauer, R. T., Yocum, R. R., Doolittle, R. F., Lewis, M., and Pabo, C. O. (1982). *Nature (London)* **298**, 447–451.
8. Anderson, J., Ptashne, M., and Harrison, S. C. (1984). *Proc. Natl. Acad. Sci. U.S.A.* **81**, 1307–1311.
9. Anderson, J. E., Ptashne, M., and Harrison, S. C. (1985). *Nature (London)* **316**, 596–601.
10. Bushman, F. D., Anderson, J. E., Harrison, S. C., and Ptashne, M. (1985). *Nature (London)* **316**, 651–653.
11. Ollis, D. L., Brick, P., Hamlin, R., Xuong, N. G., and Steitz, T. A. (1985). *Nature (London)* **313**, 762–766.
12. Richmond, T. J., Finch, J. T., Rushton, B., Rhodes, D., and Klug, A. (1984). *Nature (London)* **311**, 532–537.
13. Bentley, G. A., and Lewit-Bentley, A (1984). *J. Mol. Biol.* **176**, 55–75.
14. Burlingame, R. W., Love, W. E., Wang, B.-C., Hamlin, R., Xuong, N. H., and Moudrianankis, E. N. (1985). *Science* **228**, 546–553.
15. Tanaka, I., Appelt, K., Dij, K. L., White, S. W., and Wilson, K. S. (1984). *Nature (London)* **310**, 376–381.
16. Youderian, P., Vershon, A., Bouvier, S., Sauer, R. T., and Susskind, M. M. (1983). *Cell (Cambridge, Mass.)* **35**, 779–783.

17. Rich, A., Seeman, N. C., and Rosenberg, J. M. (1977). In "Nucleic Acid–Protein Recognition" (H. J. Vogel, ed.), pp. 361–374. Academic Press, New York.
18. Rosenberg, J. M., Seeman, N. C., Kim, J. P., Suddath, F. L., Nicholas, H. B., and Rich, A. (1973). Nature (London) **243**, 150–154.
19. Seeman, N. C., Rosenberg, J. M., Suddath, F. L., Kim, J. P., and Rich, A. (1976). J. Mol. Biol. **104**, 109–144.
20. Dickerson, R. E., and Drew, H. R. (1981). J. Mol. Biol. **149**, 761–786.
21. Dickerson, R. E., and Drew, H. R. (1981). Proc. Natl. Acad. Sci. U.S.A. **78**, 7318–7322.
22. Dickerson, R. E. (1983). J. Mol. Biol. **166**, 419–441.
23. Frederick, C. A., Grable, J., Melia, M., Samudzi, C., Jen-Jacobson, L., Wang, B.-C., Greene, P. J., Boyer, H. W., and Rosenberg, J. M. (1984). Nature (London) **309**, 327–331.
24. McClarin, J. A., Frederick, C. A., Grable, J., Samudzi, C. T., Wang, B.-C., Greene, P., Boyer, H. W., and Rosenberg, J. M. (1986). Proc. Conf. Conversation Biomol. Stereodyn., 4th, 45–68.
25. Rosenberg, J. M., McClarin, J. A., Frederick, C. A., Wang, B.-C., Boyer, H. B., and Greene, P. (1986). Chemica Scripta **263**, 147–157.
26. Greene, P. J., Gupta, M., Boyer., H. W., Brown, W. E., and Rosenberg, J. M. (1981). J. Biol. Chem. **256**, 2143–2153.
27. Newman, A. K., Rubin, R. A., Kim, S.-H., and Modrich, P. (1981). J. Biol Chem. **256**, 2131–2139.
28. Connolly, B. A., Eckstein, F., and Pigoud, A. (1984). J. Biol. Chem. **259**, 10760–10763.
29. Modrich, P., and Zabel, D. (1976). J. Biol. Chem. **251**, 5866–5874.
30. Jen-Jacobson, L., Kurpiewski, M., Lesser, D., Grable, J., Boyer, H. W., Rosenberg, J. M., and Greene, P. J. (1983). J. Biol. Chem. **258**, 14638–14646.
31. Modrich, P. (1979). Q. Rev. Biophys. **12**, 315–369.
32. Halford, S. E., and Johnson, N. P. (1980). Biochem. J. **191**, 593–604.
33. Rosenberg, J. M., Boyer, H. W., and Greene, P. J. (1981). In "Gene Amplification and Analysis" (J. G. Chirikjian, ed.), Vol. 1, pp. 131–164. Elsevier/North-Holland, New York.
34. Jack, W. E., Rubin, R. A., Newman, A., and Modrich, P. (1981). In "Gene Amplification and Analysis" (J. G. Chirikjian, ed.), Vol. 1, pp. 165–179. Elsevier/North-Holland, New York.
35. Grable, J., Frederick, C. A., Samudzi, C., Jen-Jacobson, L., Lesser, D., Greene, P. J., Boyer, H. W., Itakura, K., and Rosenberg, J. M. (1984). J. Biomol. Struct. Dyn. **1**, 1149–1160.
36. Wang, B.-C. (1985). In "Methods in Enzymology" (H. Wyckoff, C. H. W. Hirs, and S. N. Timasheff, eds.), Vol. 115, pp. 90–112. Academic Press, New York.
37. Rossmann, M. G. (1979). J. Appl. Crystallogr. **12**, 225–238.
38. Rossmann, M. G., Leslie, A. G. W., Abdel-Meguid, S. S., and Tsukihara, T. (1979). J. Appl. Crystallogr. **12**, 570–581.
39. Jones, T. A. (1978). J. Appl. Crystallogr. **11**, 268–272,
40. Jones, T. A. (1982). In "Computational Crystallography" (D. Sayre, ed.), pp. 303–317. Oxford Univ. Press (Clarendon), London and New York.
41. Pflugrath, J. W., Saper, M. A., and Quiocho, F. A., (1983). Pap. Int. Summer Sch. Crystallogr. Comput., 1983.
42. Rossmann, M. G., Liljas, A., Branden, C. I., and Banaszak, L. J. (1975). In "The Enzymes" (P. D. Boyer, ed.); Vol. 11, pp. 61–102. Academic Press, New York.

43. Richardson, J. S. (1981). *Adv. Protein Chem.* **34**, 167–339.
44. Hol, W. G. S. (1985). *Prog. Biophys. Mol. Biol.* **45**, 149–195.
45. Chothia, C. (1973). *J. Mol. Biol.* **75**, 295–302.
46. Quiocho, F. A., Gilliland, G. L., and Philligs, G. N. (1977). *J. Biol. Chem.* **252**, 5142–5149.
47. Shaw, P. S., and Muirhead, H. (1977). *J. Mol. Biol.* **109**, 475–485.
48. Weatherford, D. W., and Salemme, F. R. (1979). *Proc. Natl. Acad. Sci. U.S.A.* **76**, 19–23.
49. Schulz, G. E., Elzinga, M., Marx, F., and Schirmer, R. H. (1974). *Nature (London)* **250**, 120–123.
49a. Jen-Jacobson, L. *et al.* (1986). *Cell* **45**, 619.
50. Kim, R., Modrich, P., and Kim, S.-H. (1984). *Nucleic Acids Res.* **12**, 7285–7292.
51. Crick, F. H. C., and Klug, A. (1975). *Nature (London)* **255**, 530–533.
52. Rosenberg, J. M., Seeman, N. C., Day, R. O., and Rich, A. (1976). *Biochem. Biophys. Res. Commun.* **69**, 979–987.
53. Richmond, T. J., Finch, J. T., Rushton, B., Rhodes, D., and Klug, A. (1985). *Nature (London)* **311**, 532–537.
54. Seeman, N. C., Rosenberg, J. M., and Rich, A. (1976). *Proc. Natl. Acad. Sci. U.S.A.* **73**, 804–808.
55. Thomas, M., and Davis, R. W. (1975). *J. Mol. Biol.* **91**, 315–328.
56. Alves, J., Pingoud, A., Haupt, W., Langowski, J., Peters, F., Maass, G., and Wolff, C. (1984). *Eur. J. Biochem.* **140**, 83–92.
57. Jack, W. E., Terry, B. J., and Modrich, P. (1982). *Proc. Natl. Acad. Sci. U.S.A.* **79**, 4010–4014.
57a. Picone, J. (1987). In preparation.
58. Lillehaug, J. R., Kleppe, R. K., and Kleppe, K. (1976). *Biochemistry* **15**, 1858–1865.
59. Greene, P. J., Poonian, M. S., Nussbaum, A. L., Tobias, L., Garfin, D. E., Boyer, H. W., and Goodman, H. M. (1975). *J. Mol. Biol.* **99**, 237–261.
60. Lu, A-L., Jack, W. E., and Modrich, P. (1981). *J. Biol. Chem.* **256**, 13200–13206.
61. Polisky, B., Greene, P., Garfin, D. E., McCarthy, B. J., Goodman, H. M., and Boyer, H. W. (1975). *Proc. Natl. Acad. Sci. U.S.A.* **72**, 3310–3314.
62. Hsu, M., and Berg, P. (1978). *Biochemistry* **17**, 131–138.
63. Woodbury, C. P., Jr., Downey, R. L., and von Hippel, P. H. (1980). *J. Biol. Chem.* **255**, 11526–11533.
64. Malyguine, E., Vannier, P., and Yot, P. (1980). *Gene* **8**, 163–177.
65. Woodhead, J. L., Bhave, N., and Malcolm, A. D. B. (1981). *Eur. J. Biochem.* **115**, 293–296.
66. Rosenberg, J. M., and Greene, P. J. (1982). *DNA* **1**, 117–124.
67. Gardner, R. C., Howarth, A. J., Messing, J., and Shepherd, R. J. (1982). *DNA* **1**, 109–115.
68. Goodman, H. M., Greene, P. J., Garfin, D. E., and Boyer, H. W. (1977). *In* "Nucleic Acid–Protein Recognition" (H. J. Vogel, ed.), pp. 239–259. Academic Press, New York.
69. Smith, H. O. (1979). *Science* **205**, 455–462.

3

The Klenow Fragment Structure Suggests Mechanisms for Fidelity and Processivity of DNA Polymerase I

T. A. STEITZ
Department of Molecular Biophysics and Biochemistry
Yale University
New Haven, Connecticut 06511

INTRODUCTION

For more than 25 years a central question in molecular biology has concerned the mechanism of DNA replication by DNA polymerases. Among the many issues of interest to us are (1) the role that the enzyme plays in assuring the fidelity of template-directed DNA synthesis, (2) the mechanism of the enzyme's processivity, i.e., the successive incorporation of nucleotides without enzyme dissociation, and (3) the relationship between polymerization and editing activities. DNA polymerase I (Pol I) of *Escherichia coli* is the first template-directed DNA synthesizing enzyme for which high-resolution structural information has been obtained (1), and, as such, constitutes an excellent model for investigating the molecular details of replication. Pol I has a molecular weight of 103,000 and has three enzymatic activities: DNA polymerase, a 3'-5' exonuclease thought to edit out mismatched terminal nucleotides, and a 5'-3' exonuclease that removes DNA ahead of the growing point of a DNA chain (2). Pol I has separate binding sites for deoxynucleoside monophosphate and deoxynucleoside triphosphate; the binding of one does not affect the binding of the other. Limited proteolysis removes the 35,000-Da N-terminus domain that contains the 5'-3' exonuclease activity (3,4). The remaining 68,000-Da fragment (Klenow fragment) has the polymerization and editing activities.

High-resolution structural analyses of polymerases have not been possible until recently because of the difficulty of obtaining adequate quantities of material. For Pol I this problem has been solved by cloning the portion of the structural gene that codes for the Klenow fragment into an expression vector (5). Overproduction of the fragment has enabled us to obtain large quantities of protein rapidly using few purification steps and has eliminated the possibility of heterogeneity introduced by protease treatment of the intact molecule. The availability of large quantities of homogeneous protein has facilitated the crystallization and determination of the structure of Klenow fragment.

A combination of structural, biochemical, and genetic studies has led to the conclusion that Pol I has three domains, each responsible for a separate enzymatic activity (Fig. 1). As described below, the Klenow fragment structure has two domains, which appear to correspond to distinct polymerization and editing functions. The separation of these two active sites plus the structure of the DNA binding site which appears able to surround the duplex product of DNA syn-

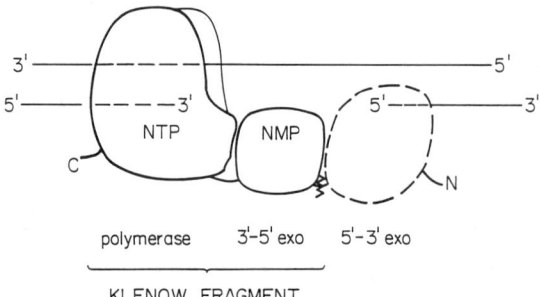

Fig. 1. A schematic drawing showing the apparent domain structure of *E. coli* DNA Pol I and the enzymatic activities associated with each domain. The solid lines show domains seen in the Klenow fragment structure, which has binding sites for deoxynucleoside monophosphate (dNMP) and deoxynucleoside triphosphate (dNTP). The small fragment produced by proteolytic cleavage of Pol I is shown in dashed lines and consists of at least one domain. The guessed domain locations of the 3' and 5' termini of a nicked DNA substrate are indicated, but their relative separation is schematic.

thesis suggest mechanisms for fidelity and processivity in this polymerase.

STRUCTURE OF THE KLENOW FRAGMENT

A model of the Klenow fragment was built by fitting its amino acid sequence (6) into a 3.3-Å-resolution electron density map using an Evans and Sutherland PS300 color graphics unit and the computer program FRODO (7). The coordinates are being refined at 2.7 Å resolution and yield a crystallographic R-factor that is currently 0.25 (L. Beese and T. Steitz, unpublished). The structure (Fig. 2) shows that the 605-amino acid polypeptide is folded into two distinct structural domains of approximately 200 and 400 amino acids.

The N-terminus one-third of the Klenow fragment [residues 324–517 in the Pol I sequence (6)] forms the smaller of the two domains.

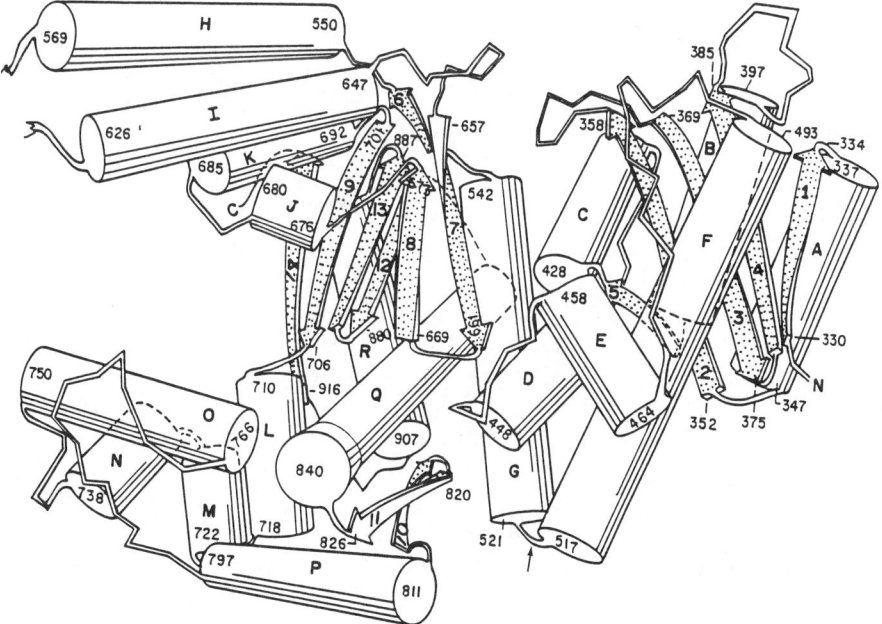

Fig. 2. Tertiary structure of the Klenow fragment. Alpha-helices are represented by tubes (lettered) and β-sheets by arrows (numbered). The broken lines on the strands between helix H and helix I shows the position of the approximately 50-residue disordered subdomain. The large and small domains are separated by the loop between helices F (residue 517) and G (residue 521).

This domain has a central core of β-pleated sheet (mostly parallel) with α-helices on both sides. In the crystal, this domain can bind one molecule of dTMP, as well as divalent metal ions that interact with both the protein and the 5' phosphate of dTMP. One metal is bound to the protein by the carboxylate groups of Asp 355, Glu 357, and Asp 501, with the 5' phosphate providing a fourth ligand. A second metal appears to lie between Asp 424 and the 5' phosphate of the dNMP.

The 400 amino acids of the larger domain (521–920) form a structure that contains a very deep cleft that is about 20–24 Å wide and 25–35 Å deep. A six-stranded antiparallel β-sheet forms the bottom of the cleft and large protrusions of α-helix form its sides. The structure is similar in shape to a right hand grasping a rod. Thus, one side of the cleft forms a wall, 50 Å long, that can be compared with the curled over fingers of a right hand. The other side of the cleft is formed primarily by two long α-helices, I and H, projecting from the protein like a thumb. At the tip of the thumblike protrusion and hanging over the crevice are 50 amino acid residues (not shown in Fig. 2) that appear to be partially disordered in the crystal.

FUNCTION OF THE SMALL DOMAIN

The main reason for suggesting that the small domain of Klenow fragment has the 3'-5' exonuclease activity is the location of the dNMP binding site on this domain (1). dNMP inhibits (8) the 3'-5' exonuclease reaction (presumably by product inhibition) and therefore probably marks the position of the DNA 3' terminus in the exonuclease active site. dNMP fails to inhibit polymerization, supporting the idea that the polymerase and exonuclease active sites are distinct from one another. Further circumstantial evidence for this location of the exonuclease active site is provided by the protein sequence homology between the small domain and epsilon, the 3'-5' exonuclease subunit of DNA polymerase III (9). The homologies, though weak, involve regions of the protein that surround the dNMP site, and residues that interact with the dNMP tend to be conserved (9). In equating the dNMP binding site with the 3'-5' exonuclease active site, we are assuming that the dNMP site observed crystallographically is the identical site responsible for the inhibition observed *in vitro*. The interactions between the dNMP molecule and the protein support this assumption. In particular, efficient inhibition of the exonuclease reaction requires a free 3' hydroxyl group (8); the structure shows that this hydroxyl interacts with the protein, forming a hydrogen bond to the backbone amide of Thr 358. This interaction also makes biological sense if we assume that the dNMP molecule marks the position of the DNA 3' terminus: a buried 3' hydroxyl group is compatible with the exonuclease reaction but would be inappropriate for a DNA 3' terminus at the polymerase active site.

FUNCTION OF THE LARGE DOMAIN

Model building suggests that the cleft in the large domain can bind double-stranded B-DNA (Fig. 3). The J and K α-helices are placed partially into a major groove and may function to fix the translational position of the DNA on the protein, so that movement of the polymerase would require the protein to spiral along the DNA. The location of the small, disordered 50-residue subdomain above the cleft suggests that it might bind to the DNA in the complex allowing the protein to completely surround the DNA substrate. This model for the

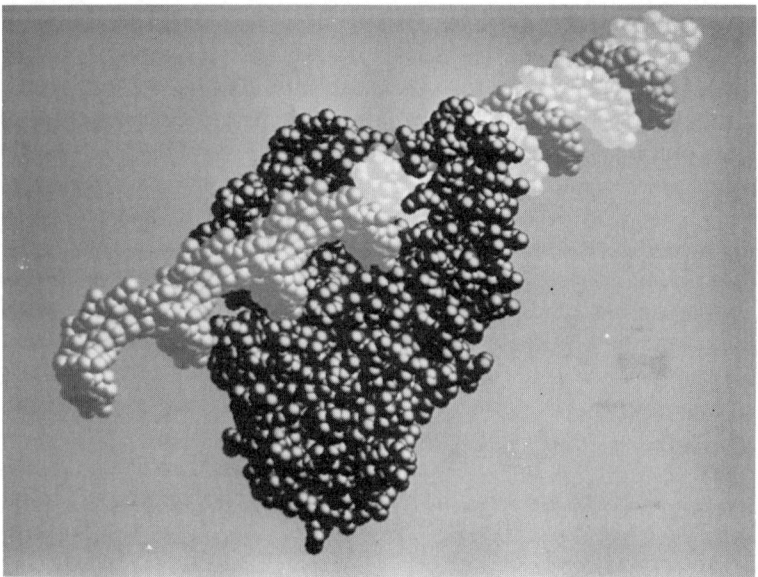

Fig. 3. Space-filling model of the Klenow fragment bound to a DNA substrate. The placement of the DNA is achieved by model building, whereas the nucleoside monophosphate position is experimentally determined. Note that the minor groove of the duplex product of DNA synthesis adjacent to the putative 3' terminus is facing the top of the cleft where the 50-residue disordered domain is located.

location of the DNA binding site is strongly supported by theoretical calculations of the electrostatic surface potential (10). Virtually all the positive electrostatic charge potential lies within the cleft and describes an approximately spiral path with a pitch of about 34 Å. The rest of the protein surface has negative electrostatic charge potential, as would be expected for an acidic protein. Additional evidence is provided by genetics. Two mutations, *polA5* and *polA6*, which affect the enzyme's interaction with DNA, are both in amino acids within the proposed binding cleft (1,11). DNA footprinting with DNAse I has been used to measure the length of DNA covered by Klenow fragment bound at a primer terminus (12,12a). Combining this length measurement with the model building described above has allowed placement of a probable location for the primer terminus and thus the polymerase active site (Fig. 3).

An important conclusion from this work is that the polymerase active site is likely to be located on the large domain; there is no way the primer terminus can reach the small domain unless the protein undergoes a large conformational change on binding DNA. To confirm that the large domain contains the polymerase active site, the DNA encoding the large domain has been cloned in a high-expression vector and the protein product purified. The purified large domain has DNA polymerase activity (although at about 50-fold lower specific activity than the intact Klenow fragment), but does not show any measurable 3'-5' exonuclease activity (12b).

A more precise indication of the polymerase active site location is provided by photoaffinity labeling. Rush and Konigsberg have cross-linked the dNTP analogue, 8-azido-dATP, to Klenow fragment and have isolated and sequenced the labeled peptide (12). The site of cross-linking has been tentatively identified as Tyr 776, located at the end of helix O, pointing toward the proposed DNA binding cleft. Allowing for some uncertainties in the interpretation of the data, both the photoaffinity labeling and the footprinting data suggest that the polymerase active site lies somewhere in the vicinity of the N-terminus part of helix Q. It is not excluded that the 50 residue disordered subdomain may form some or all of the polymerase active site.

CO-CRYSTALLIZATION OF KLENOW FRAGMENT WITH DNA

In order to determine the location and nature of the polymerase active site (or perhaps even to form it) we must, of course, determine the crystal structures of the binary complex between Klenow fragment and DNA as well as the ternary complex with both the DNA and dNTP substrate. Crystals now have been grown of the Klenow fragment complexed with an eight base pair duplex DNA containing a three base, single stranded 5' overhang (P. Freemont, J. Friedman, and T. Steitz, unpublished). These crystals diffract to better than 2.8 Å in all directions and have the same space group and unit cell dimensions as structure that is solved but show a change in the diffraction intensities. HPLC analysis of the crystals shows that one molecule of duplex DNA is bound per molecule of protein. Thus, the structure of the binary complex will be possible by measuring one high resolution data set from the complex crystals and calculating a difference electron density map. A similar approach should work with complexes containing a mismatched base pair.

RELATION TO OTHER POLYMERASES

Extensive amino acid sequence homologies have been found between the large domain of Klenow fragment and phage T7 DNA polymerase (13). The strongest homologies involve residues that form the putative DNA binding cleft. This is consistent with the notion that both Pol I and T7 DNA polymerase evolved from a common ancestor and that the cleft shares a common and important function, presumably DNA binding and polymerization activity. No sequence homologies between Pol I or T7 DNA polymerase and the DNA polymerases from adenovirus, Epstein Barr virus, and phage T4 have been seen (D. Ollis, and T.A. Steitz, unpublished observation).

PROCESSIVITY

Pol I, like other polymerases, is a processive enzyme, that is, it incorporates about 20 nucleotides into a growing DNA chain before dissociating (14). The structure of the Klenow fragment suggests a mechanism for this processivity (1). The 50-residue subdomain that is flexibly attached to the tip of the I and H helices could close off the cleft after DNA binds. If the protein does envelop the DNA substrate, the rate of dissociation of the DNA product may be very slow since it would require a protein conformational change. If the dissociation rate is slower than the rate of polymerization, processive polymerization would result. Instead of dissociating from the product, the enzyme would simply slide to the new 3' terminus between polymerization steps.

FIDELITY OF DNA SYNTHESIS

Clearly, it is of utmost importance that DNA polymerases make as few errors as possible in copying DNA. *In vitro* experiments indicate that Pol I has an error rate of around 1 in 10^6 when copying a natural DNA template (15). How is high fidelity achieved? The enzyme must play a role in enhancing fidelity both at the step of incorporating dNTP and by editing out incorrectly incorporated base pairs. It appears that the 3'-5' exonuclease activity functions to excise incorrectly incorporated bases (2). However, for this editing activity to take place the formation of a mismatched base pair must first be detected at the polymerase active site since the polymerase and exonuclease active sites appear to be separated by 20–30 Å. How is this done? A detailed answer to this question must await the determination of the structure of the Klenow fragment complexed to DNA containing a mismatched base pair at the primer terminus; for the present, the structure of the native enzyme provides some clues as to how this might be achieved. At the present time, the structure provides no insight into how the enzyme enhances the incorporation of correct dNTP's.

As we described above, the binding cleft, together with the flexible subdomain, could allow the enzyme to completely surround its DNA substrate, forming a tight orifice through which the duplex product of DNA synthesis must pass. Thus the enzyme could, in principle, detect the mismatched base pair by scanning either the sugar–phosphate backbone, the major groove, or the minor groove.

It is proposed that the enzyme contains a "reading head" that can detect mismatched base pairs in the minor groove. Pol I cannot be scanning the major groove since substitutions at the 5 position of pyrimidine do not affect polymerization (16). The enzyme is not detecting mismatched base pairs by scanning the DNA backbone structure since the crystal structures of two mismatched base pairs in a B-DNA duplex T·G (17) and A·G (18) show little deviation in the sugar–phosphate backbone conformation. However, there are substantial changes, for example, in the locations of the guanine N-3 and thymine O-2 atoms in the minor groove (17,18). Since all four arrangements of the Watson–Crick base pairs present the same pattern of hydrogen bond acceptors in the minor groove (19), scanning in the minor groove could be sequence independent; only the mismatched base pair would be different. Thus, an enzyme "reading head" scanning the minor groove could detect the mismatched base pair and reduce the rate of translocation of the newly formed 3' terminus from the triphosphate site to the primer terminus site (Fig. 4).

That the incorporation of a mismatched base pair probably stops or greatly slows the translocation step (Fig. 4) is concluded from the following considerations. Since the crystal structures of a duplex DNA containing AG and GT base pairs show identical conformations for the sugar–phosphate backbone of the mismatched base pairs, it follows

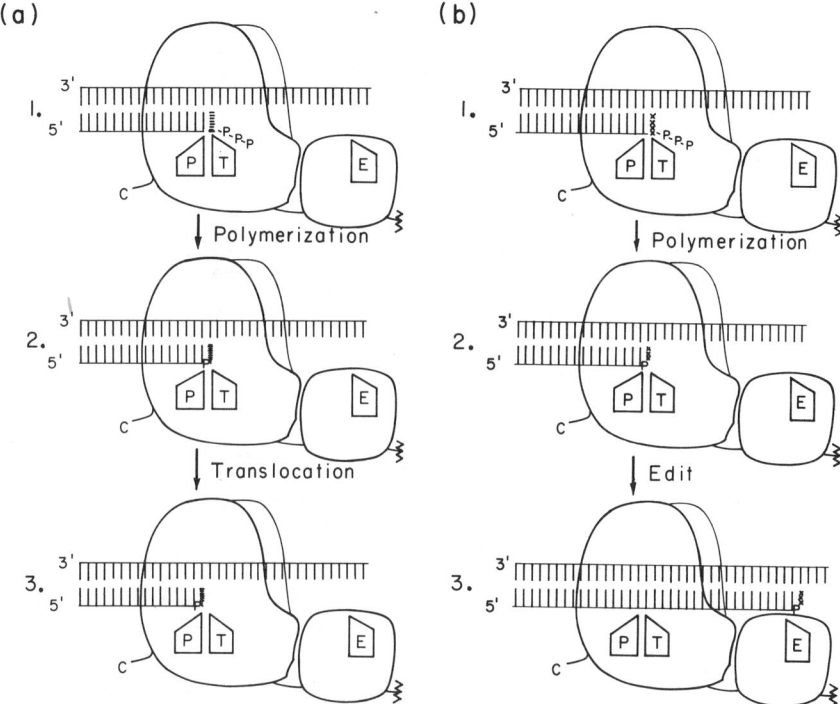

Fig. 4. Schematic drawing of proposed processive polymerization reaction and excision of mismatched base pair. (a) The binding site for the nucleoside triphosphate (T) and the 3' primer terminus (P) lies on the large domain of the Klenow fragment whereas the 3'-5' editing exonuclease site (E) lies on the small domain. After polymerization, a correctly incorporated nucleotide translocates by sliding to the P site. Processivity in the enzyme presumably results from the rate of sliding between the T and P sites being faster than dissociation, because of the protein surrounding the DNA. (b) Incorporation of a mismatched base pair is proposed to prevent translocation of the new primer terminus from the T site to the P site. Excision of the mismatched base at the E site requires sliding of the DNA terminus to the E site.

that the 3' OH of such a mismatched base pair at the primer terminus would be in the same place as with a correct base pair. Thus, if this mismatched base pair were translocated from the triphosphate site to the primer terminus site, there might be little or nothing to prevent a subsequent polymerization step. (More complex mechanisms involving a conformational interaction between the "reading head" and the correctly assembled catalytic site also are possible.)

Removal of the mismatched base pair could be acheived by pyrophosphorolysis at the triphosphate side (reversal of the polymerase step) or, more likely, removal of the mismatched nucleoside monophosphate at the 3'-5' exonuclease site. Given the 20- to 30-Å separation between the proposed polymerase and exonuclease active sites in the native structure, editing of a mismatched base pair would require substantial movement of the DNA to the 3'-5' exonuclease site. Since the cocrystals with DNA are isomorphous with the solved structure, it is unlikely that the 3'-5' exonuclease domain moves relative to the polymerase domain when DNA is bound; in any case, it is not sterically possible to move the monophosphate binding site (3'-5' exonuclease active site) to the same location as the presumed polymerase active site. Thus, for excision of the mismatched base pair to occur, the DNA must slide from the polymerase to the exonuclease active site. Presumably, with correctly incorporated nucleotides, the rates of translocation and polymerization are faster than the sliding of DNA to the exonuclease site, whereas with mismatched base pairs the rate of translocation is so slow that sliding to the exonuclease site followed by editing hydrolysis is the preferred pathway.

This steric blocking model for polymerase fidelity makes a testable prediction, namely, that a mismatched base at the 3' terminus of a DNA substrate will bind to the triphosphate site of the large (polymerase) domain of the Klenow fragment. If true it would then block the binding of nucleoside triphosphates to that site. Such experiments to test the model are in progress.

ACKNOWLEDGMENTS

I wish to thank and acknowledge my collaborators on the research reviewed here: Lorena Beese, Peter Brick, Paul Freemont, Jonathan Friedman, Nigel Grindley, William Konigsberg, Connie Kline, Cathy Joyce, David Ollis, John Rush, and Jim Warwicker. Research was supported by American Cancer Grant NP-421 and USPHS Grant GM-22778.

REFERENCES

1. Ollis, D. L., Brick, P., Hamlin, R., Xuong, N. G., and Steitz, T. A. (1985). *Nature (London)* **313**, 762–766.
2. Kornberg, A. (1980). "DNA Replication." Freeman, San Francisco, California.
3. Klenow, H., and Henningsen, I. (1970). *Proc. Natl. Acad. Sci. U.S.A.* **65**, 168–175.
4. Brutlag, D., Atkinson, M. R., Setlow, P., and Kornberg, A. (1968). *Biochem. Biophys. Res. Commun.* **37**, 982–989.
5. Joyce, C. M., and Grindley, N. D. F. (1983). *Proc. Natl. Acad. Sci. U.S.A.* **80**, 1830–1834.
6. Joyce, C. M., Kelley, W. S., and Grindley, N. D. F. (1982). *J. Biol. Chem.* **257**, 1958–1964.
7. Jones, A. T. (1978). *J. Appl. Crystallogr.* **11**, 268–272.
8. Que, B. G., Downey, K. M., and So, A. (1978). *Biochemistry* **17**, 1603–1606.
9. Scheuermann, R. H., and Echols, H. (1984). *Proc. Natl. Acad. Sci. U.S.A.* **81**, 7747–7751.
10. Warwicker, J., Ollis, D. L., Richards, F. M., and Steitz, T. A. (1985). *J. Mol. Biol.* **186**, 645–649.
11. Joyce, C. M., Fujii, D. N., Laks, H. F., Hughes, C. M., and Grindley, N. D. F. (1985). *J. Mol. Biol.* **186**, 283–297.
12. Joyce, C. M., Ollis, D. L., Rush, J., Steitz, T. A., Konigsberg, W. H., and Grindley, N. D. F. (1986). *In* "Protein Structure, Folding and Design" (D. Oxender, ed.), UCLA Symp. Mol. Cell. Biol. pp. 197–205. Alan R. Liss, Inc., New York.
12a. Joyce, C. M. (1987). In preparation.
12b. Fremont, P. S., Ollis, D. L., Steitz, T. A., and Joyce, C. M. (1986). *Proteins* **1**, 66–73.
13. Ollis, D. L., Kline, C., and Steitz, T. A. (1985). *Nature (London)* **313**, 818–819.
14. Bambara, R. A., Uyenura, D., and Choi, T. (1978). *J. Biol. Chem.* **253**, 413–423.
15. Kunkel, T. A., and Loeb, L. A. (1980). *J. Biol. Chem.* **255**, 9961–9966.
16. Dale, R. M. K., and Ward, D. C. (1975). *Biochemistry* **14**, 2458–2469.
17. Brown, T., Kennard, O., Kneale, G., and Rabinovich, D. (1985). *Nature (London)* **315**, 604–606.
18. Kennard, O. (1985). *J. Biomol. Struct. Dyn.* **3**, 205–226.
19. Seeman, N. C., Rosenberg, J. M., and Rich, A. (1976). *Proc. Natl. Acad. Sci. U.S.A.* **73**, 804–808.

4
Structural Studies of DNA–Protein Recognition

R. G. BRENNAN,* H. M. HOLDEN,*,† Y. TAKEDA,‡ AND
B. W. MATTHEWS*

Institute of Molecular Biology and Department of Physics
University of Oregon
Eugene, Oregon 97403
and
† Department of Biochemistry
University of Arizona
Tucson, Arizona 85721
and
‡ Chemistry Department
University of Maryland
Baltimore County
Catonsville, Maryland 21228

The structures of several proteins that regulate gene expression have been determined recently and suggest that these proteins interact with their specific DNA recognition sites by having α-helices of the protein penetrate the major grooves of the DNA, thereby forming a network of complementary hydrogen bonds between side chains of the protein and the exposed parts of the DNA base pairs. The proteins contain a substructure consisting of a helix-turn-helix unit that is virtually identical in each case. Structural and amino acid sequence comparisons suggest that this DNA-binding helix-turn-helix unit occurs in a number of proteins that regulate gene expression. As a step toward confirming these inferences concerning DNA–protein recognition, crystals have been obtained of a Cro repressor: operator complex and also of the cII gene activator protein from bacteriophage λ.

INTRODUCTION

The control of the genetic information encoded in DNA is of critical importance in all living systems. It has been known for some time that, at least in simple organisms, this control is achieved by proteins that recognize and bind to specific sites on the DNA. Until recently, little was known of the structures of these molecules or the way in which they recognize their target sites. However, within the past few years, the three-dimensional structures of three DNA-binding proteins ("Cro," "CAP," and "λ repressor") have been determined and have suggested how these proteins bind to their specific recognition sites on the DNA. As discussed elsewhere in this volume these results have recently been extended by the determinations of the structures of an endonuclease–DNA complex and a repressor–DNA complex.

In this chapter we briefly review the structures of Cro, CAP, and λ repressor and discuss the modes of DNA–protein interaction that are suggested. As will become apparent, there are similarities between the three proteins, but there are also striking differences as well. For additional background and more detailed information, reference can be made to the reviews of Ohlendorf and Matthews (1), Takeda et al. (2), and Pabo and Sauer (3).

THREE DNA-BINDING PROTEINS

Cro, λ repressor, and CAP are all dimeric DNA-binding proteins but have substantial differences in their overall structures. For Cro the 66-amino acid polypeptide chain forms a single domain whereas the respective polypeptide chains of CAP and λ repressor fold into two domains. In λ repressor, the amino-terminus domain binds to the DNA; in CAP it is the carboxyl-terminus part of the molecule that has this function.

A sketch of the structures of Cro (4), the amino-terminus domain of λ repressor (5), and the carboxyl-terminus domain of CAP (6) as determined from the respective crystal structures is shown in Fig. 1.

MODELS FOR DNA BINDING

The dimeric Cro protein displays a 34Å spacing between the two-fold-related α_3-helices, which, together with their angle of tilt (Fig. 1), strongly suggested that these α-helices bind within successive major

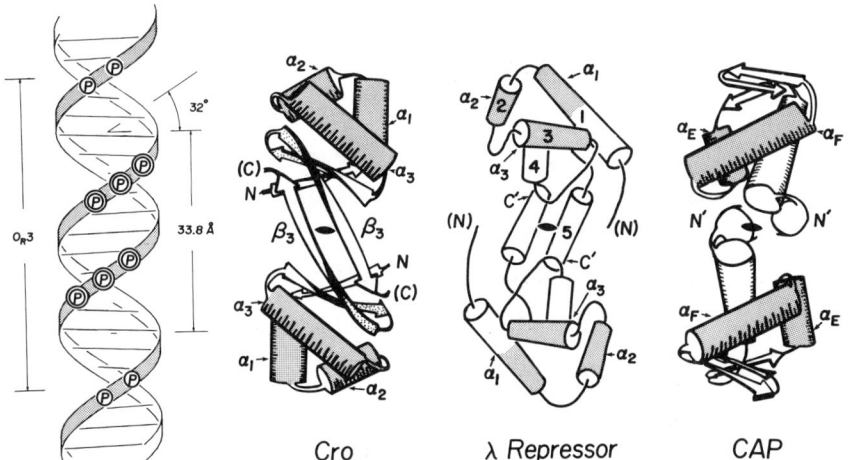

Fig. 1. Schematic drawing of a segment of Watson–Crick B-form DNA together with dimers of Cro, λ repressor amino-terminus domains, and CAP carboxyl-terminus domains viewed down their respective twofold symmetry axes. The corresponding α_2-α_3 (or α_E-α_F) helices are shaded. DNA phosphates whose ethylation interferes with binding of both λ repressor and Cro are indicated by the letter P within a double circle. Phosphates whose ethylation affects λ repressor (and also P22 repressor) binding, but not Cro, are indicated by a P in a single circle (after refs. 1,2).

grooves of right-handed Watson–Crick DNA as illustrated in Fig. 2. It is presumed that the flexible carboxyl-terminus residues of Cro participate in DNA binding by lying along the minor groove. A characteristic feature of the model is the match between the twofold symmetry of the protein and the (approximate) twofold sequence and spatial symmetry of the DNA binding site (4,7,8). Recognition of specific base sequences on the DNA is thought to be due in large part to a multiple network of specific hydrogen bonds between the side chains of the protein and the parts of the DNA base pairs exposed within the grooves of the DNA.

A similar mode of DNA binding has been proposed for λ repressor. Here also there is a pair of twofold-related α_3-helices (Fig. 1) that are presumed to bind within successive major grooves of Watson–Crick DNA. Furthermore, the amino-terminus residues of λ repressor form two long "arms" with flexible ends that "wrap around" the DNA when the protein binds. In this case the "arms" contact the major groove of the DNA (9). Current structural models (10,11) suggest that CAP also binds to DNA in a manner similar to that proposed for Cro and λ repressor.

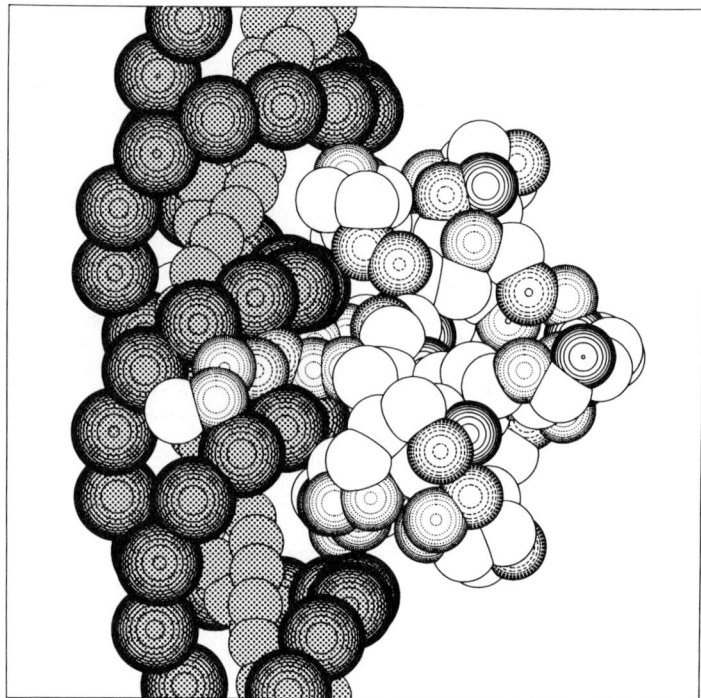

Fig. 2. Stylized drawing showing the complementarity between the structure of Cro repressor protein and DNA. In the presumed sequence-specific complex, the protein is assumed to move closer to the DNA, with the α_3-helices penetrating further into the major grooves of the DNA than is shown in the figure. The carboxyl-terminus residues of the protein are presumed to bind in the vicinity of the minor groove of the DNA. To maximize the contacts between Cro and DNA, the protein may undergo a hinge-bending motion and/or the DNA may bend (as shown), although these are not essential features of the model. The DNA is represented stylistically by large dotted spheres centered at the phosphate positions and small dotted spheres that follow the bottom of the major groove. In the protein, each residue is represented by a single sphere. Acidic residues have solid concentric circle shading, basic residues have broken circle shading, uncharged hydrophilic residues have dotted circle shading, and hydrophobic residues have no shading. Reproduced with permission from the *Annual Review of Biophysics and Bioengineering*, Vol. 12. © 1983 by Annual Reviews Inc.

A COMMON HELIX-TURN-HELIX DNA-BINDING MOTIF

Following the structure determinations of Cro, CAP, and λ repressor, it has become apparent that these three proteins have features in common which extend to a number of other DNA-binding proteins.

The suggestion that several DNA-binding proteins might have structural similarities first came from comparisons of their amino acid sequences. In some cases, such as Cro and λ repressor, the sequence homology is poor, and was not apparent on first inspection. However, the addition of other available sequences made the overall homology quite convincing (Fig. 3). The sequence homology includes not only repressors and activator proteins from different phages, but also other DNA-binding proteins such as the *lac* and *trp* repressors from *Escherichia coli* and MatA1 protein from yeast (12–17).

The region of best sequence homology occurs within the parts of the sequences that align with the α_2- and α_3-helices of Cro and of λ repressor, i.e., within the part of the respective proteins that are assumed to interact with the DNA. Thus, it is reasonable to infer that the homologous proteins contain an α-helical DNA-binding supersecondary structure similar to the α_2-α_3 fold seen in Cro and λ repressor.

As well as the above sequence relationships, it was found that Cro and CAP (18) and Cro and λ repressor (19) have a striking structural correspondence in their presumed DNA binding regions. This correspondence includes the α_2-α_3 helix-turn-helix unit in Cro and λ repressor and the α_E-α_F unit in CAP (Fig. 1). These helix-turn-helix units have virtually identical configurations in the three proteins (18,19).

The amino acid sequence comparisons and the structural comparisons both point to a special role for the two-helical "α_2-α_3" unit in DNA recognition and binding. The mode of interaction of this unit with DNA, as inferred from the structure of Cro, is sketched in Fig. 4. The α_3-helix occupies the major groove of the DNA with its amino acid side chains positioned so as to make sequence-specific interactions with the exposed parts of the DNA base pairs. Side chains of the α_2-helix are also presumed to contact the DNA, these interactions being primarily to the phosphate backbone.

It is reasonable to anticipate that similar although not necessarily identical modes of DNA binding will be found for a number of other gene regulatory proteins whose sequences have been shown to be homologous with Cro, λ repressor, and CAP. Indeed the presence of the helix-turn-helix unit in *lac* repressor has been confirmed by NMR

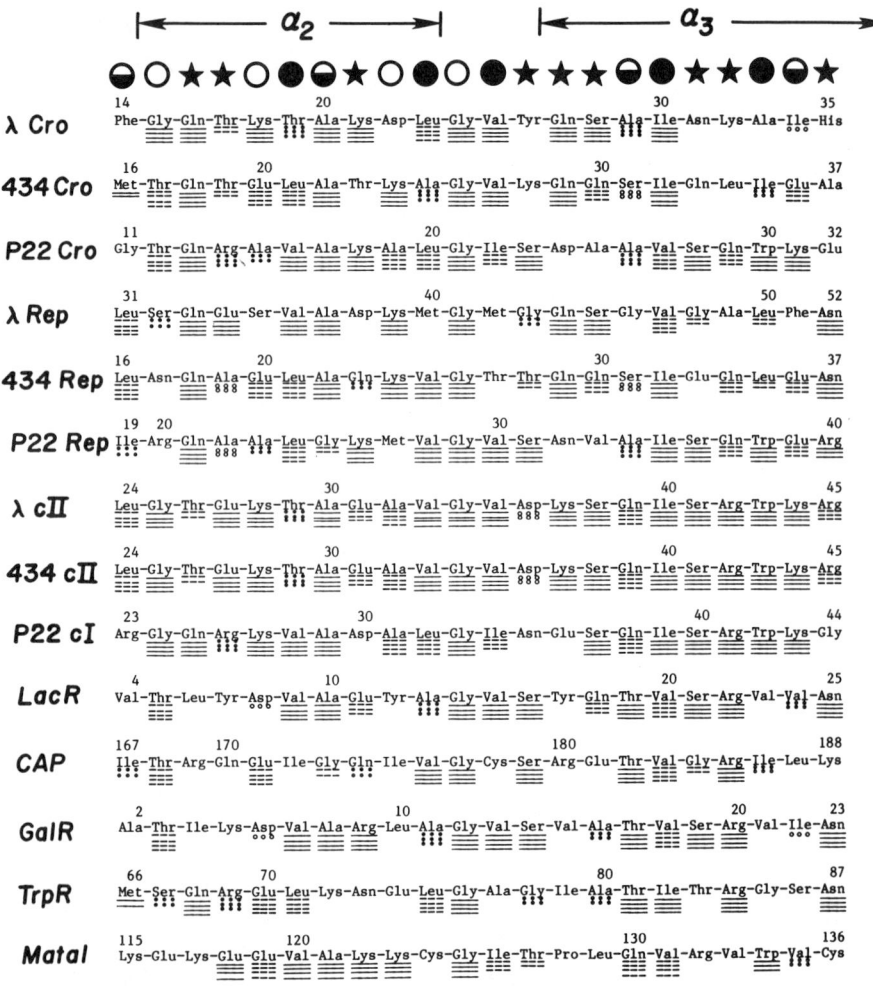

Fig. 3. Segments of the amino acid sequences of a number of gene-regulatory proteins that appear to be homologous with the helix-turn-helix (α_2-α_3) unit of Cro, λ repressor, and CAP. Amino acids that are identical in two or more sequences are underlined. The symbols at the top of the figure indicate the locations of the residues in the helix-turn-helix unit in Cro; open circles indicate full exposure to solvent, half-open circles indicate part exposure, and solid circles indicate buried residues. Stars indicate presumed DNA-contact residues. Reproduced with permission from the *Annual Review of Biophysics and Bioengineering*, Vol. 12. © 1983 by Annual Reviews Inc.

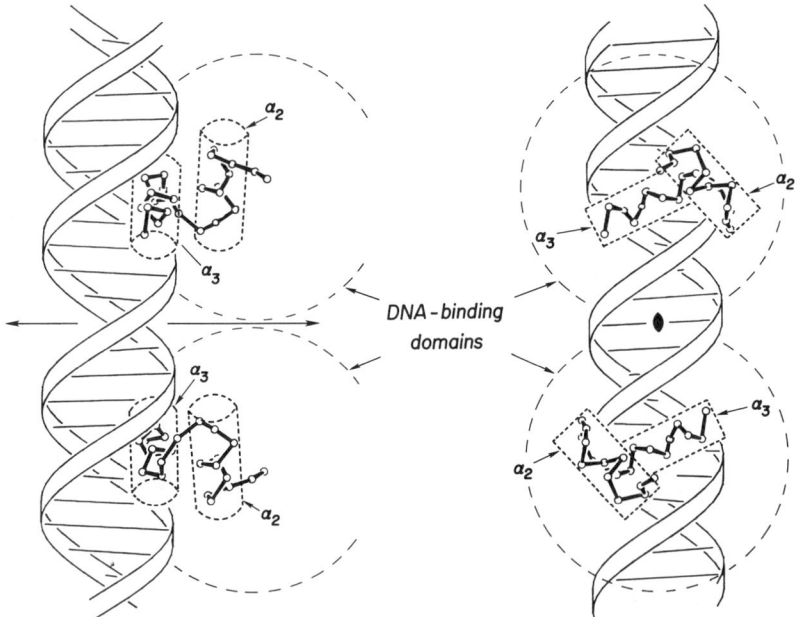

Fig. 4. The figure illustrates the general nature of the interaction presumed to occur in many DNA-regulatory proteins between a common α_2-α_3 helical unit and right-handed B-form DNA. At left is a "side view" with the twofold axis of symmetry (arrowed) extending from left to right. On the right the view is "face on" (after ref. 15).

(20,21) and in *trp* repressor by mutant studies (22) and by the recent determination of the crystal structure of the protein (23).

STUDIES IN PROGRESS

There is considerable evidence in support of the general features of the proposed complex between Cro and its specific recognition sequence (1,2,23–30). However, it is very desirable that the structures of Cro:DNA complexes be determined directly by X-ray crystallography. Complexes of Cro with 6-base pair and 9-base pair DNA duplexes have been described (31). Also we have recently obtained crystals of Cro complexed with a synthetic 17-base pair DNA duplex, which has a sequence similar to the tight-binding $O_{R}3$ operator (32). Subsequent analysis suggests that the crystals have space group $P3_1$ (or $P3_2$) with cell dimensions $a = b = 154.8$ Å, $c = 86.0$ Å. The DNA duplexes in these crystals may be stacked end to end to form essentially continuous DNA strands running parallel to the a and b crystallographic axes. There is some indication that the DNA is not exactly straight but is bent or curved possibly into the form of a superhelix.

Also, in collaboration with Dr. H. Echols at Berkeley, we have undertaken a crystallographic analysis of the "cII" gene activator protein from bacteriophage λ. This protein is one of those expected to have the "helix-turn-helix" DNA-binding motif (see Fig. 3). Crystals have been obtained in space group $P2_12_12_1$ with cell dimensions $a = 71.7$, $b = 105.8$, and $c = 128.2$ Å (unpublished results).

ACKNOWLEDGMENTS

This work was supported in part by grants from the NIH, the NSF, and the M.J. Murdock Charitable Trust. R.G.B. is supported in part by an NIH postdoctoral fellowship (GM10476) and H.M.H. by a Damon Runyon–Walter Winchell Cancer Fund fellowship.

REFERENCES

1. Ohlendorf, D. H., and Matthews, B. W. (1983). *Annu. Rev. Biophy. Bioeng.* **12**, 259.
2. Takeda, Y., Ohlendorf, D. H., Anderson, W. F., and Matthews, B. W. (1983). *Science* **221**, 1020.
3. Pabo, C. O., and Sauer, R. T. (1984). *Annu. Rev. Biochem.* **53**, 293–321.
4. Anderson, W. F., Ohlendorf, D. H., Takeda, Y., and Matthews, B. W. (1981). *Nature (London)* **290**, 654.
5. Pabo, C. O., and Lewis, M. (1982). *Nature (London)* **298**, 443.
6. McKay, D. B., and Steitz, T. A. (1981). *Nature (London)* **290**, 744.
7. Matthews, B. W., Ohlendorf, D. H., Anderson, W. F., Fisher, R. G., and Takeda, Y. (1983). *Cold Spring Harbor Symp. Quant. Biol.* **47**, 427.
8. Ohlendorf, D. H., Anderson, W. F., Fisher, R. G., Takeda, Y., and Matthews, B. W. (1982). *Nature (London)* **298**, 718.
9. Pabo, C. O., Krovatin, W., Jeffrey, A., and Sauer, R. T. (1982). *Nature (London)* **298**, 441.
10. Steitz, T. A., Weber, I. T., Ollis, D., and Brick, P. (1983a). *J. Biomol. Struct. Dyn.* **1**, 1023–1037.
11. Weber, I. T., and Steitz, T. A. (1984). *Proc. Natl. Acad. Sci. U.S.A.* **81**, 3973–3977.
12. Anderson, W. F., Takeda, Y., Ohlendorf, D. H., and Matthews, B. W. (1982). *J. Mol. Biol.* **159**, 745.
13. Matthews, B. W., Ohlendorf, D. H., Anderson, W. F., and Takeda, Y. (1982). *Proc. Natl. Acad. Sci. U.S.A.* **79**, 1428.
14. Sauer, R. T., Yocum, R. R., Doolittle, R. F., Lewis, M., and Pabo, C. O. (1982). *Nature (London)* **298**, 447.

15. Ohlendorf, D. H., Anderson, W. F., and Matthews, B. W. (1983b). *J. Mol. Evol.* **19**, 109.
16. Weber, I. T., McKay, D. B., and Steitz, T. A. (1982). *Nucleic Acids Res.* **10**, 5085.
17. Brennan, R. G., Weaver, L. H., and Matthews, B. W. (1986). *Chemics Scripta* **2673**, 251.
18. Steitz, T. A., Ohlendorf, D. H., McKay, D. B., Anderson, W. F., and Matthews, B. W. (1982). *Proc. Natl. Acad. Sci. U.S.A.* **79**, 3097.
19. Ohlendorf, D. H., Anderson, W. F., Lewis, M., Pabo, C. O., and Matthews, B. W. (1983). *J. Mol. Biol.* **169**, 757.
20. Zuiderweg, E. R. P., Billeter, M., Boelens, R., Scheek, R. M., Wüthrich, K., and Kaptein, R. (1984). *FEBS Lett.* **174**, 243–247.
21. Zuiderweg, E. R. P., Kaptein, R., and Wüthrich, K. (1983). *Proc. Natl. Acad. Sci. U.S.A.* **80**, 5837.
22. Kelley, R. L., and Yanofsky, C. (1985). *Proc. Natl. Acad. Sci. U.S.A.* **82**, 483–487.
23. Schevitz, R. W., Otinowski, Z., Joachimiak, A., Lawson, C. L., and Sigler, P. B. (1985). *Nature (London)* **317**, 782.
24. Arndt, K. T., Boschelli, F., Cook, J., Takeda, Y., Tecza, E., and Lu, P. (1983). *J. Biol. Chem.* **258**, 4177.
25. Kirpichnikov, M. P., Kurochkin, A. V., Chernov, B. K., and Skryabin, K. G. (1984). *FEBS Lett.* **175**, 317–320.
26. Metzler, W. J., Arndt, K., Teeza, E., Wasilewski, J., and Lu, P. (1985). *Biochemistry* **24**, 1418–1424.
27. Ohlendorf, D. H., Anderson, W. F., Takeda, Y., and Matthews, B. W. (1983). *J. Biomol. Struct. Dyn.* **1**, 553–563.
28. Matthew, J. B., and Ohlendorf, D. H. (1985). *J. Biol. Chem.* **260**, 5860–5862.
29. Takeda, Y., Kim, J., Caday, C. G., Steers, E., Jr., Ohlendorf, D. H., Anderson, W. F., and Matthews, B. W. (1986). J. Biol. Chem. **261**, 8608–8616.
30. Eisenbeis, S. J., Nasoff, M. S., Noble, S. A., Bracco, L. P., Dodds, D. R., and Caruthers, M. H. (1985). *Proc. Natl. Acad. Sci. U.S.A.* **82**, 1084–1088.
31. Anderson, W. F., Cygler, M., Vandonselaar, M., Ohlendorf, D. H., Matthews, B. W., Kim, J., and Takeda, Y. (1983). *J. Mol. Biol.* **168**, 903.
32. Brennan, R. G., Takeda, Y., Kim, J., Anderson, W. F., and Matthews, B. W. (1986). *J. Mol. Biol.* **188**, 115–118.

PART II

VIRUSES

5
Refinement of the Structure of Tobacco Mosaic Virus

GERALD STUBBS
Department of Molecular Biology
Vanderbilt University
Nashville, Tennessee 37235

INTRODUCTION

Tobacco mosaic virus (TMV) has served for many years as a classic system for the study of protein–nucleic acid interactions and, in particular, the assembly of systems of interacting proteins and nucleic acids. Its assembly properties have been reviewed by Caspar (1), Stubbs (2), and Butler (3). Although some insight into the molecular basis of these interactions was obtained from the determination of the structure at a nominal resolution of 4 Å by fiber diffraction methods (4), a proper understanding is only now beginning to emerge, with an improved determination at 3.6 Å, including structure refinement (5), and the extension of resolution to 3.0 Å. In this chapter we outline the process of structure determination and develop some of the speculations about TMV assembly that the structure has made possible.

The virus is rod-shaped, 3000 Å long, and 180 Å in diameter. There are about 2100 identical coat protein subunits of MW ~18,000, forming a right-handed helix with 49 subunits in three turns. A single strand of RNA follows the basic helix, with three nucleotides binding to each protein subunit (Fig. 1). Although the virus does not form crystals suitable for high-resolution analysis, it does form very well oriented gels (7,8), which give X-ray fiber diffraction patterns of extremely high quality (5,9). Analysis of these patterns at 4 Å resolution showed the general fold of the protein chain and the conformation of the RNA (4). The coat protein of TMV can be isolated and crystallized

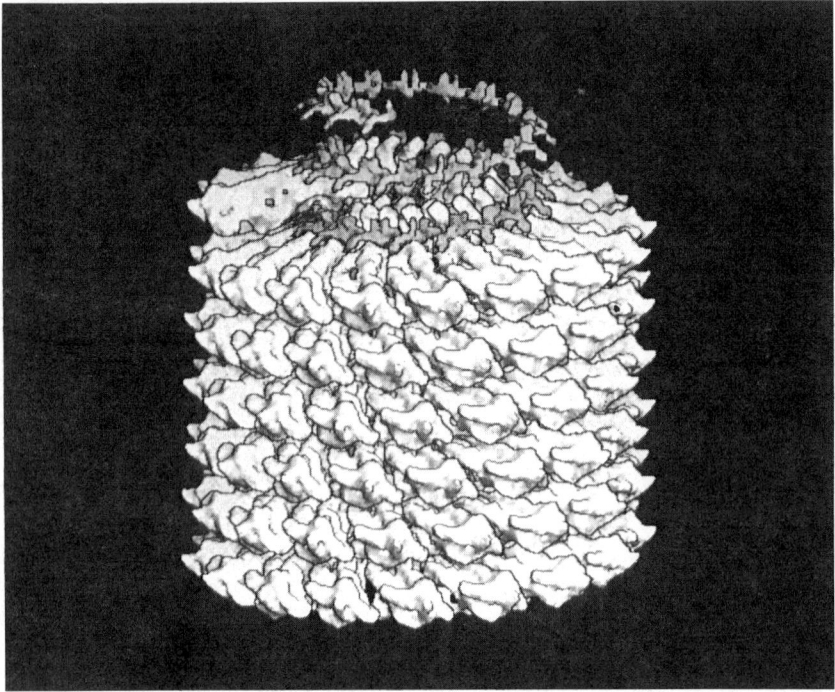

Fig. 1. Computer graphics representation of about one-twentieth of the TMV particle. Protein subunits are shown in light gray and RNA in dark gray. The protein was reconstructed from atomic coordinates to 10 Å resolution; the RNA (which is shown for clarity as extending beyond the protein) to 4 Å resolution. I thank Dr. K. Namba for this figure and for Fig. 2. The computer graphics system used for these two figures has been described by Namba *et al.* (6).

(10) in the form of a disk with two layers of 17 subunits each, both facing the same way. The structure of the disk has been determined at 2.8 Å resolution (11).

The two structures have many features in common. A core of four approximately parallel α-helices (Fig. 2) runs radially from about 45 to 65 Å, with the parallel α-helix packing extending to neighboring subunits. At the inner end of this core is the RNA binding site. In the disk crystals, which do not contain RNA, the protein inside this radius is disordered, so that there is a flexible peptide loop of indeterminate structure between residues 90 and 113. This flexibility is maintained in solution at neutral pH, as has been shown by nuclear magnetic resonance experiments (12), but in the virus the loop is ordered (4), forming an inner wall 20 Å from the virus axis. At the outer end of the

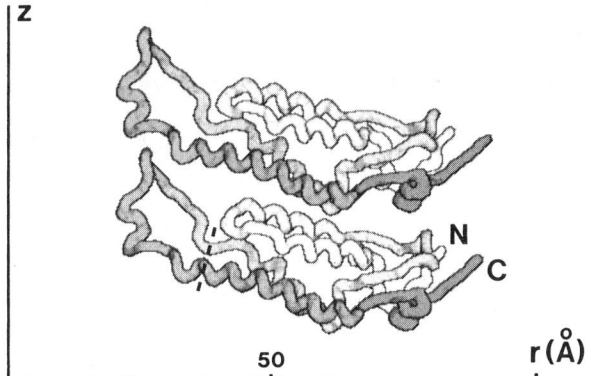

Fig. 2. Protein backbone of two subunits of TMV. The z axis is the virus axis. Shading is from the N terminus (light) to the C terminus (dark). The peptide inside the dashed line is disordered in the disk crystals. The four core α-helices form two layers; those in the top layer are referred to as the left (toward the reader) and right (away) slewed helices and those in the bottom layer are the left and right radial helices.

core an irregular four-stranded β-sheet lies at right angles to the helices, and beyond this sheet lie most of the aromatic residues, in a region that has been called the "hydrophobic girdle" (11). The N and C termini are both on the outer surface of the virus.

The packing of the subunits is quite different in the virus and the disk. Whereas in the virus each subunit is above and to the left of its lower neighbor by about one-third of a subunit (Fig. 1), the interface between the layers in the disk is dislocated relative to the virus so that each top subunit is about one-fifth of a subunit to the right of its lower neighbor. Thus the top-to-bottom contacts in the two assemblies are completely different. The side-to-side contacts are quite similar, but even here the 3.6-Å virus structure shows that there are significant differences. Although the specific side chain interactions tend to be the same, major side chain conformational changes are required for this to be the case.

Most of the protein backbone structure in the virus and the disk is very similar (rms difference less than 1 Å), but the notable exception is the disordered inner loop. This loop includes six carboxyl groups and two arginines, most of which form a large concentration of charge [the "carboxyl cage" (4,13)] and play a major part in the assembly of the virus. It has been known for many years (1,14) that TMV contains at least two anomalously titrating protons, believed to be from pairs of carboxyl groups forced into proximity by the molecular structure (1). At least one of these carboxyl pairs is part of the low-radius charge cluster (4). The concentration of carboxyl groups provides a sensitive switch (15), active at physiological pH, which enables the coat protein to respond to its environment and change conformation as part of the assembly and disassembly of the virus.

The present model of the virus allows us to describe the molecular structure of this charge cluster in detail, to locate another charge cluster, which may be the site of the second anomalously titrating proton, and to describe many of the details of the protein–RNA interaction. It is possible to identify both base-specific and nonspecific aspects of the nucleic acid binding.

STRUCTURE DETERMINATION

X-RAY METHODS

Data were collected from oriented specimens using an Elliot rotating anode X-ray generator, with a point-focused beam from either two bent quartz or germanium crystals or two bent mirrors. Diffraction patterns were recorded on Ilford Industrial G or Kodirex film, digitized with an Optronics Photoscan densitometer, converted from film space to reciprocal space (16), and corrected for nonlinear response and geometric factors. A modified version of Makowski's angular deconvolution procedure (17) was used to determine intensities and layer-line positions.

THE PHASE PROBLEM IN FIBER DIFFRACTION

The data available from a fiber diffraction experiment are two-dimensional, in contrast to the three-dimensional data available from crystals. This is because the particles in an oriented gel (or fiber) are randomly oriented about their long axes, so that the diffraction pattern is cylindrically averaged. The diffracted intensity at a point is then

$$I = \Sigma\ G\ G^*$$

(18,19), where G is a complex Fourier–Bessel structure factor (20), analogous to the crystallographic F. To calculate an electron density map, it is necessary to know both the phases and the magnitudes of the G terms; thus the phase problem in fiber diffraction is multidimensional.

A solution to this problem based on an extension into multiple dimensions of crystallographic isomorphous replacement was presented by Stubbs and Diamond (21). This approach required large numbers of heavy-atom derivatives, two for each **G** term to be separated. The number can be reduced by making use of the fine splitting of layer-lines (22), caused by the fact that like many helical diffracting systems, TMV does not repeat perfectly in a small number of turns, having in fact 49.02 subunits in three turns. Phases determined by multidimensional isomorphous replacement with layer-line splitting were refined by calculating an electron density map and using a solvent flattening procedure, which included further refinement of heavy-atom parameters. Eventually a model was built using the program FRODO (23) in an Evans and Sutherland computer graphics system, and this model was refined iteratively with the phases, using the model to calculate the separation of the **G** terms, and calculating the phase of each **G** term by isomorphous replacement. The complete process has been described in detail by Namba and Stubbs (5). At this point the model was based on data extending to 3.6 Å resolution.

Further refinement and extension of the phases to beyond 3.0 Å resolution were carried out using the restrained least-squares technique originally developed for protein crystallography by Hendrickson and Konnert (24), adapted for use in fiber diffraction (25). The atomic coordinates were refined against the diffraction data by least-squares methods, using restraints on bond lengths, bond angles, and van der Waals distances. The R-factor of the starting model for data between 10 and 3.6 Å was 0.31, and this was reduced to 0.17 after 30 cycles of refinement of atomic coordinates. A further 15 cycles, in which individual temperature factors (restrained to correlate with the temperature factors of neighboring atoms) were refined, reduced the R-factor to 0.14. Phases were extended to data at higher resolution in four steps, typically requiring about 15 cycles of refinement before the next shell of data was added. This process is continuing; at present the R-factor between 10 and 2.94 Å resolution is 0.13, with the outermost shell of data also having an R-factor of 0.13.

STRUCTURAL DETAILS

PROTEIN STRUCTURE

Many of the most interesting aspects of the structure of TMV concern the interactions between the protein subunits, in particular the electrostatic interactions. These are mostly found in two charge clusters: an extensive one near the inner surface of the virus, between 20

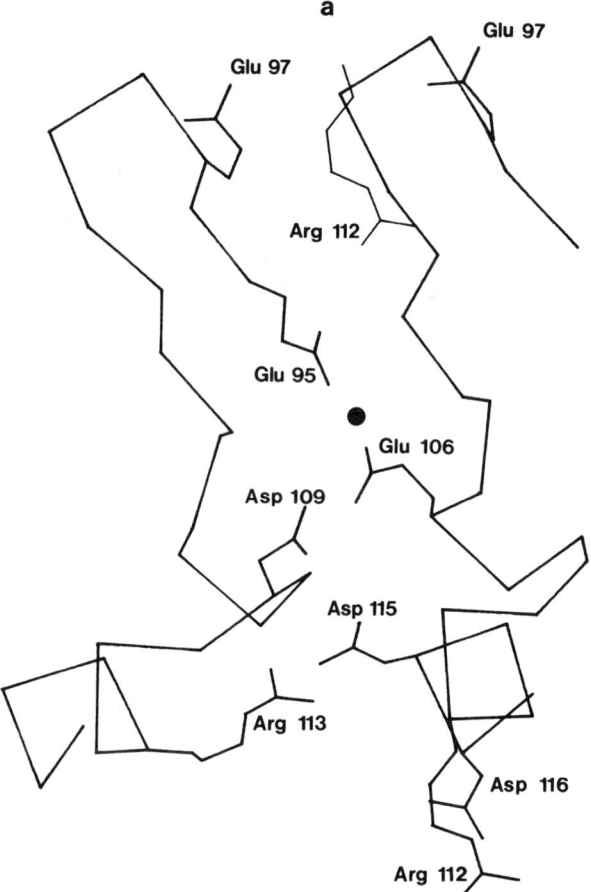

Fig. 3. Skeletal model of the subunit interface near the inner surface of the virus, including the inner charge cluster. Residues 95 to 115 are shown for two adjacent subunits. Cα atoms are connected by straight lines. Charged side chains extending into the interface are shown. Arg 112 is shown extending from the subunit above, as well as in one of the subunits shown. The dot represents the position of the uranyl ion, which is believed to replace a proton or a calcium ion. (a) Viewed toward the virus axis. (b) Viewed orthogonal to (a). Light lines represent the subunit away from the reader.

and 50 Å from the axis, and a smaller one between 55 and 60 Å from the axis. The first of these (Fig. 3) includes a number of carboxyl groups in close proximity, which has been described as a "carboxyl cage" (4). Such intersubunit carboxyl clusters have now been found to be a characteristic feature of many virus structures (2,26). All the side chains contributing to this inner charge concentration are in the pep-

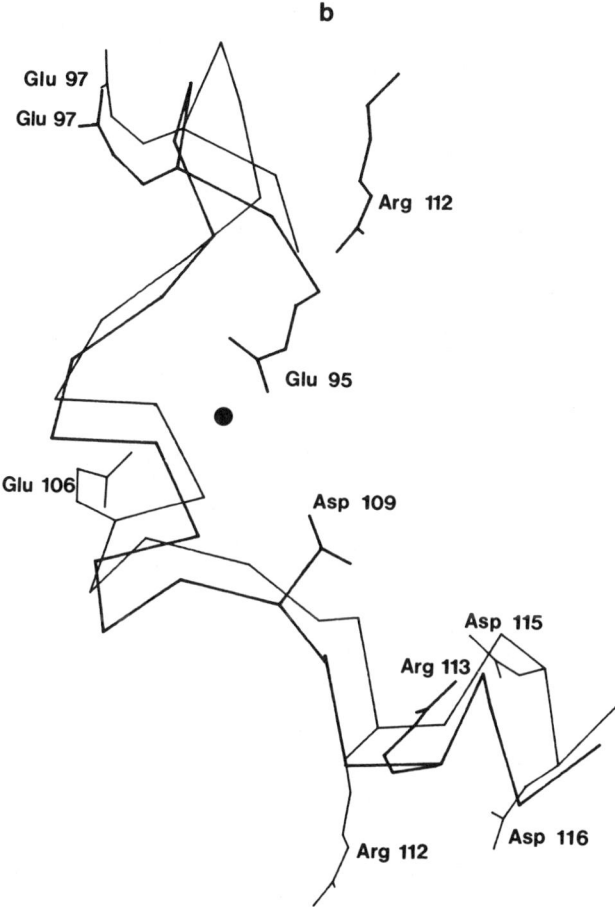

Fig. 3. *Continued*

tide loop between residues 88 and 122, with many between 90 and 113, the disordered loop of the protein disk crystals. Three carboxyl groups, Glu 95 and Asp 109 from one subunit and Glu 106 from its side-to-side (1-start helical) neighbor, are close together, about 25 Å from the virus axis. These groups probably form the binding site for lead and uranyl ions in heavy-atom derivatives, together with Asn 101 and perhaps one or more main chain carbonyl groups. At slightly higher radius are Arg 112 (from the subunit above) and Arg 113, and still further out Asp 115 and Asp 116. Arg 113 and Asp 115 form an intersubunit salt bridge, as do Asp 88 and Arg 122 at still higher radius. This latter bridge has also been found in the protein disk (11), although the side chain conformations are somewhat altered because of small but significant differences in the side-to-side packing of viral subunits and disk subunits. The phosphate groups are close to Asp 115 and Asp 116, forming another concentration of negative charge.

In the outer charge cluster, four subunits interact, through the right slewed α-helix of subunit 1, the left slewed helix of 2, the right radial helix of 16, and the left radial helix of 17 (Fig. 4). A second uranyl binding site is located in this cluster, and a series of alternating positive and negative charges, including the bound metal ion, provides most of the electrostatic interactions. Arg 71 from subunit 16 forms a salt bridge with Asp 77 in the same subunit, which is within 4 Å of Glu

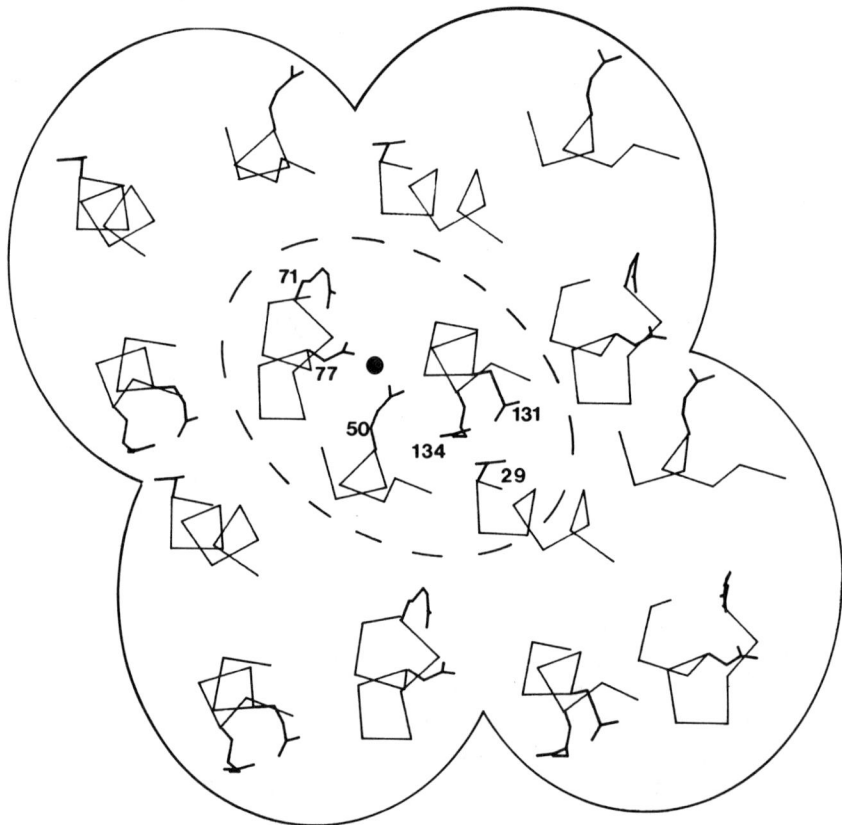

Fig. 4. Skeletal model of a cylindrical section through four subunits of the virus, viewed toward the virus axis, and showing the outer charge cluster. Residues 20 to 29, 47 to 53, 71 to 80, and 128 to 135 are included. Cα atoms are connected by straight lines. Side chains are shown for the groups in the electrostatic interaction: Arg 71, Asp 77, the uranyl binding site (dot), Glu 50, Arg 134, Glu 131, and Asn 29.

50 from subunit 1. This Glu 50 forms an ion pair with Arg 134 from subunit 17, and Arg 134 also forms a pair with Glu 131 from the same subunit. It appears that Asn 29 from subunit 2 may form hydrogen bonds with both Arg 134 and Glu 131, and that hydrogen bonds may also be formed between Arg 71 and Thr 81 within a subunit and between Tyr 72 and Thr 28 across the 1-start helix intersubunit boundary.

It is also of interest to consider the interactions between the aromatic side chains. Most of these are clustered between the β-sheet and the outer surface of the virus, forming the 'hydrophobic girdle" together with a number of prolines and aliphatic groups. Three other aromatics, Trp 52, Phe 87, and Phe 35, are part of the hydrophobic interface between the core α-helices. Tyr 72 is in the subunit interface, close to Trp 52 and Tyr 70. All these side chains form a network of interacting aromatic rings, with the planes of neighboring pairs approximately at right angles. This is a striking example of the type of aromatic interaction found by Burley and Petsko (27) in a number of globular proteins and proposed by them as a general type of stabilizing structure. The conformation of these side chains is generally very similar in the virus and the crystalline disk, but even when it is not, the right-angled interaction is conserved. For example, Phe 10 and Phe 67 interact across the side-to-side intersubunit boundary and have very different conformations in the disk and the virus; the differences are in fact greater than for any other side chains except those binding the RNA. Nonetheless, although the rings move about 4 Å relative to the bulk of the structure, their interaction remains approximately the same.

PROTEIN–NUCLEIC ACID INTERACTIONS

The protein–RNA interaction in TMV must combine elements of base specificity (with which to recognize the viral RNA) with nonspecific elements (for the entire genome to be coated). TMV protein reconstitutes only in poor yield with unrelated viral RNA (2). It was shown by Zimmern and Wilson (28) that protein binding is initiated at a site about one-sixth of the way along the molecule that includes the sequence AAGAAG (29) and has G in six consecutive equivalent positions in the trinucleotide protein binding unit. Steckert and Schuster (30) found relatively strong binding of the trinucleotide AAG to TMV protein, with GAA, GAG, AUG, and CAG also binding strongly. They concluded that the nucleotide binding site at the 3′ end of the trimer favored purines, whereas the site at the 5′ end favored purines and amino substituents away from the ribose ring (that is, adenine and cytosine). The complete sequence of TMV–RNA (31) shows that the favored GAA sequence is a feature only of the nucleation site.

The conformation of the RNA was determined by Stubbs and Stauffacher (32) by fitting models to the 4-Å-resolution map (4). The model has changed very little since then, the rms difference being only 0.7 Å, but we are now able to say a great deal more about the interactions with the protein. The RNA bases lie against the left radial helix, so that the flat base surfaces make non-base-specific van der Waals contacts with both main chain and side chain atoms. Two of the bases point up into a cavity between the left radial helices of adjacent subunits, stacking together, one in the anti conformation and one syn. The third extends along the helix toward higher radius, lying on the bottom of the subunit, between the left radial helix and the peptide connecting the left slewed and right slewed helices in the subunit below. The electron density suggests two possible conformations, both anti, for this base, which may reflect the fact that we see only electron density averaged over the four possible bases.

The sugar rings appear to make only van der Waals contacts with the protein; somewhat unexpectedly, this model does not contain ribose hydroxyl hydrogen bonds, although the bonding potential could be satisfied by water molecules not yet located.

The phosphate groups interact with the protein through ion pairs and hydrogen bonds. Numbering from the 5' end, phosphate 1 is neutralized by Arg 90 and is also close to Arg 41. Phosphate 2 is neutralized by Arg 92. Phosphate 3 does not appear to be neutralized by a close positive charge, but does form a hydrogen bond with Thr 37. It is also possible that Arg 112 is part of the general charge interaction, but the electron density for this residue is not clearly interpretable. There is a close approach between phosphate 2 and Asp 116, which could bind a proton with an unusually high pK, as suggested by Durham and Hendry (33).

It appears that base specificity is achieved through a number of protein–RNA hydrogen bonds. These structural features must be considered tentative at the present stage of resolution and refinement, but they correlate well with the known specificities of TMV protein. Base 1, in the anti conformation, lies between two side-to-side intersubunit salt bridges, Arg 122–Asp 88 and Asp 115–Arg 113. If this base is guanine, it could form hydrogen bonds with Arg 122 through O-6 and with Asp 115 through N-2. Base 3 could form adenine-specific hydrogen bonds with the main chain carbonyl groups of Ala 86 and Thr 89, through the amino substituent atom N-6. Base 2 is close to Asn 127, placed so that it could form hydrogen bonds through N-1 and N-6 (for

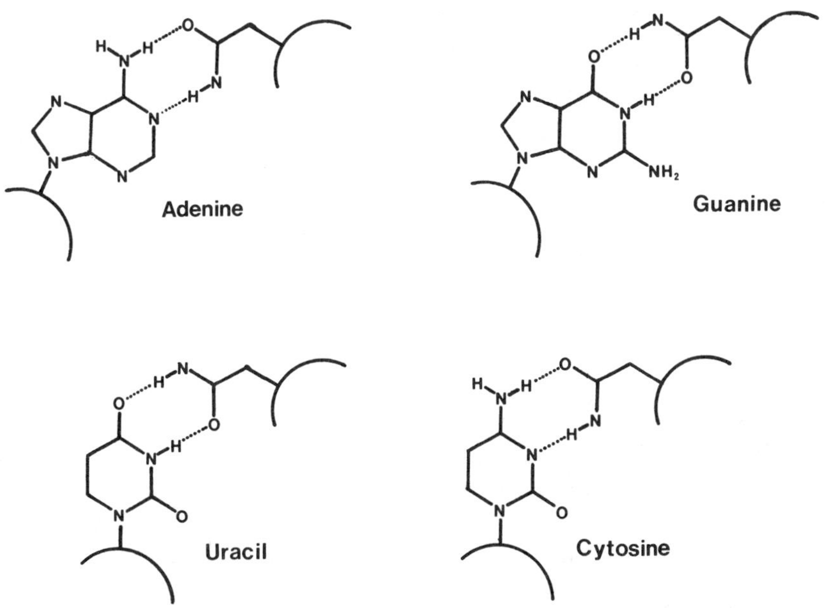

Fig. 5. Asparagine can, if appropriately placed, form hydrogen bonds with any RNA base, requiring only small conformational changes in the protein or the RNA. In this figure, the amide group and the base are shown as coplanar; if they are not, binding of particular bases can be favored.

adenine). Because of the capacity of the amide group to rotate about the $C\beta$–$C\gamma$ bond (Fig. 5), this need not be a base-specific interaction. However, if the idealized planar base–amide complex were to be distorted, such an interaction could become base-specific. Although we cannot exclude the possibility that further refinement at higher resolution might produce changes, the present model is distorted in such a way as to favor adenine binding. If this is the case, cytosine binding might also be expected (Fig. 5), which would be in agreement with the observations of Steckert and Schuster (30) on trinucleotide binding.

TMV ASSEMBLY

The broad outline of TMV assembly was described by Butler *et al.* (34) and Lebeurier *et al.* (35) in 1977. TMV protein exists in solution in a variety of aggregated forms; under the standard conditions for assembly, pH 7, 20°C, and low ionic strength, the predominant form is a 20 S aggregate. Smaller aggregates (A-protein, typically about 4 S, but almost always present as a mixture of oligomers) are also present (36,37). On the basis of electron microscopic observations, the 20 S aggregate has usually been identified as the 34-subunit double disk (3), but this identification will be discussed further below. The assembly initiation sequence is believed to fold in a stable hairpin loop (29), which is inserted into the center of the 20 S aggregate. RNA binds to the protein, and assembly proceeds by further addition of 20 S aggregates (38,39), with smaller aggregates also being incorporated (40). The partially assembled virus contains RNA bound to the protein helix, with the 3' tail free at one end and the 5' tail looped back from the other end, through the central hole of the virus, and thus appearing at the same end of the rod as the 3' tail. Assembly proceeds by pulling the 5' tail through the hole.

To describe this process at the molecular level, it is necessary to describe more accurately the 20 S aggregate. Three lines of evidence converge to suggest that it is not the double disk. Circular dichroism experiments (41,42) show that the protein conformation under the assembly conditions described above is different from that at the higher pH and ionic strength required for crystal formation. Sedimentation equilibrium experiments (43) show that the number of subunits in the aggregate is 39 ± 2, rather than the 34 of the disk. Our structural results, taken together with titration data, suggest that the aggregate is in fact a short helix. It has been known for many years (1,44) that, during assembly, TMV protein binds two protons, and these have been attributed to carboxyl–carboxylate pairs (1). In the 20 S aggregate, half a proton is bound to each protein monomer (44). Our structure contains two proton binding sites which could have pK's near 7: the two carboxyl pairs described above. The identification of the outer pair is strengthened by the fact that such a top-to-bottom pair would indeed bind only half a proton per subunit (44); indeed, Shalaby and Lauffer correctly predicted that one site would be in a side-to-side subunit interface and one top-to-bottom. Neither pair exists in the disk structure (11); the inner pair is part of the disordered loop, whereas the outer carboxyls are too far apart (11 Å) to form an anomalous proton binding site. It thus appears that the disk does not exist in significant concentration under the virus nucleating conditions.

Consideration of the structures of the 20 S aggregate and the assembled virus leads to a plausible molecular theory of assembly. The array of arginine groups on the top surface of the aggregate could bind the phosphate groups of free RNA, although not in a base-specific way. When the AAG-rich assembly initiating sequence is encountered, the high affinity of the protein for this sequence would lead to further binding of the RNA, either by intercalation or by bases binding to the bottom surface of the aggregate.

In speculating on the possible structure of the protein–RNA complex, it is particularly helpful to consider the role of the disordered inner peptide loop. While it is not possible to directly determine the structure of this protein–RNA complex (referred to below as P'R), there is considerable evidence that it is not particularly stable. RNA protection studies (45,46) suggest that the stable nucleation complex binds 250 nucleotides, which, in the case of stoichiometric binding, would correspond to 83 subunits, that is, approximately two 20 S aggregates. Kinetic studies (47) suggest a similar-sized nucleus having 88 subunits. Considering the energetically favorable contributions of RNA binding and protein–protein binding, set against the unfavorable (because of the mutual repulsion of the carboxyl groups) contribution of the folding of the inner loop, it is reasonable to conclude that the inner loop is not folded in P'R. Addition of another 20 S aggregate, however, would require folding to sterically permit protein–protein interactions. Indeed, we have suggested (48) that one function of the unfolded inner loop is prevention of premature protein aggregation. The folding would be highly cooperative and would lead to a stable nucleating complex derived from one RNA molecule and two 20 S protein aggregates.

In summary, if P represents a short helical aggregate of about 39 subunits, P' represents such an aggregate bound to RNA, but still with a disordered inner loop, H represents the bound aggregate with the loop folded, and R represents the RNA:

$$P + R \rightleftharpoons P'R$$
$$P'R + P \rightleftharpoons P'_2R \rightarrow H_2R$$

H_2R is the stable nucleation complex.

It is now over 50 years since the first X-ray diffraction experiments (49) with TMV were carried out. We have progressed from there to a knowledge of the molecular details of the protein–nucleic acid interaction and the assembly process. Refinement of the TMV structure is continuing, and even a small further improvement in resolution will improve our confidence in these details and perhaps reveal more about the structure. In view of the importance of the anomalously titrating carboxyl groups, it will be necessary to repeat the structure determination at low pH, where the carboxyl–carboxylate interactions can be visualized directly. Strains of TMV exist in which the electrostatic interactions and therefore the control of viral assembly differ from those in TMV vulgare. These systems are under investigation and will all contribute to a greater understanding of the relationship between macromolecular structure and assembly.

ACKNOWLEDGMENTS

Most of the results described here come from work carried out with Dr. K. Namba. Many others have contributed to TMV structural work in recent years, in particular Drs. S. G. Warren, C. V. Stauffacher, and L. Makowski. I began this work in the laboratory of Dr. K. C. Holmes, where it was already an established project, and continued it until 1983 in the laboratory of Dr. D. L. D. Caspar, receiving a great deal of encouragement from both. The work is currently supported at Vanderbilt by NIH Grant GM33265.

REFERENCES

1. Caspar, D. L. D. (1963). *Adv. Protein Chem.* **18**, 37–121.
2. Stubbs, G. (1984). *In* "Biological Macromolecules and Assemblies" (F. A. Jurnak and A. McPherson, eds.), Vol.1, pp. 149–202. Wiley, New York.
3. Butler, P. J. G. (1984). *J. Gen. Virol.* **65**, 253–279.
4. Stubbs, G., Warren, S., and Holmes, K. (1977). *Nature (London)* **267**, 216–221.
5. Namba, K., and Stubbs, G. (1985). *Acta Crystallogr., Sect. A* **A41**, 252–262.
6. Namba, K., Caspar, D. L. D., and Stubbs, G. (1985). *Science* **227**, 773–776.
7. Bernal, J. D., and Fankuchen, I. (1941). *J. Gen. Physiol.* **28**, 111–165.
8. Gregory, J., and Holmes, K. C. (1965). *J. Mol. Biol.* **13**, 796–801.
9. Holmes, K. C., Stubbs, G., Mandelkow, E., and Gallwitz, U. (1975). *Nature (London)* **254**, 192–196.
10. Finch, J. T., Leberman, R., Chang, Y. S., and Klug, A. (1966). *Nature (London)* **212**, 349–350.
11. Bloomer, A. C., Champness, J. N., Bricogne, G., Staden, R., and Klug, A. (1978). *Nature (London)* **276**, 362–368.
12. Jardetzky, O., Akasaka, K., Vogel, D., Morris, S., and Holmes, K. C. (1978). *Nature (London)* **273**, 564–566.
13. Stubbs, G., Warren, S., and Mandelkow, E. (1979). *J. Supramol. Struct.* **12**, 177–183.
14. Fraenkel-Conrat, H., and Narita, K. (1958). *In* "Symposium on Protein Structure" (A. Neuberger, ed.), pp. 249–261. Wiley, New York.
15. Caspar, D. L. D. (1976). *In* "Structure–Function Relationships of Proteins" (R. Markham and R. W. Horne, eds.), pp. 85–99. North-Holland Publ., Amsterdam.
16. Fraser, R. D. B., Macrae, T. P., Miller, A., and Rowlands, R. J. (1976). *J. Appl. Crystallogr.* **9**, 81–94.
17. Makowski, L. (1978). *J. Appl. Crystallogr.* **11**, 273–283.
18. Waser, J. (1955). *Acta Crystallogr.* **8**, 142–150.
19. Franklin, R. E., and Klug, A. (1955). *Acta Crystallogr.* **8**, 777–780.
20. Klug, A., Crick, F. H. C., and Wyckoff, H. W. (1958). *Acta Crystallogr.* **11**, 199–213.
21. Stubbs, G., and Diamond, R. (1975). *Acta Crystallogr., Sect. A* **A31**, 709–718.

22. Stubbs, G., and Makowski, L. (1982). *Acta Crystallogr., Sect. A* **A38**, 417–425.
23. Jones, T. A. (1982). *In* "Computational Crystallography" (D. Sayre, ed.), pp. 303–317. Oxford Univ. Press, London and New York.
24. Hendrickson, W. A., and Konnert, J. H. (1980). *In* "Computing in Crystallography" (R. Diamond, S. Ramaseshan, and K. Venkatesan, eds.), pp. 13.01–13.26. Indian Acad. Sci., Bangalore.
25. Stubbs, G., Namba, K., and Makowski, L. (1986). *Biophys. J.* **49**, 58–60.
26. Abdel-Meguid, S. S., Yamane, T., Fukuyama, K., and Rossmann, M. G. (1981). *Virology* **114**, 81–85.
27. Burley, S. K., and Petsko, G. A. (1985). *Science* **229**, 23–28.
28. Zimmern, D., and Wilson, T. M. A. (1976). *FEBS Lett.* **71**, 294–298.
29. Zimmern, D. (1977). *Cell (Cambridge, Mass.)* **11**, 455–462.
30. Steckert, J. J., and Schuster, T. M. (1982). *Nature (London)* **299**, 32–36.
31. Goelet, P., Lomonossoff, G. P., Butler, P. J. G., Akam, M. E., Gait, M. J., and Karn, J. (1982). *Proc. Natl. Acad. Sci. U.S.A.* **79**, 5818–5822.
32. Stubbs, G., and Stauffacher, C. V. (1981). *J. Mol. Biol.* **152**, 387–396.
33. Durham, A. C. H., and Hendry, D. A. (1977). *Virology* **77**, 510–519.
34. Butler, P. J. G., Finch, J. T., and Zimmern, D. (1977). *Nature (London)* **265**, 217–219.
35. Lebeurier, G., Nicolaieff, A., and Richards, K. E. (1977). *Proc. Natl. Acad. Sci. U.S.A.* **74**, 149–153.
36. Durham, A. C. H. (1972). *J. Mol. Biol.* **67**, 289–305.
37. Durham, A. C. H., Finch, J. T., and Klug, A. (1971). *Nature (London), New Biol.* **229**, 37–42.
38. Butler, P. J. G., and Klug, A. (1971). *Nature (London), New Biol.* **229**, 47–50.
39. Ohno, T., Yamaura, R., Kuriyama, K., Inoue, H., and Okada, Y. (1972). *Virology* **50**, 76–83.
40. Shire, S. J., Steckert, J. J., and Schuster, T. M. (1981). *Proc. Natl. Acad. Sci. U.S.A.* **78**, 256–260.
41. Vogel, D., and Jaenicke, R. (1976). *Eur. J. Biochem.* **61**, 423–431.
42. Raghavendra, K., Adams, M. L., and Schuster, T. M. (1985). *Biochemistry* **24**, 3298–3304.
43. Correia, J. J., Shire, S., Yphantis, D. A., and Schuster, T. M. (1985). *Biochemistry* **24**, 3292–3297.
44. Shalaby, R. A. F., and Lauffer, M. A. (1977). *J. Mol. Biol.* **116**, 709–725.
45. Zimmern, D. (1976). *Philos. Trans. R. Soc. London, Ser. B* **276**, 189–204.
46. Zimmern, D., and Butler, P. J. G. (1977). *Cell (Cambridge, Mass.)* **11**, 455–462.
47. Shire, S. J., Steckert, J. J., Adams, M. L., and Schuster, T. M. (1979). *Proc. Natl. Acad. Sci. U.S.A.* **76**, 2745–2749.
48. Namba, K., and Stubbs, G. (1986). *Science* **231**, 1401–1406.
49. Bawden, F. C., Pirie, N. W., Bernal, J. D., and Fankuchen, I. (1936). *Nature (London)* **138**, 1051–1053.

6

The Structure of a Human Common Cold Virus (Rhinovirus 14) and Its Functional Relations to Other Picornaviruses

MICHAEL G. ROSSMANN,* EDWARD ARNOLD,* JOHN W. ERICKSON,*,[1] ELIZABETH A. FRANKENBERGER,*,[2] JAMES P. GRIFFITH,* HANS-JÜRGEN HECHT,*,[3] JOHN E. JOHNSON,* GREG KAMER,* MING LUO,* ANNE G. MOSSER,† ROLAND R. RUECKERT,† BARBARA SHERRY,† AND GERRIT VRIEND*

*Department of Biological Sciences
Purdue University
West Lafayette, Indiana 47907
and
†Biophysics Laboratory
University of Wisconsin
Madison, Wisconsin 53706

We report here the first atomic resolution structure of any animal virus, namely, that of human rhinovirus 14. The structure has been solved to 3.0 Å resolution using primarily a technique dependent on

---

[1] Present address: Department of Physical Biochemistry, AP-9A D-47E, Abbott Laboratories, Abbott Park, North Chicago, Illinois 60064.
[2] Present address: Department of Agronomy, Purdue University, West Lafayette, Indiana 47907.
[3] Present address: F. G. Roentgenstrukturanalyse, Universitaet Wuerzburg, Zentralbau Chemie, Am Hubland, D-8700 Wuerzburg, West Germany.

the viral symmetry rather than isomorphous replacement. The course of all four capsid polypeptides has been traced and correlated with the known amino acid sequences. The tertiary structures of the three larger proteins are each strikingly similar to those of the known icosahedral plant RNA viruses, as is also their quaternary organization in the virus coat. Four neutralizing immunogenic regions have been identified by sequencing mutants selected for their ability to survive in the presence of neutralizing antibodies. The altered amino acids, as well as corresponding antigenic sequences in the homologous polio and foot-and-mouth disease viruses, reside on protrusions. A large cleft, spanning the center of each icosahedral face, is most probably the host cell receptor binding site.

INTRODUCTION

Picornaviruses are associated with serious diseases in humans and other animals, and they comprise one of the largest families of viral pathogens. For example, the common cold, poliomyelitis, foot-and-mouth disease, and hepatitis can be caused by these viruses. They are among the smallest RNA-containing animal viruses (1–3). Their molecular weight is around 8.5×10^6 and they contain about 30% by weight RNA. Their external diameter is roughly 300 Å and they form icosahedral shells. Picornaviridae have been subdivided into four genera on the basis of their buoyant density, pH stability, and sedimentation coefficients: enterovirus (e.g., polio and Coxsackie viruses), cardiovirus (e.g., encephalomyocarditis and Mengo viruses), aphthovirus (e.g., foot-and-mouth disease virus), and rhinovirus. They differ also in the number of known serotypes. For instance, there are three known serotypes for polioviruses, seven for foot-and-mouth disease viruses (FMDV), and at least 89 for human rhinoviruses (HRV). Accordingly, it has been possible to produce effective vaccines for poliomyelitis and, with greater difficulty, for foot-and-mouth disease, but not for the common cold.

Picornavirions contain 60 protomers (4), each composed of four structural proteins VP1, VP2, VP3, and VP4 corresponding to genes *1D*, *1B*, *1C*, and *1A*, respectively (for nomenclature, see ref. 5). Their molecular weights in HRV14 are 32,000, 29,000, 26,000, and 7000. The capsid protein VP0, corresponding to gene *1AB*, is cleaved into its components VP4 and VP2 only in the final stages of assembly (4,6,7). The cleavage occurs at an Asn–Ser peptide in HRV14, and hence is not effected by the viral protease 3C, whose specificity is for

Gln–Gly peptides. The assembly of the capsid occurs in a series of steps culminating in the insertion of RNA into capsids to produce mature virions with the concomitant cleavage of VP0 (3,8–11).

The virions contain a single, positive strand of RNA with a protein, VPg, covalently attached to the 5′ end (12–17). The RNA, containing one long, open reading frame, is translated into a polyprotein which is processed into its component proteins in a series of steps (18). The gene order is essentially the same for HRV, FMDV, poliovirus, and encephalomyocarditis virus (EMCV). The initial cleavage of capsid precursor protein from the polyprotein is probably by a host cell protease, but subsequent cleavages are mostly dependent on the release of viral protease excised from the polypeptide (19).

Considerable effort has been devoted to mapping topological relationships among VP1, VP2, VP3, and VP4 within the capsid (cf. ref. 1) using chemical labeling of the surface of intact particles (20–22), treatment with cross-linking reagents (23–25), reaction with specific antibodies, and cross-linking with UV light (26). The consensus is that VP1 is the most external and immunodominant protein, whereas VP4 is inaccessible from the outside but can be cross-linked with RNA on the interior. Heat treatment or mild denaturing agents cause a conformational change to the capsid, thus altering the response to antisera. Surprisingly, the internal capsid protein VP4 can dissociate and escape from the capsid during antigenic conversion (3).

The RNA, and hence by inference the polyprotein, has been sequenced for all three strains of poliovirus (27–30), for various FMDV strains (31–34), for two strains of rhinovirus (35–37), for EMCV (38), and for hepatitis A virus (39,40). Protein-to-protein comparisons between HRV, poliovirus, EMCV, FMDV, and hepatitis A virus sequences show that HRV and poliovirus are closely related, whereas sequence homology of HRV to EMCV, FMDV, or hepatitis A virus is not immediately obvious, particularly for the structural proteins. Homologies for nonstructural proteins between HRV and poliovirus range from 44% in the protease gene *3C* to 65% in the RNA polymerase gene *3D* (35–37). Comparison of the structural proteins of HRV14 with poliovirus shows a 60% homology for VP4 and VP2, whereas VP3 and VP1 show only 47 and 44% conservation of amino acids. The polymerase, which is among the most conserved picornavirus proteins, has been shown to be homologous to the polymerase of the cowpea mosaic plant virus (41,42) as well as those of other plant and animal viruses (43–46).

An animal's immunological response to a virus is one of the major defenses against disease. Antibodies can bind to viruses but they do not necessarily neutralize infectivity (cf. ref. 47). In spite of the 60-fold equivalence of each potential binding site on the virus, as few as four neutralizing antibodies per virion can be sufficient to inhibit infectivity of poliovirus (48). Neutralizing antibodies usually change the isoelectric point of the picornavirions (49,50), indicating that a conformational change frequently accompanies neutralization. Antibodies may neutralize by interfering with cell attachment, membrane penetration, or virus uncoating (51,52). Antibodies that bind to poliovirus may require bivalent attachment for neutralization of the virus (48,53). Extensive studies have been reported (see below) on mapping the antigenic surfaces of HRV14 (54,55), polio, and FDMV. Four major immunogenic sites have been identified for HRV14. One of the HRV14 immunogenic sites on VP1 coincides with the dominant immunogen of polio.

In spite of the sequence and surface similarities of picornaviruses, they have different host and tissue specificity. Abraham and Colonno (56,57) have shown, using 24 rhinovirus serotypes in competition binding assays, that, while the majority recognize one receptor, a second, smaller group recognizes a different receptor. HRV14 [sequenced by Callahan et al. (36) and by Stanway et al. (35)] belongs to the larger receptor group and HRV2 [sequenced by Skern et al. (37)] belongs to the smaller one. Minor et al. (58) have been able to produce monoclonal antibodies that block cellular receptors of poliovirus as have Colonno and co-workers for the large rhinovirus receptor group (57). Krah and Crowell (59) have characterized some properties of HeLa cell receptors for group B Coxsackie viruses. They found that concanavalin A and other lectins adsorbed to receptors and inhibited virus attachment, a finding similar to that of Lonberg-Holm (60).

X-ray diffraction studies of crystalline picornaviruses have been limited. Coxsackie virus crystals (61) and poliovirus crystals (62) were reported a long time ago and rhinovirus strain 1A crystals a little later (63). Preliminary examination of the polio crystals (64) showed the particles to possess icosahedral symmetry. However, the technical problems involved in a complete structure determination were not solved until the elucidation of the small plant RNA viruses tomato bushy stunt virus (TBSV) (65), southern bean mosaic virus (SBMV) (66) and satellite tobacco necrosis virus (STNV) (67). This encouraged the renewal of the crystallographic study of poliovirus (68) and stimulated work on rhinovirus (69,70) as well as Mengo virus (71). It was shown that rhinovirus and poliovirus crystals can be roughly isomorphous, suggesting close similarities of the viral capsids, a result later supported by sequence homologies. Comparison of the X-ray diffraction patterns of rhinovirus and Mengo virus crystals also indicated significant structural homologies (71).

STRUCTURE DETERMINATION

HRV14 crystals were prepared as described by Arnold *et al.* (70). The crystals are cubic with $a = 445.1$ Å belonging to space group $P2_13$. There are four particles per crystal cell with each virion situated on a crystallographic threefold axis along the body diagonal. Thus, one-third of the virus, or 20 icosahedral asymmetric units, is in the crystallographic asymmetric unit. Packing considerations showed that the particle had to be near (0,0,0) or $(\frac{1}{4},\frac{1}{4},\frac{1}{4})$ given the choice of origin as defined in International Tables A (72).

Initial 5-Å-resolution data were collected on rotating anode X-ray generators and these were used to compute a rotation function (73) which determined the particle orientation in the cell (70). However, the crystals suffered appreciable radiation damage and high-resolution data were not practicably attainable. This was solved by using X-ray radiation from a synchrotron source where 0.3° oscillation photographs extended to 2.6 Å resolution. Some experiments were performed at the DESY synchrotron in Hamburg, but the data used for the results given here were collected at CHESS (Cornell High Energy Synchrotron Source). A new crystal was used for every exposure. A total of 83 film packs were eventually included in the native data (Table I). Surveys for suitable isomorphous heavy atom derivatives eventually produced two related compounds, namely, 1 mM KAu(CN)$_2$ and 5mM KAu(CN)$_2$. Full high-resolution data were collected of the former, but only partial low resolution (0.6 to 0.8° oscillation angles) of the latter.

The mean differences between heavy atom derivative and native data sets were not encouraging (Table II). Not only were the differences rather small in proportion to the size of the native amplitudes [8% for 1 mM KAu(CN)$_2$ and 12% for 5 mM KAu(CN)$_2$], but also the size of the differences increased at a resolution greater than 7 Å. Thus, there appeared to be little substitution and some concern about lack of isomorphism, particularly at higher resolution.

In spite of the lack of confidence in the available heavy atom derivative data, systematic search procedures (74,75) were applied to 6-Å difference Pattersons. The first three-dimensional searches were based on locating the self-Patterson vectors within an icosahedral virion given its known orientation. Trial heavy atom sites were used

TABLE I
Data Used in Structure Determination[a]

	Number of unique observations $[F^2 > 1\sigma(F^2)]$ for each data set and percentage of possible total					
	Native		1 mM KAu(CN)$_2$		5 mM KAu(CN)$_2$	
Resolution range (Å)	Number	%	Number	%	Number	%
−30	358	63	331	57	231	40
30–15	3,445	86	2,847	70	1,850	46
15–10	9,109	84	7,757	70	5,029	46
10–7.5	17,532	83	14,641	68	9,486	45
7.5–5.0	70,043	81	57,577	65	34,046	39
5.0–3.5	184,864	78	138,477	58		
3.5–3.0	133,101	63	83,852	39		
3.0–2.75	60,629	36	25,271	15		
2.75–2.6	26,834	11	8,175	3		
Total	509,915	58	338,928	39	50,641	41
Number of film packs		83		48		11
R-factor (%)		11.0		12.9		12.0

[a] $R = \dfrac{\sum_h \sum_i |(I_h - I_{hi})|}{\sum_h \sum_i I_h} \times 100$

where I_h is the mean of the I_{hi} observations of reflection h.

TABLE II
Distribution of Heavy Atom Differences

Resolution (Å)	Mean native [F]	Estimated error on native	Mean differences	
			1 mM Au	5 mM Au
11.6	307	12	29	36
7.6	269	15	23	37
6.0	261	17	24	45
5.2	282	19	28	50
4.6	313	21	32	—
4.2	292	25	35	—
3.9	266	26	37	—
3.6	246	27	39	—
3.4	223	28	40	—
3.2	206	30	42	—
3.1	186	31	42	—
2.9	170	36	44	—

systematically to explore all possibilities in an icosahedral asymmetric unit between 80- and 160-Å radius. Plausible sites were then explored further using partial four-dimensional searches in which the virion position was slightly varied and cross-Patterson vectors between particles were also considered. The results showed clearly a single site, A, on the icosahedral threefold axes at a radius of 111.0 Å for the 1mM KAu(CN)$_2$ compound and a possible substitution at the same site in the 5 mM KAu(CN)$_2$ derivative. The best particle position was found to be at $x = y = z = 0.0006$. Three-dimensional searches were then conducted for a second site assuming the particle position as well as the A site, yielding site B in the 5 mM but not in the 1 mM KAu(CN)$_2$ derivative. These sites were checked with a reciprocal space "feedback" technique (76). The heavy atom parameters for both heavy atom data sets were used for determining phases to 6 Å resolution. The resultant electron density map, averaged over the 20 different noncrystallographic asymmetric units, showed clearly the viral protein envelope between about 100- and 150-Å radius.

The 6-Å map was improved with two cycles of molecular replacement averaging and back-transformation (77–79). A difference map of the two Au heavy atom data sets with respect to the native data, based on the improved phases, showed an unambiguous third site, C, in the 5 mM KAu(CN)$_2$ derivative. In contrast, difference maps of two other heavy atom data sets (KUO$_2$F$_5$ and mercurochrome) showed no interpretable sites. The heavy atom parameters are shown in Table III. These were used to compute phases from 25 to 5 Å resolution. The resultant map was the starting point for five cycles of molecular replacement. The electron density map was exceptionally clear, showing that there were probably three β-barrels per icosahedral asymmetric unit.

Phase extension beyond 5 Å, to eventually 3.0 Å resolution, was performed in 20 steps. In each step, the resolution was extended by three reciprocal lattice points. Two to three molecular replacement real-space averaging cycles were usually sufficient to improve the R-factor to less than 30% and correlation coefficient to greater than 0.5 in the current outermost resolution shell. Both R and C measure the agreement of the observed structure amplitudes and those calculated by Fourier back-transformation of the averaged cell (Fig. 1). Attempts at increasing the resolution in larger steps led to erroneous phases, probably the result of satisfying the noncrystallographic symmetry constraints in isolated outer regions of reciprocal space independently of the established phases at lesser resolution. The particle envelope was defined by an external and internal sphere of 163- and 104-Å radii,

TABLE III
Heavy Atom Parameters Used for Double Isomorphous Replacement Phasing

Derivative	Site	Relative occupancy	Number of sites per virion	Fractional cell coordinates			Orthogonal coordinates (Å) with respect to icosahedron			Polar coordinates with respect to icosahedron			Chemical site of attachment
				x	y	z	P	Q	R	R (Å)	ψ (Å)	ϕ (Å)	
1 mM KAu(CN)$_2$	A	80.4	20	0.2389	−0.0186	−0.0739	0.0	39.8	104.2	111.5	69.1	−90.0	VP2 Cys 7
5 mM KAu(CN)$_2$	A	64.2	20	0.2368	−0.0185	−0.0732	1.0	38.7	103.5	110.5	69.5	−89.5	VP2 Cys 7
	B	33.7	60	0.2637	0.0823	−0.0206	43.7	13.9	113.0	122.0	83.0	−69.0	VP1 Cys 69
	C	24.7	60	0.2606	0.0004	0.0389	6.1	−10.2	116.8	117.0	95.0	−87.0	VP2 Cys 248

6. THE STRUCTURE OF A HUMAN COMMON COLD VIRUS

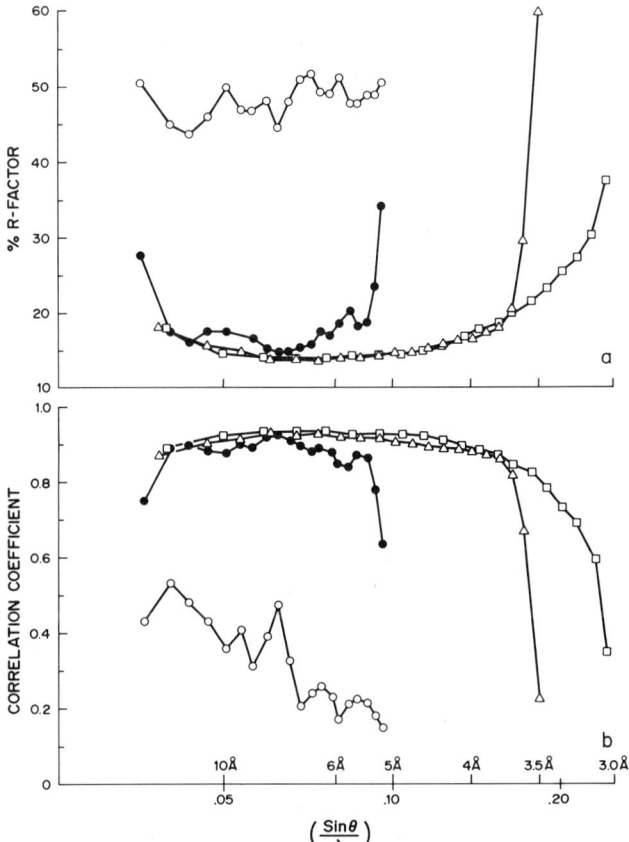

Fig. 1. Phase determination from 5 to 3 Å resolution by extension from 5 Å in small incremental steps of resolution. Shown are the correlation coefficients and R-factors obtained from the initial 5-Å DIR determination (open circles), the 5-Å-resolution refined phases, cycle 5 (solid circles), the 3.5-Å-resolution extension, cycle 40 (open triangles), and the final 3.0-Å-resolution results, cycle 60 (open squares). (a) R-factor. (b) Correlation coefficient, C, where

$$C = \frac{\Sigma(\langle F_h \rangle - |F_o|)(\langle F_h \rangle - |F_c|)}{[\Sigma(\langle F_h \rangle - |F_o|)^2 (\langle F_h \rangle - |F_c|)^2]^{1/2}}$$

$$R = \frac{\Sigma|(|F_o| - |F_c|)|}{|F_o|}$$

Here $|F_o|$ and $|F_c|$ are observed and calculated structure amplitudes placed on the same relative scale for each local resolution shell, and $\langle F_h \rangle$ is the mean observed amplitude in that shell. Structure factors were calculated by Fourier back-transformation of the averaged map.

respectively. Density outside the envelope was set to zero. The external sphere was truncated by planes, tangential to the interparticle contacts, at a radial distance of 157 Å to avoid particle overlap, while allowing all the protein density to be included within the envelope. Electron densities used for averaging were computed on grids with spacing less than one-fifth of the resolution of the data.

The final electron density map was averaged and displayed in sections at 1-Å intervals perpendicular to an icosahedral twofold axis. A minimap of the 3.5-Å-resolution electron density was used for the original chain tracing and amino acid identification. An atomic model was then built into the final 3.0-Å-resolution map using an Evans and Sutherland PS300 computer graphics system and the FRODO program written by Alwyn Jones (80), modified for the PS300 at Rice and Purdue Universities.

Phase determination based on noncrystallographic symmetry was first suggested by Rossmann and Blow (73,81). It was used to compute very low resolution maps of SBMV (82) and polyoma virus (83) and also was used for phase extension in the determination of STNV (67) and hemocyanin (84). However, there was considerable trepidation that the molecular replacement method might fail in extending the phases all the way from low resolution (5 Å) to high resolution (3 Å). This is the first time that such phase extension has been used successfully.

The high quality of the map reflects the power and exactness of the noncrystallographic constraints as opposed to the approximations required in an isomorphous replacement phase determination. The noncrystallographic constraints, however, are only meaningful in the presence of precise data and the availability of a sufficiently powerful computer to perform the large number of iterations during phase extension. Hence, our results were possible only because of the use of the CHESS synchrotron and the Cyber 205 supercomputer. The success of these techniques holds promise for a rapid series of structural determinations of viruses and other biological assemblies of high symmetries, by reducing the dependence on finding satisfactory isomorphous heavy atom derivatives.

STRUCTURE

The particle consists of an icosahedral protein shell (Fig. 2a) surrounding an RNA core. The lack of visible structure in the central cavity results from the random orientation of the asymmetric RNA

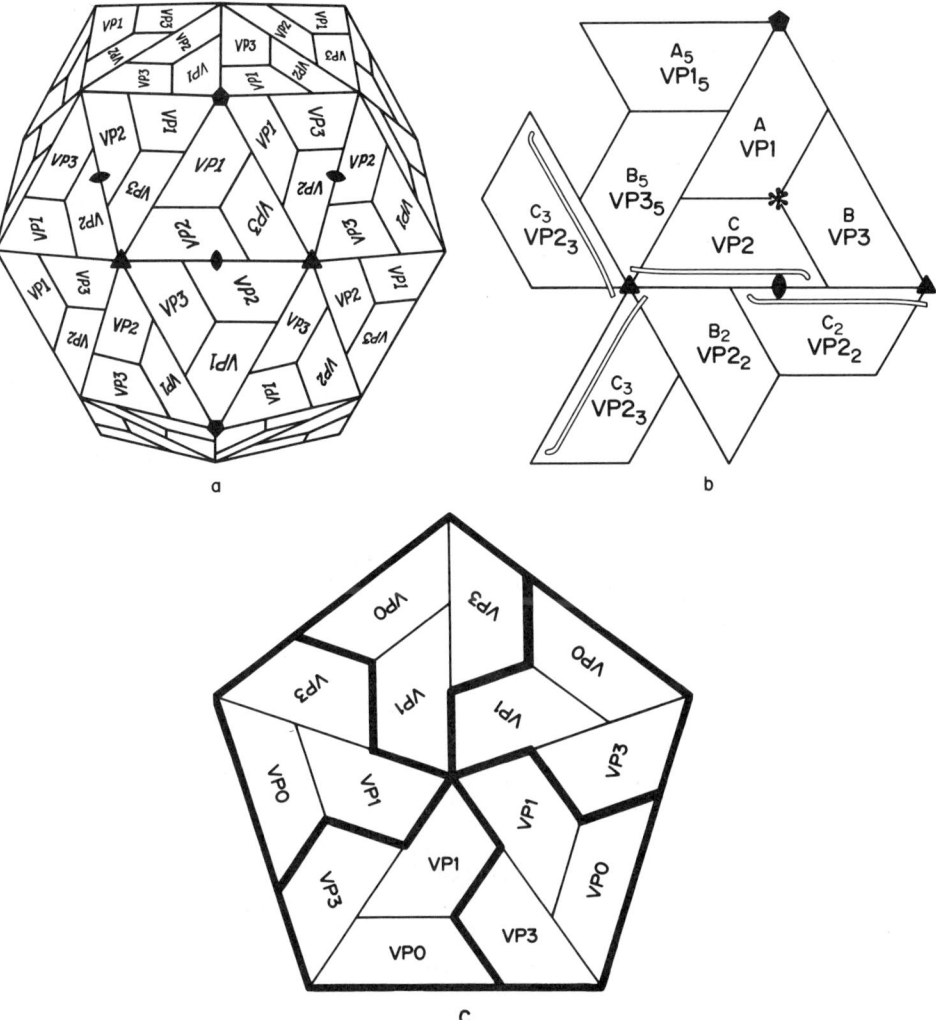

Fig. 2. Relation of the pseudoequivalent VP1, VP2, and VP3 subunits to the quasi-equivalent subunits A, C, and B in TBSV and SBMV. (a) Icosahedral capsid. (b) Icosahedral asymmetric unit, showing the ordered amino-terminus arm βA present only in the C subunit of the plant viruses and VP2 of HRV14. The amino end of the arm interacts with two other VP2 arms across the threefold axis, whereas the carboxy end of the arm interacts with a VP2 across the twofold axis. An asterisk indicates the position of the quasi-threefold axis in SBMV and TBSV analogous to the pseudothreefold axis in HRV14. Subscripts designate the symmetry operation required to obtain the given subunit from the basic triangle. (c) The thickly outlined VP1, VP3, VP0 unit corresponds to the 6 S protomer, and the 15-mer cap to the 14 S pentamer observed in assembly experiments.

molecule. Both the tertiary fold of the VP1, VP2, and VP3 polypeptide chains and their quaternary organization within the HRV14 capsid are closely similar to the two published high-resolution structures of $T = 3$ (180 identical subunits per capsid) RNA plant viruses, TBSV (65) and SBMV (66). Although the subunit organization within the icosahedron is somewhat different (85), a similar tertiary structure has also been found for $T = 1$ (60 subunits per capsid) STNV (67). The radial position and orientation of structurally equivalent atoms of HRV14 and SBMV generally agree to better than 3 Å relative to the icosahedral symmetry axes. In the plant viruses, the three quasi-equivalent subunits A, B, and C have the same amino acid sequence but cannot have identical geometrical environments (86). For SBMV, there are 260 amino acids per subunit but the first 63 residues are associated with the RNA and, therefore, do not have icosahedral symmetry. These are said to be in the "random" domain. In the C subunit, unlike the A and B subunits, residues 38 to 63 are ordered with their ends forming a "β-annulus" about the icosahedral threefold axes (Fig. 2b). Residues 64 to 260 of SBMV are referred to as the "shell" domain and form an eight-stranded antiparallel β-barrel (Fig. 3a). The β-barrels in HRV14, like those in SBMV, are wedge-shaped with the thin end (left in Fig. 3) pointing toward the five- or threefold (quasi-sixfold) axes. There are four excursions of the polypeptide chain toward the wedge-shaped end. Each excursion makes a sharp bend or "corner." The most exterior (top left in Fig. 3) corner is formed between the β-sheets βB and βC, the second corner down is formed between βH and βI, the third corner is between βD and $\beta E1$, and the most internal corner connects βF and βG. The $\beta F-\beta G$ corner is the site of a 25-residue insertion in SBMV, including the α-helix αC, that is not present in any of the viral proteins of picornaviruses, nor in TBSV or STNV. The alignment of the VP1, VP2, and VP3 chains of several picornaviruses is shown relative to the capsid protein of SBMV in Fig. 4.

The three larger capsid proteins VP1, VP2, and VP3 in HRV14 are oriented and situated at essentially the same radius and position as the A, C, and B subunits in SBMV, respectively. The capsid proteins of HRV14 are related by a pseudothreefold axis, analogous to the quasi-threefold axis in SBMV and TBSV, at the asterisk in Fig. 2b. However, VP1 and VP3 have additions at their amino and carboxy termini. The amino ends, as in plant viruses, are in contact with the RNA, but unlike plant viruses, they are more acidic than basic and have only a very few amino-terminus residues "disordered" (lacking icosahedral symmetry). The first 64 of the 73 amino-terminus residues of VP1 reside under VP3, whereas the first 42 of the 71 amino-terminus resi-

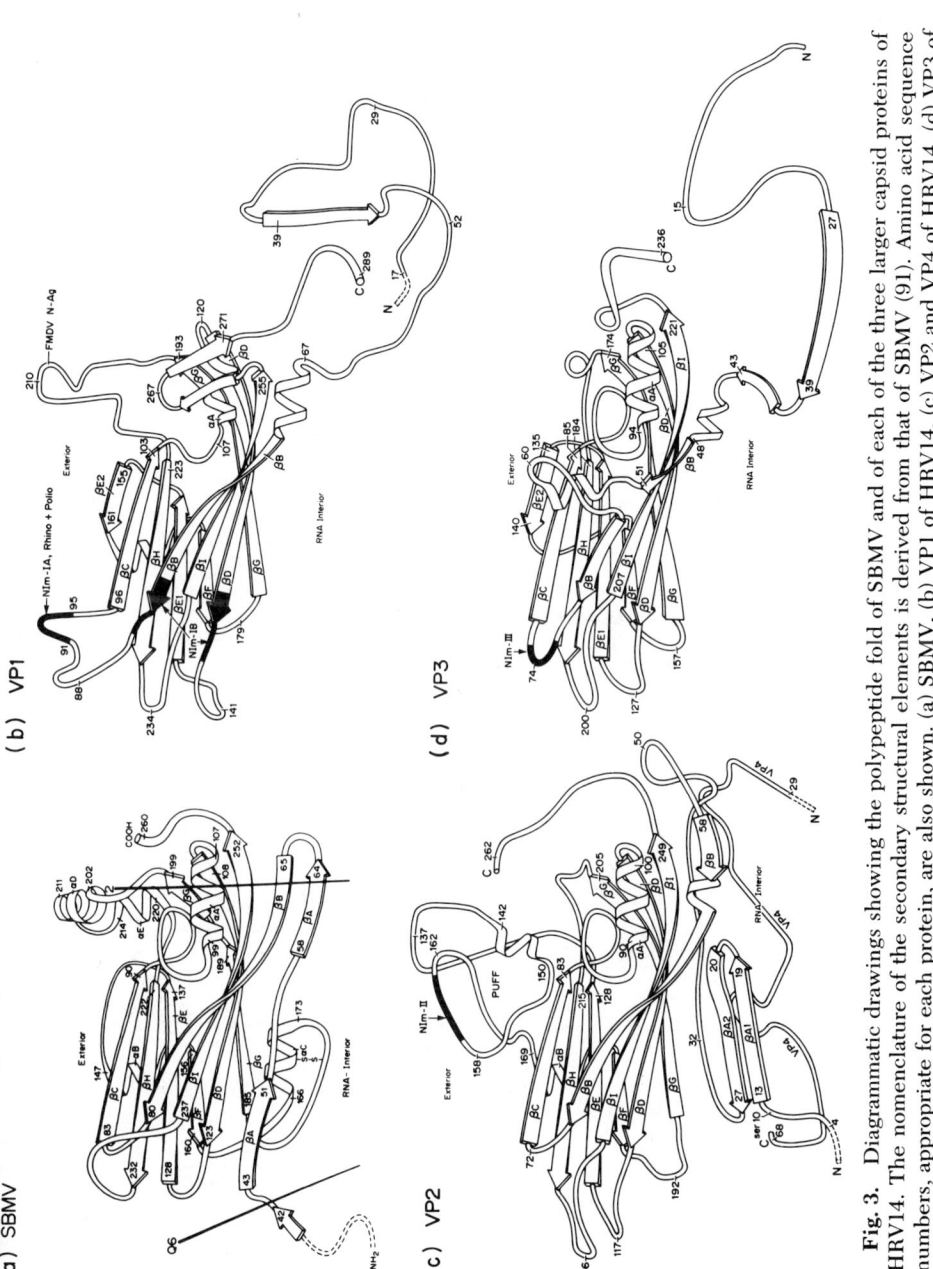

Fig. 3. Diagrammatic drawings showing the polypeptide fold of SBMV and of each of the three larger capsid proteins of HRV14. The nomenclature of the secondary structural elements is derived from that of SBMV (91). Amino acid sequence numbers, appropriate for each protein, are also shown. (a) SBMV, (b) VP1 of HRV14, (c) VP2 and VP4 of HRV14, (d) VP3 of HRV14. (Adapted from a drawing of SBMV by Jane Richardson.)

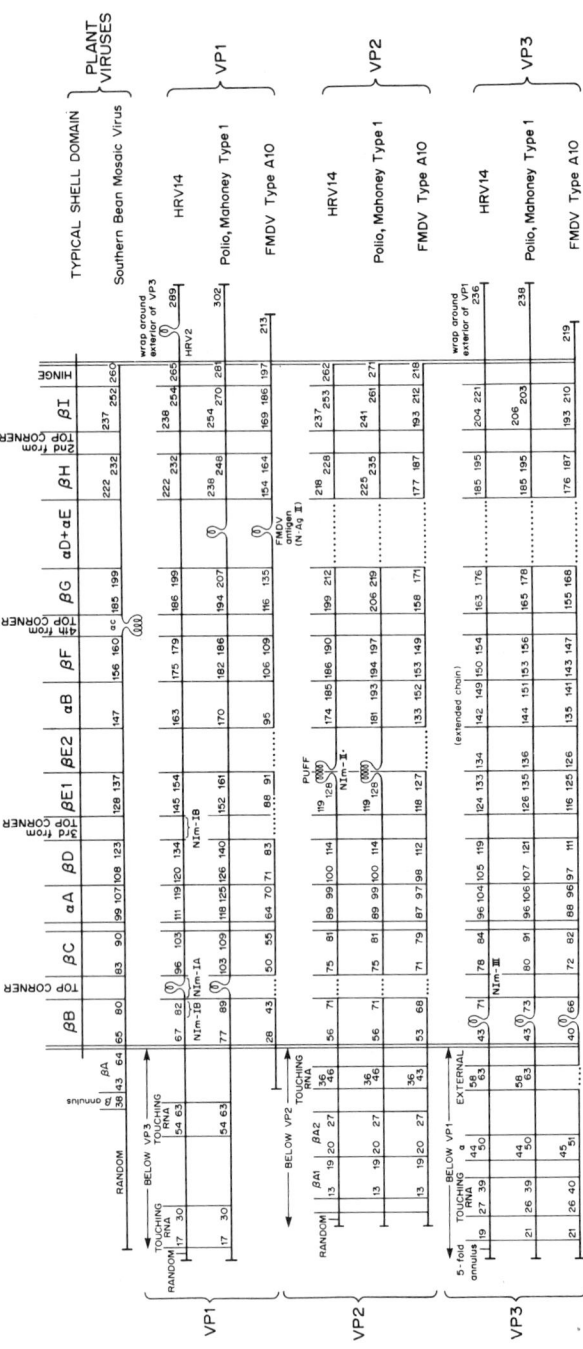

Fig. 4. Alignment of the three larger viral proteins VP1, VP2, and VP3 with a typical plant viral capsid protein as, for instance, the SBMV coat protein. The loop symbols and dot symbols represent insertions and deletions with respect to the typical plant virus capsid protein. Immunogenic sites found in HRV14 are also shown. Three different picornaviruses are considered. The assignments of HRV14 are derived from the electron density map and the known sequence of its RNA. The alignment of polio type 1 Mahoney was deduced from the amino acid homology. The alignment of FMDV type A10, determined from amino acid homology with all available picorna sequences, is uncertain from residue 1 to 90 in VP1.

dues of VP3 are under VP1. Thus, the predominant positions of VP1 and VP3 at the RNA-protein interface are exchanged relative to their positions at the exterior surface.

The first 25 of the 69 residues of the internal structural protein VP4 are not seen in the electron density map, implying that they lack icosahedral symmetry. VP4 is positioned in part below VP1 and VP2 with its visible amino end surrounding the fivefold axis. The carboxy ends of VP1 and VP3 are external and function in part to associate proteins within a protomer (Fig. 2c).

Large sequence insertions relative to the typical shell domain (Fig. 4) form protrusions on VP1, VP2, and VP3 and create a deep cleft or "canyon wall" on the viral surface. The canyon separates the major part of five VP1 subunits (in the "North") clustered about a pentamer axis from the surrounding VP2 and VP3 subunits (in the "South"), thus forming a moat around the VP1 protrusions on the fivefold axis. The South canyon walls are lined with the carboxy-terminus ends of VP1 and a large sequence insertion in VP1 corresponding to helices αD and αE in the equivalent SBMV capsid protein. The North canyon wall is partially lined with the carboxy terminus of VP3. VP2 is hardly associated at all with the canyon, whereas VP1 is the major contributor to the residues lining the canyon. The canyon is ~25 Å deep and 12 to 30 Å wide.

Because of the additional elaborations which VP1 has on the surface relative to VP2 and VP3, its overall shape is that of a kidney, with the depression forming a large part of the canyon. The first 16 residues of VP1 (Fig. 3b) are not seen in the electron density map. The shell domain of VP1 in HRV14 starts at residue 74. The small sequence insertion between βB and βC in rhinovirus and poliovirus is not found in FMDV. This loop forms a major immunogen in HRV14 (NIm-IA to be discussed below) and poliovirus. Five carbonyl oxygens of fivefold-related Tyr 142 residues converge on the vertex to chelate a presumably cationic ligand on the fivefold axis. The residues in VP1 of HRV14, which are analogous to αD and αE helices in SMBV, protrude to the surface and form part of the South rim of the canyon, but do not form helices in HRV14. There is an 8-residue insertion in polio and FMDV relative to HRV14 (Fig. 4) at the most external portion of this segment. These additional residues contain the major antigenic site of FMDV and have been predicted to form an α-helix (87). The carboxy-terminus 23 residues of VP1 line the South rim of the canyon in HRV14. There is little conservation of these residues among picornaviruses.

The first three residues of VP2 (Fig. 3b) are not seen in the electron density. The proximity of serine 10 in VP2 to the carboxy terminus of VP4 suggests that the VP0 cleavage, which occurs during the virus maturation step, may be autoproteolytic. There is no histidine in the immediate vicinity of Ser 10 and the carboxy end of VP4. However, nucleotide bases of the RNA might act as proton acceptors in the autocatalysis. Thus the insertion of RNA into the growing capsids might trigger the change of VP0 into VP2 and VP4 [cf. serine proteases (88,89)]. The first nine residues of VP2 are likely to have been rearranged after cleavage. Portions of the region between residues 10 and 72 are involved in contacts between twofold-related VP2 polypeptide chains (Fig. 2b). The αD and αE helices are absent in VP2 (Fig. 3b). There is a large 43-residue insertion, in the VP2 position corresponding to βE2 of VP1 and VP3, forming an external mushroom-shaped "puff." This is positioned adjacent to the VP1 elaborations, associated with the major antigenic site in FMDV, which line the South canyon wall. The most external residues of this puff correspond to NIm-II of HRV14. In contrast to VP1 and VP3, the carboxy terminus of VP2 has no extension beyond the shell domain.

All residues of VP3 (Fig. 3d) can be seen in the electron density. The 26 amino-terminus residues form a fivefold β-barrel about the pentamer axis analogous to the β-annulus (65) about trimer axes in SBMV and TBSV. This fivefold annulus extends down into the RNA to a radius of 111 Å. The polypeptide emerges from the β-annulus, circles around the base of the VP1 shell domain while making extensive contact with the RNA, emerges on the viral surface near residue 61, and then enters the shell domain at residue 72. The top corner of the VP3 shell domain, between βB and βC, is the NIm-III site of HRV14, structurally equivalent to NIm-IA in VP1. Helix αB of SBMV is replaced by an extended chain in VP3 and the external helices αD and αE of SBMV are absent as in VP2. When the shell domains of VP3 and VP1 are superimposed, much of the amino- and carboxy-terminus axes are about structurally equivalent.

IMMUNOGENIC SITES ON HRV14

Amino acid residues within the major neutralization immunogens of HRV14 have been identified by Sherry and Rueckert (54) and by Sherry et al. (55). They isolated mouse hybridoma lines which secrete monoclonal antibodies that neutralize HRV14. Each was then used to select several viral mutants resistant to neutralization by that antibody. Finally, every monoclonal antibody was assayed for its ability to neutralize the mutants. The results revealed four major immunogenic neutralization sites. Each immunogenic site was composed of overlapping epitopes where a given mutant was resistant to many or all of the antibodies directed against that site. The amino acid residues defining the four immunogens are summarized in Table IV. These results were obtained without knowledge of the three-dimensional structure. When, however, the electron density map became available, it was immediately clear that the substitutions that could confer resistance to neutralization, regardless of their location in the amino acid sequences, were localized into four distinct areas corresponding exactly to the proposed immunogens. Moreover, these residues invariably

TABLE IV
Amino Acids Associated with Each of the Four Major Immunogenic Sites in HRV14[a]

	Amino acid number	Wild type		Observed mutations
NIm-IA				
VP1	91	D	→	A F G H N V Y
VP1	95	E	→	G K
NIm-IB				
VP1	83	Q	→	H
VP1	85	K	→	N
VP1	138	D	→	E G
VP1	139	S	→	P
NIm-II				
VP2	158	S	→	F
VP2	159	A	→	V
VP2	161	E	→	D V[b] D[b] K
VP2	162	V	→	M A[b]
VP1	210	E	→	D[b]
VP2	136	E	→	G
NIm-III				
VP3	72	N	→	I
VP3	75	R	→	G K M
VP3	78	E	→	K V
VP1	287	K	→	I
VP3	203	G	→	D

[a] Substitutions in mutants selected for resistance to neutralization by monoclonal antibodies.

[b] A double mutation to V161 and A162 occurred in two cases; a double mutation to D161 and D210 occurred in a third case.

faced outward toward the viral exterior. The possible effects which a given mutation might have on the capacity of the antibody to neutralize the virus have not yet been fully scrutinized in light of the structure. For a fuller understanding of these phenomena, it will be necessary to differentiate between mutants that resist neutralization while still maintaining their ability to bind antibodies and those mutants which entirely fail to bind antibodies.

The immunogenic site NIm-IA is an insertion between βB and βC at the topmost corner of the VP1 wedge. It is the most external portion of the complete virion, being 160 Å from the center. Residues 91 and 95, associated with NIm-IA, are on the extreme external portion of the loop.

Residues associated with site NIm-IB on VP1 are on the carboxy ends of βB and βD, which are situated on either side of the amino end of strand βI. No mutations conferring resistance have yet been found on the βI strand. The NIm-IB site is close to the fivefold axis, a little below the canyon rim.

NIm-II on VP2 is at the extreme outside of the puff, 155 Å from the viral center. This immunogenic puff is adjacent to the external loop formed by the sequence insertion in VP1 corresponding to αD and αE of SBMV. One of the residues (210E on VP1) that is associated with NIm-II is, indeed, on VP1 (Table IV).

NIm-III on VP3 is 149 Å from the viral center, to some extent in the shadow of a larger protrusion of VP3 (residues 58 to 63) and the carboxy end of VP1 (282–286). NIm-III is in a position on VP3 corresponding to NIm-IA on VP1, that is, in the loop between βB and βC. Residue K287 on VP1, associated with this site (Table IV), points directly toward and is adjacent to the other residues on VP3 associated with NIm-III. The large protrusion near NIm-III, consisting of residues 282–286 in VP1 and 58–63 in VP3, has not been shown to be antigenic for HRV14 but in FMDV residues associated with the carboxy-terminus end of VP1 have shown antigenicity (Table V) (90–99). It is possible that this structural component lacks antigenicity in HRV14 because of greater rigidity (100,101), and this hypothesis will be checked when the HRV14 structure is refined and temperature factors have been determined.

ANTIGENIC SITES IN POLIOVIRUS AND FMDV

Table V summarizes results obtained in mapping the immunogenic and antigenic surfaces of other picornaviruses. There is a dominant immunogen in poliovirus corresponding to NIm-IA of rhinovirus. Ab-

TABLE V
Evidence for Antigenic Regions in Poliovirus and FMDV

	Virus	Coat protein	Amino acid number	HRV14[a] amino acid equivalent	Serotype[b]	Ref.	Method[c]	Position in HRV14 structure
1.	Polio	VP1	11–17	Deleted	1M	50	3	Within RNA
2.	Polio	VP1	24–40	14–30	1M	90	5	Faces RNA
3.	Polio	VP1	70–75	60–65	1M	50	3,6	Faces RNA
4.	Polio	VP1	61–80	51–70	1M	90	2	Faces RNA
5.	Polio	VP1	86–103	77–97	1M	90	5	NIm-IB + NIm-IA
6.	Polio	VP1	91–109	84–103	1M	90	2	NIm-IB + NIm-IA
7.	Polio	VP1	100–109	93–103	1M	90	2	Part of NIm-IA
8.	Polio	VP1	93–103	86–97	1M	50	2,3,6	NIm-IB(?) + NIm-IA[d]
9.	Polio	VP1	93–104	86–98	1	91	4	NIm-IB(?) + NIm-IA[d]
10.	Polio	VP1	93–100	86–93	3L	92,93	1	NIm-IB(?) + part of NIm-IA[d]
11.	Polio	VP1	141–147	135–141	1M,S	94	3	NIm-IB
12.	Polio	VP1	161–181	154–174	1M	90	5	Partly exposed on canyon
13.	Polio	VP1	182–201	175–193	1M	90	2	Buried on 4th corner down
14.	Polio	VP1	222–241	212–225	1M	90	2	NIm-II
15.	FMDV	VP1	141–160	210–228	O,K	95	2	NIm-II
16.	FMDV	VP1	144–159	211–227	O,K	87	2	NIm-II
17.	FMDV	VP1	145–168	211–236	A12	96	8	NIm-II
18.	FMDV	VP1	146–154	211–223	O,K	97	7	NIm-II
19.	FMDV	VP1	169–179	237–247	A12	96,98	8	NIm-IB central strand
20.	Polio	VP1	270–287	254–272	1M	90	5	Partly exposed on canyon wall
21.	FMDV	VP1	200–213	268–294	O,K	95	2	NIm-III(?)
22.	FMDV	VP1	200–213	268–294	O,K	97	7	NIm-III(?)
23.	Polio	VP2	162–173	161–170	1M	99	2,3	NIm-II
24.	Polio	VP3	71–82	70–81	1M	99	6	NIm-III

[a] Aligned by eye and by fitting to the shell domain structure. [b] 1M (type 1, Mahoney), S (Sabin), 3L (type 3, Leon), O_1K (type O_1, strain Kaufbeuern), A12 (type A, subtype 12). [c] Method key: (1) Monoclonal antibodies raised against intact virus select for resistant mutations. (2) Synthetic peptides induce neutralizing antibodies. (3) Synthetic peptides prime for high titer neutralizing response. (4) Synthetic peptide competes with monoclonal antibody to inhibit neutralization. (5) Synthetic peptides induce antibodies which bind virus but neutralize poorly. (6) Neutralizing antisera or monoclonal antibodies bind peptide in ELISA. (7) Deduced from ability or inability of protein fragments to induce neutralizing antibodies. (8) Deduced from ability or inability of neutralizing monoclonal antibodies to bind protein fragments. [d] Uncertainty as a result of somewhat arbitrary nature of computer alignments.

sence of a corresponding immunogen in FMDV is explained by deletion of this loop from its VP1.

The dominant antigenic site in FMDV resides in a region homologous to NIm-II in HRV14 (Table V). The VP1 contribution to this antigen has also been shown for HRV14 (E210 of NIm-II in Table IV) and for poliovirus (residues 222–241, Table V). In FMDV, however, the immunogenic puff on VP2 is absent and eight amino acids are inserted at the extreme surface of the $\alpha D-\alpha E$ region of VP1. The resulting protrusion of NIm-II in FMDV would be unsupported by the puff. This protrusion may, therefore, occupy the space which in HRV14 is occupied by the puff. The inability of FMDV VP1 to elicit neutralizing antibodies (102) is consistent with the lack of structural support for this NIm-II protrusion in the absence of the neighboring VP2.

Poliovirus can also be neutralized by antibodies that bind to NIm-II or NIm-III (Table V). Thus, there is an overall consistency in identifying at least three of the major HRV14 neutralization sites in poliovirus.

Many of the methods used to determine antibody binding sites depended on the use of synthetic peptides as antigens (Table V). Peptides associated with neutralizing antigenic regions do, in general, elicit antibodies that can neutralize the intact virus. In a significant number of cases (lines 1, 2, 3, 4, 8, and 13 of Table V), however, the sequence in question lies far below the viral surface or even buried in the RNA. This suggests that some peptides can elicit antibodies which subsequently bind to totally unrelated portions of the native virus. Alternatively, some of the results might be accounted for by conformational changes occurring during the isoelectric transition of the virus (103).

The correspondence of sequence variability with antigenic sites has frequently been pointed out in the study of picornaviruses. This is certainly apparent on comparing the two available rhinovirus sequences of HRV14 (35,36) and HRV2 (37), as well as in comparative studies of polio (27) and FMDV (31,104) sequences. Nevertheless, the surface protrusion caused by HRV14 VP3 58–63 together with VP1 282–286 is equally variable, but has not yet been associated with an antigenic site in any picornaviruses.

THE CANYON AS RECEPTOR BINDING SITE

The 25-Å-deep canyon, circulating around each of the 12 pentamer vertices, suggests that this is the site for cell receptor binding. An antibody molecule would have difficulty reaching the canyon floor, its entrance being blocked by the canyon rim. Thus, the residues in the deeper recesses of the canyon would not be under immune selection and could remain constant, permitting the virus to retain its ability to seek out the same cell receptors. A similar situation of a protected sialic acid binding site exists in the neuraminidase spike of influenza virus (105,106).

While retention of the canyon structure for all picornaviruses is to be expected, variation in the residues lining the canyon should be anticipated between viruses that attach themselves to different host cell receptors. That is, FMDV, polio, HRV14, and HRV2, all of which recognize different receptors, should exhibit some variation in the residues lining the canyon wall. It is thus noteworthy that not only are those parts of the carboxy-terminus ends of VP1 and VP3 which line the canyon walls some of the least conserved amino acids among picornaviruses, but also there is a deletion in HRV2 relative to HRV14 in βB at NIm-IB which is possibly associated with cell receptor carbohydrate recognition (see below).

Since the topology of the canyon should be retained, the highly conserved sequence (MYVPPGAPNP starting at 151 of VP1 and AYTPPGARGP starting at 130 of VP3 for HRV14) in rhino, polio, and FMD viruses situated in the middle of the floor of the canyon may be significant.

FMDV can be treated with trypsin, causing cleavage at residues between 138 and 154 of VP1. This causes the virus to lose its ability to attach to cells and its ability to stimulate neutralizing antibodies (97,98,107,108). The enzymatic cleavages occur in NIm-II on the αD–αE protrusion of VP1, a large loop which also forms part of the presumed host receptor binding site.

Concanavalin A is a lectin extracted from jack beans. It has been shown to compete with solubilized Coxsackie B3 virus for receptors on HeLa cells (59). The lectin also interferes with infection of HRV and poliovirus (60), suggesting that there might be common structural features between concanavalin A and these viruses. Argos et al. (109) have shown that there is a remarkable similarity in the folding topology of concanavalin A and the shell domain of TBSV (and hence also to VP1 of HRV14), thus providing some rationalization of the competition experiments at the molecular level. Many animal and bacterial viruses recognize their host cells by virtue of a specific polysaccharide (110). The latter may be a part of the virus or of the host cell receptor. Concanavalin A can bind D-glucose and D-mannose pyranosides. The functional sugar binding site of concanavalin A is in a position equivalent to the amino end of βI, that is, roughly the NIm-IB site. If this comparison is meaningful, it might be anticipated that the NIm-IB site, which lies on the rim of the receptor binding cavity, would be associated with polysaccharide recognition, a hypothesis that should be readily testable using crystallography.

NEUTRALIZATION AND SEROTYPES

Some poliovirus neutralizing antibodies that bind to VP1 require bivalent attachment (52,53) associated with a change in isoelectric point from pI=6 to pI=4. Cleavage with papain of the bound antibody restores infectivity and the original isoelectric point. Icenogle et al. (48) have shown that only about four copies per virion of a particular, bivalently bound, monoclonal antibody were needed to neutralize poliovirus, without altering the pI. Both these observations suggest that neutralization may involve conformational changes that interfere with cell attachment, penetration, or uncoating.

Since an antibody itself has a twofold axis, bivalent attachment could occur across icosahedral twofold axes. The distance between the nearest twofold-related immunogenic sites IA, IB, II, and III is 120, 120, 50, and 60 Å, respectively. All other symmetrical counterparts of these antigens related by twofold axes are at least 170 Å apart. Lower and upper limits of the distance of the two antibody binding sites on an immunoglobulin molecule are not well known but probably lie in the range of 50 to 180 Å (111–113).

Polyclonal antisera raised against virus are likely to contain antibodies against all immunogenic regions on the viral surface. Thus, different serotypes must express at least one mutation at each immunogenic site. If, say, four separate mutational events (one for each immunogenic site in HRV14) occurred in the corresponding limited region of the RNA, it is highly probable that mutations also will have occurred at other regions. Apparently, then, there could be some constraint on additional mutations accumulating over time which limits the known number of polio serotypes, while permitting numerous serotypes for rhinoviruses. Constraints limiting the diversity of viruses might be the cell receptor specificity (mediated perhaps by the nature of the carbohydrate attachment near NIm-IB), RNA structure providing hot spots for mutations, a reduced number of generations in particularly virulent pathogens, and the particular viral habitat.

ASSEMBLY

Assembly of picornaviruses (1–3,8–11) proceeds from 6 S protomers of VP1, VP3, and VP0, via 14 S pentamers of five 6 S protomers, to mature virions. The final step involves inclusion of the RNA into empty capsids or partially assembled shells with simultaneous cleavage of VP0 into VP2 and VP4. Conversely, *in vitro* disassembly, produced by mild denaturation, proceeds via the expulsion of VP4 followed by the RNA (3).

Both the amino and carboxy ends of VP1 and VP3 are intertwined with each other. Furthermore, if VP4 and VP2 are considered as VP0, then VP0 is also intertwined with VP1 and VP3. This strongly suggests that the 6 S protomer is as shown in Fig. 2c. These protomers are themselves intertwined by virtue of the fivefold β-annulus formed by the amino ends of the VP3's and the proximity of the observed amino ends of VP4's to the fivefold axis. Thus, the 14 S pentamers closely correlate with the observed structure shown diagrammatically in Fig. 2c. Such an assembly sequence matches that observed in plant viruses, in particular that of SBMV, where the building blocks are dimers corresponding to VP1 and VP3 and where the formation of intermediates with fivefold symmetry is considered to be a critical stage in the formation of $T = 1$ and $T = 3$ capsids (114). Indeed, it has now been shown (115) that the 15-protein cluster, corresponding to the one shown in Fig. 2c, is conserved between $T = 1$ and $T = 3$ SBMV particles.

VP2, once cleaved from VP4, is globular and does not contact the other proteins extensively. There are large solvent-accessible regions between VP2 and the surrounding proteins. This, as well as the extraordinarily internal heavy atom sites on VP2 (Table III), is consistent with the loose binding of VP2 to the capsid. (The ability of the heavy atoms to penetrate deeply into the shell suggests cautious interpretation of data based on the use of small chemical labels for mapping viral surfaces.) Disruption of pentamer–pentamer contacts, mediated by a slight reorientation of VP2 or its complete removal, could provide a port by which the VP4 and RNA exit. Binding of a cell receptor in the canyon adjacent to VP3 could facilitate this process, possibly accompanied by an isoelectric change.

It is remarkable how, in both plant and animal RNA viruses, there is a β-annulus type structure between the amino ends of some subunits. It is equally remarkable how the amino ends of the capsid proteins are invariably associated with the RNA. In TBSV, SBMV, and STNV they are basic, whereas in HRV14 they are slightly acidic. These properties may be significant for the initial events of assembly.

ACKNOWLEDGMENTS

This paper was adapted from a manuscript submitted to *Nature*. We are most grateful to Sharon Wilder for her outstanding assistance in the organization of the Purdue structural laboratory over many years and in the preparation of this manuscript. We wish to thank Kathy Shuster, Bill Boyle, and Jun Tsao for preparation of the figures. The data used in this study were collected at the Cornell High Energy Synchrotron Source (CHESS). We are, indeed, most grateful for all the dedicated help we have received from Keith Moffat, Wilfred Schildkamp, Robert Hunt, Don Bilderbeck, Boris Batterman, Aggie Sirrine, and all the CHESS staff and operators. We would also like to thank Dr. Gale Rhodes and Diana Delatore for helping us on our visits to CHESS. The work was also aided by a number of visits to the EMBO Outstation at the DESY synchrotron in Hamburg, where we were instructed in the use of the equipment by Hans Bartunik and Klaus Bartels. We are most grateful to the Purdue University Computer Center staff (in particular Saul Rosen, John Steele, Tom Putnam, and Paul Townsend) for their generous help and encouragement of our use of the Cyber 205 supercomputer and to those at Purdue University like Don Brown and Struther Arnott who have made the supercomputer and other resources available to us. We have benefited greatly by the helpful and stimulating discussions with our other colleagues in the Purdue structural groups, including Jeffrey Bolin, Abelardo Silva, Ignacio Fita, Celerino Abad-Zapatero, R. Usha, M. V. Hosur, Cynthia Stauffacher, Patrick Argos, and J. K. Mohana Rao. We thank Richard Colonno (Merck Sharp & Dohme Co.) for his interest in the work and for providing the HRV14-RNA sequence prior to publication and also thank Ann Palmenberg (University of Wisconsin) for providing her sequence alignments of picornaviruses. We also thank Tim Schmidt for maintaining the X-ray equipment at Purdue University. The work was supported by an NIH grant to M.G.R., and NSF grant for supercomputer time to M.G.R., and an ACS grant to R.R.R. CHESS is supported by an NSF grant to Boris Batterman and the macromolecular diffraction facility at CHESS by an NIH grant to Keith Moffat. A postdoctoral Walter Winchell–Damon Runyon fellowship supported E.A. for some of the time and a predoctoral NIH training grant supported B.S. A Purdue University Showalter Foundation grant was awarded to M.G.R. to equip the cell culture laboratory and a recent grant from the Merck Sharp & Dohme Co. contributed to the salary of one technician involved in the virus propagation.

REFERENCES

1. Putnak, J. R., and Phillips, B. A. (1981). *Microbiol. Rev.* **45**, 287–315.
2. Sangar, D. V. (1979). *J. Gen. Virol.* **45**, 1–13.
3. Rueckert, R. R. (1976). *In* "Comprehensive Virology" (H. Fraenkel-Conrat and R. R. Wagner, eds.), Vol. 6, pp. 131–213. Plenum, New York.
4. Rueckert, R. R., Dunker, A. K., and Stoltzfus, C. M. (1969). *Proc. Natl. Acad. Sci. U.S.A.* **62**, 912–919.
5. Rueckert, R. R., and Wimmer, E. (1984). *J. Virol.* **50**, 957–959.
6. McGregor, S., Hall, L., and Rueckert, R. R. (1975). *J. Virol.* **15**, 1107–1120.

7. Jacobson, M. F., Asso, J., and Baltimore, D. (1970). *J. Mol. Biol.* **49**, 657–669.
8. McGregor, S., and Rueckert, R. R. (1977). *J. Virol.* **21**, 548–553.
9. Fernandez-Tomas, C. B., Guttman, N., and Baltimore, D. (1973). *J. Virol.* **12**, 1181–1183.
10. Jacobson, M. F., and Baltimore, D. (1968). *J. Mol. Biol.* **33**, 369–378.
11. Fernandez-Tomas, C. B., and Baltimore, D. (1973). *J. Virol.* **12**, 1122–1130.
12. Lee, Y. F., Nomoto, A., Detjen, B. M., and Wimmer, E. (1977). *Proc. Natl. Acad. Sci. U.S.A.* **74**, 59–63.
13. Sangar, D. V., Rowlands, D. J., Harris, T. J. R., and Brown, F. (1977). *Nature (London)* **268**, 648–650.
14. Ambros, V., and Baltimore, D. (1978). *J. Biol. Chem.* **253**, 5263–5266.
15. Golini, F., Semler, B. L., Dorner, A. J., and Wimmer, E. (1980). *Nature (London)* **287**, 600–603.
16. Vartapetian, A. B., Koonin, E. V., Agol, V. I., and Bogdanov, A. A. (1984). *EMBO J.* **3**, 2593–2598.
17. Wimmer, E. (1982). *Cell (Cambridge, Mass.)* **28**, 199–201.
18. Pallansch, M. A., Kew, O. M., Semler, B. L., Omilianowski, D. R., Anderson, C. W., Wimmer, E., and Rueckert, R. R. (1984). *J. Virol.* **49**, 873–880.
19. Hanecak, R., Semler, B. L., Anderson, C. W., and Wimmer, E. (1982). *Proc. Natl. Acad. Sci. U.S.A.* **79**, 3973–3977.
20. Carthew, P., and Martin, S. J. (1974). *J. Gen. Virol.* **24**, 525–534.
21. Lonberg-Holm, K., and Butterworth, B. E. (1976). *Virology* **71**, 207–216.
22. Beneke, T. W., Habermehl, K. O., Diefenthal, W., and Buchholz, M. (1977). *J. Gen. Virol.* **34**, 387–390.
23. Lund, G. A., Ziola, B. R., Salmi, A., and Scraba, D. G. (1977). *Virology* **78**, 35–44.
24. Wetz, K., and Habermehl, K. O. (1979). *J. Gen. Virol.* **44**, 525–534.
25. Hordern, J. S., Leonard, J. D., and Scraba, D. G. (1979). *Virology* **97**, 131–140.
26. Wetz, K., and Habermehl, K. O. (1982). *J. Gen. Virol.* **59**, 397–401.
27. Toyoda, H., Kohara, M., Kataoka, Y., Suganuma, T., Omata, T., Imura, N., and Nomoto, A. (1984). *J. Mol. Biol.* **174**, 561–585.
28. Kitamura, N., Semler, B. L., Rothberg, P. G., Larsen, G. R., Adler, C. J., Dorner, A. J., Emini, E. A., Hanecak, R., Lee, J. J., Can der Werf, S., Anderson, C. W., and Wimmer, E. (1981). *Nature (London)* **291**, 547–553.
29. Stanway, G., Cann, A. J., Hauptmann, R., Hughes, P., Clarke, L. D., Mountford, R. C., Minor, P. D., Schild, G. C., and Almond, J. W. (1983). *Nucleic Acids Res.* **11**, 5629–5643.
30. Racaniello, V. R., and Baltimore, D. (1981). *Proc. Natl. Acad. Sci. U.S.A.* **78**, 4887–4891.
31. Makoff, A. J., Paynter, C. A., Rowlands, D. J., and Boothroyd, J. C. (1982). *Nucleic Acids Res.* **10**, 8285–8295.
32. Carroll, A. R., Rowlands, D. J., and Clarke, B. E. (1984). *Nucleic Acids Res.* **12**, 2461–2472.
33. Forss, S., Strebel, K., Beck, E., and Schaller, H. (1984). *Nucleic Acids Res.* **12**, 6587–6601.
34. Boothroyd, J. C., Highfield, P. E., Cross, G. A. M., Rowlands, D. J., Lowe, P. A., Brown, F., and Harris, T. J. R. (1981). *Nature (London)* **290**, 800–802.
35. Stanway, G., Hughes, P. J., Mountford, R. C., Minor, P. D., and Almond, J. W. (1984). *Nucleic Acids Res.* **12**, 7859–7875.
36. Callahan, P. L., Mizutani, S., and Colonno, R. J. (1985). *Proc. Natl. Acad. Sci. U.S.A.* **82**, 732–736.

37. Skern, T., Sommergruber, W., Blaas, D., Gruendler, P., Fraundorfer, F., Pieler, C., Fogy, I., and Kuechler, E. (1985). *Nucleic Acids Res.* **13**, 2111–2126.
38. Palmenberg, A. C., Kirby, E. M., Janda, M. R., Drake, N. L., Duke, G. M., Potratz, K. F., and Collett, M. S. (1984). *Nucleic Acids Res.* **12**, 2969–2985.
39. Najarian, R., Caput, D., Gee, W., Potter, S. J., Renard, A., Merryweather, J., Van Nest, G., and Dina, D. (1985). *Proc. Natl. Acad. Sci. U.S.A.* **82**, 2627–2631.
40. Linemeyer, D. L., Menke, J. G., Martin-Gallardo, A., Hughes, J. V., Young, A., and Mitra, S. W. (1985). *J. Virol.* **54**, 247–255.
41. Argos, P., Kamer, G., Nicklin, M. J. H., and Wimmer, E. (1984). *Nucleic Acids Res.* **12**, 7251–7267.
42. Franssen, H., Leunissen, J., Goldbach, R., Lomonossoff, G., and Zimmern, D. (1984). *EMBO J.* **3**, 855–861.
43. Kamer, G., and Argos, P. (1984). *Nucleic Acids Res.* **12**, 7269–7282.
44. Haseloff, J., Goelet, P., Zimmern, D., Ahlquist, P., Dasgupta, R., and Kaesberg, P. (1984). *Proc. Natl. Acad. Sci. U.S.A.* **81**, 4358–4362.
45. Ahlquist, P., Strauss, E. G., Rice, C. M., Strauss, J. H., Haseloff, J., and Zimmern, D. (1985). *J. Virol.* **53**, 536–542.
46. Rezaian, M. A., Williams, R. H. V., Gordon, K. H. J., Gould, A. R., and Symons, R. H. (1984). *Eur. J. Biochem.* **143**, 277–284.
47. Dimmock, N. J. (1984). *J. Gen. Virol.* **65**, 1015–1022.
48. Icenogle, J., Shiwen, H., Duke, G., Gilbert, S., Rueckert, R., and Anderegg, J. (1983). *Virology* **127**, 412–425.
49. Mandel, B. (1976). *Virology* **69**, 500–510.
50. Emini, E. A., Jameson, B. A., and Wimmer, E. (1983). *Nature (London)* **304**, 699–703.
51. Schrom, M., Laffin, J. A., Evans, B., McSharry, J. J., and Caliguiri, L. A. (1982). *Virology* **122**, 492–497.
52. Emini, E. A., Kao, S., Lewis, A. J., Crainic, R., and Wimmer, E. (1983). *J. Virol.* **46**, 466–474.
53. Emini, E. A., Ostapchuk, P., and Wimmer, E. (1983). *J. Virol.* **48**, 547–550.
54. Sherry, B., and Rueckert, R. (1985). *J. Virol.* **53**, 137–143.
55. Sherry, B., Mosser, A. G., Colonno, R. J., and Rueckert, R. R. (1986). *J. Virol.* **57**, 246–257.
56. Abraham, G., and Colonno, R. J. (1984). *J. Virol.* **51**, 340–345.
57. Colonno, R. J., personal communication.
58. Minor, P. D., Pipkin, P. A., Hockley, D., Schild, G. C., and Almond, J. W. (1984). *Virus Res.* **1**, 203–212.
59. Krah, D. L., and Crowell, R. L. (1985). *J. Virol.* **53**, 867–870.
60. Lonberg-Holm, K. (1975). *J. Gen. Virol.* **28**, 313–327.
61. Mattern, C. F. T., and duBuy, H. G. (1956). *Science* **123**, 1037–1038.
62. Schaffer, F. L., and Schwerdt, C. E. (1955). *Proc. Natl. Acad. Sci. U.S.A.* **41**, 1020–1023.
63. Korant, B. D., and Stasny, J. T. (1973). *Virology* **55**, 410–417.
64. Finch, J. T., and Klug, A. (1959). *Nature (London)* **183**, 1709–1714.
65. Harrison, S. C., Olson, A. J., Schutt, C. E., Winkler, F. K., and Bricogne, G. (1978). *Nature (London)* **276**, 368–373.
66. Abad-Zapatero, C., Abdel-Meguid, S. S., Johnson, J. E., Leslie, A. G. W., Rayment, I., Rossmann, M. G., Suck, D., and Tsukihara, T. (1980). *Nature (London)* **286**, 33–39.

67. Liljas, L., Unge, T., Jones, T. A., Fridborg, K., Lövgren, S., Skoglund, U., and Strandberg, B. (1982). *J. Mol. Biol.* **159**, 93–108.
68. Hogle, J. M. (1982). *J. Mol. Biol.* **160**, 663–668.
69. Erickson, J. W., Frankenberger, E. A., Rossmann, M. G., Fout, G. S., Medappa, K. C., and Rueckert, R. R. (1983). *Proc. Natl. Acad. Sci. U.S.A.* **80**, 931–934.
70. Arnold, E., Erickson, J. W., Fout, G. S., Frankenberger, E. A., Hecht, H. J., Luo, M., Rossmann, M. G., and Rueckert, R. R. (1984). *J. Mol. Biol.* **177**, 417–430.
71. Luo, M., Arnold, E., Erickson, J. W., Rossmann, M. G., Boege, U., and Scraba, D. G. (1984). *J. Mol. Biol.* **180**, 703–714.
72. Hahn, T., ed. (1983). "International Tables for Crystallography" Vol. A. Reidel Publ. Dordrecht, Netherlands.
73. Rossmann, M. G., and Blow, D. M. (1962). *Acta Crystallogr.* **15**, 24–31.
74. Argos, P., and Rossmann, M. G. (1976). *Acta Crystallogr., Sect. B* **B32**, 2975–2979.
75. Argos, P., and Rossmann, M. G. (1974). *Acta Crystallogr., Sect. A* **A30**, 672–677.
76. Arnold, E., and Rossmann, M. G. (1986). *Proc. Natl. Acad. Sci. U.S.A.* **83**, 5489–5493.
77. Argos, P., Ford, G. C., and Rossmann, M. G. (1975). *Acta Crystallogr., Sect. A* **A31**, 499–506.
78. Bricogne, G. (1976). *Acta Crystallogr., Sect. A* **A32**, 832–847.
79. Johnson, J. E. (1978). *Acta Crystallogr. Sect. B* **B34**, 576–577.
80. Jones, T. A. (1978). *J. Appl. Crystallogr.* **11**, 268–272.
81. Rossmann, M. G., and Blow, D. M. (1963). *Acta Crystallogr.* **16**, 39–45.
82. Johnson, J. E., Akimoto, T., Suck, D., Rayment, I., and Rossmann, M. G. (1976). *Virology* **75**, 394–400.
83. Rayment, I., Baker, T. S., Caspar, D. L. D., and Murakami, W. T. (1982). *Nature (London)* **295**, 110–115.
84. Gaykema, W. P. J., Hol, W. G. J., Vereijken, J. M., Soeter, N. M., Bak, H. J., and Beintema, J. J. (1984). *Nature (London)* **309**, 23–29.
85. Rossmann, M. G., Abad-Zapatero, C., Murthy, M. R. N., Liljas, L., Jones, T. A., and Strandberg, B. (1983). *J. Mol. Biol.* **165**, 711–736.
86. Caspar, D. L. D., and Klug, A. (1962). *Cold Spring Harbor Symp. Quant. Biol.* **27**, 1–24.
87. Pfaff, E., Mussgay, M., Böhm, H. O., Schulz, G. E., and Schaller, H. (1982). *EMBO J.* **1**, 869–874.
88. Neurath, H. (1984). *Science* **224**, 350–357.
89. Steitz, T. A., and Shulman, R. G. (1982). *Annu. Rev. Biophys. Bioeng.* **11**, 419–444.
90. Chow, M., Yabrov, R., Bittle, J., Hogle, J., and Baltimore, D. (1985). *Proc. Natl. Acad. Sci. U.S.A.* **82**, 910–914.
91. Wychowski, C., van der Werf, S., Siffert, O., Crainic, R., Bruneau, P., and Girard, M. (1983). *EMBO J.* **2**, 2019–2024.
92. Evans, D. M. A., Minor, P. D., Schild, G. S., and Almond, J. W. (1983). *Nature (London)* **304**, 459–462.
93. Minor, P. D., Schild, G. C., Bootman, J., Evans, D. M. A., Ferguson, M., Reeve, P., Spitz, M., Stanway, G., Cann, A. J., Hauptmann, R., Clarke, L. D., Mountford, R. C., and Almond, J. W. (1983). *Nature (London)* **301**, 674–679.
94. van der Werf, S., Wychowski, C., Bruneau, P., Blondel, B., Crainic, R., Horodniceanu, F., and Girard, M. (1983). *Proc. Natl. Acad. Sci. U.S.A.* **80**, 5080–5084.
95. Bittle, J. L., Houghten, R. A., Alexander, H., Shinnick, T. M., Sutcliffe, J. G.,

Lerner, R. A., Rowlands, D. J., and Brown, F. (1982). *Nature (London)* **298**, 30–33.
96. Robertson, B. H., Morgan, D. O., and Moore, D. M. (1984). *Virus Res.* **1**, 489–500.
97. Strohmaier, K., Franze, R., and Adam, K. H. (1982). *J. Gen. Virol.* **59**, 295–306.
98. Baxt, B., Morgan, D. O., Robertson, B. H., and Timpone, C. A. (1984). *J. Virol.* **51**, 298–305.
99. Emini, E. A., Jameson, B. A., and Wimmer, E. (1984). *In* "Modern Approaches to Vaccines" (R. M. Chanock and R. A. Lerner, eds.), pp. 65–75. Cold Spring Harbor Lab., Cold Spring Harbor, New York.
100. Westhof, E., Altschuh, D., Moras, D., Bloomer, A. C., Mondragon, A., Klug, A., and Van Regenmortel, M. H. V. (1984). *Nature (London)* **311**, 123–131.
101. Tainer, J. A., Getzoff, E. D., Alexander, H., Houghten, R. A., Olson, A. J., Lerner, R. A., and Hendrickson, W. A. (1984). *Nature (London)* **312**, 127–134.
102. Meloen, R. H., Briaire, J., Woortmeyer, R. J., and Van Zaane, D. (1983). *J. Gen. Virol.* **64**, 1193–1198.
103. Mandel, B. (1979). *In* "Comprehensive Virology" (H. Fraenkel-Conrat and R. R. Wagner, eds.), Vol. 15, pp. 37–121. Plenum, New York.
104. Beck, E., Feil, G., and Strohmaier, K. (1983). *EMBO J.* **2**, 555–559.
105. Colman, P. M., Varghese, J. N., and Laver, W. G. (1983). *Nature (London)* **303**, 41–44.
106. Varghese, J. N., Laver, W. G., and Colman, P. M. (1983). *Nature (London)* **303**, 35–40.
107. Cavanagh, D., Sangar, D. V., Rowlands, D. J., and Brown, F. (1977). *J. Gen. Virol.* **35**, 149–158.
108. Wild, T. F., Burroughs, J. N., and Brown, F. (1969). *J. Gen. Virol.* **4**, 313–320.
109. Argos, P., Tsukihara, T., and Rossmann, M. G. (1980). *J. Mol. Evol.* **15**, 169–179.
110. Lonberg-Holm, K., and Philipson, L. (1974). *Monogr. Virol.* **9**.
111. Werner, T. C., Bunting, J. R., and Cathou, R. E. (1972). *Proc. Natl. Acad. Sci. U.S.A.* **69**, 795–799.
112. Marquart, M., Deisenhofer, J., Huber, R., and Palm, W. (1980). *J. Mol. Biol.* **141**, 369–391.
113. Thomas, A. A. M., Brioen, P., and Boeye, A. (1985). *J. Virol.* **54**, 7–13.
114. Rossmann, M. G., Abad-Zapatero, C., Hermodson, M. A., and Erickson, J. W. (1983). *J. Mol. Biol.* **166**, 37–83.
115. Erickson, J. W., Silva, A. M., Murthy, M. R. N., Fita, I., and Rossmann, M. G. (1985). *Science* **229**, 625–629.

7
Adenovirus Hexon: A Novel Use of the Viral Beta-Barrel

MICHAEL M. ROBERTS AND ROGER M. BURNETT
Department of Biochemistry and Molecular Biophysics
College of Physicians and Surgeons
Columbia University
630 West 168th Street
New York, New York 10032

Adenoviruses differ from other viruses whose structures have been solved so far by X-ray crystallography. These are satellite tobacco necrosis virus (STNV) (Liljas *et al.*, 1982), tomato bushy stunt virus (TBSV) (Olson *et al.*, 1983), southern bean mosaic virus (SBMV) (Abad-Zapatero *et al.*, 1980), rhinovirus (Rossmann *et al.*, 1985), and poliovirus (Hogle *et al.*, 1985). Adenoviruses contain DNA, rather than RNA, and their construction is far more complex (Ginsberg, 1984). They contain at least ten different structural proteins and, with a particle mass of 150×10^6 Da, are 20-fold larger (van Oostrum and Burnett, 1985). The intact adenovirus virion presents a formidable structural problem, which has required a novel approach to the ultimate goal of obtaining a detailed picture of its architecture. An X-ray crystallographic investigation of the major coat protein hexon (Burnett, 1984; Burnett *et al.*, 1985) has been combined with electron microscopic studies on capsid fragments containing small arrays of hexons (van Oostrum and Burnett, 1984; Burnett, 1985; van Oostrum *et al.*, 1987).

Hexon is a trimer of three identical polypeptides (Grütter and Franklin, 1974). It crystallizes with one subunit per asymmetric unit in the cubic space group P2$_1$3, $a = 150.6$ Å (Franklin *et al.*, 1971). The polypeptide of 967 amino acids has a molecular mass of 109,077 Da (Akusjärvi *et al.*, 1984) and the 240 hexons in the virion account for

60% of its protein. The low-resolution 6-Å structure (Burnett et al., 1985) revealed that hexon consists of a pseudohexagonal "base" and a triangular "top" (Fig. 1). The base has two kinds of vertical basal faces, A and B, which alternate about the molecular threefold axis. These form A : B hexon–hexon contacts within the group of 12 hexons on a p3 net that forms each facet of the icosahedral capsid (Fig. 2). Hexons lie in the capsid with their triangular tops facing to the outside of the virion. The recent determination of the crystal structure at 2.9 Å resolution (Roberts et al., 1986) has shown how hexon forms a stable building block of the outer protective coat of the adenovirus virion. We describe here the structural function of hexon's component β-barrel domains and compare this function with that of the β-barrel in the spherical RNA viruses.

The basic construction of the hexon trimer is best described at the tertiary level. Hexon is formed from three pairs of pedestal domains, P1 and P2, in the base and three tower domains T (Figs. 1 and 4). Domains P1 and P2 are very similar, each consisting of an eight-stranded β-barrel arranged in the same "jellyroll greek key" topology (Richardson, 1981) found in the coat proteins of all the small spherical RNA viruses whose structures are known. Each tower domain is formed by loops arising from the P domains. Loops l_1 from P1 and l_4 from P2 in the adjacent subunit combine with loop l_2 from the opposite subunit (Figs. 1 and 4). The tower occurs at the interface between two subunits, where it strengthens their intersubunit bonding and links this pair to the third subunit to give the whole assembly great strength.

A distinctive feature occurs within the top of the molecule around the molecular threefold axis. The l_2 loops from each subunit bond together at the "β-constriction" (Fig. 4) to hold the three T domains together within the top portion of the trimer. The β-constriction is a six-stranded antiparallel β-barrel (Fig. 5). A different form of link is provided by loop l_3 from P2, which holds the two P domains in place within one subunit by filling the space between them. These interactions across the polypeptide chains of hexon provide for its molecular function as a highly stable building block (Burnett, 1984, 1985). The stability is demonstrated by the inability of 5 M urea to denature hexon even though its action will break up the capsid (Maizel et al., 1968). Such chemical resistance is especially remarkable as disulfide bridges are not present in hexon.

The interweaving of loops l_2 at the β-constriction in hexon can be compared to a similar structural feature in the soybean trypsin inhibitor (Sweet et al., 1974; Blow et al., 1974). It has been suggested

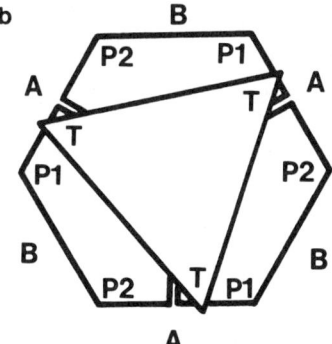

Fig. 1. (a) An envelope model of the hexon trimer at 6 Å resolution showing the pseudohexagonal base and the less massive top with its three towers forming a triangle. (b) A symbolic hexon viewed from above showing the locations of the A and B vertical basal faces, domains P1 and P2 in the base, and domain T in the top. The intersubunit boundary is approximately vertical in the base, where it is indicated, but is very complex at the top of the molecule. The vertices of the top (the towers) are rotated by 10° counterclockwise from the midpoints of the A faces.

Fig. 2. A model showing the distribution of hexons on each of the twenty $p3$ facets of the icosahedral adenovirus capsid. All other protein components are excluded for clarity. By referring to Fig. 1b, it can be seen that all twelve hexons within each $p3$ facet have A:B interactions, whereas those at the edges have A:A and B:B interactions. From Roberts *et al.* (1986). Copyright 1986 by the American Association for the Advancement of Science.

(McLachlan, 1979), on the basis of topological similarities, that this molecule evolved from a trimer of three identical polypeptides held together by a similar feature to the β-constriction. The feature remains in the present-day monomeric form of the molecule in which its function is still to stabilize the assembly. By contrast, the interweaving of the polypeptide chains about the threefold axes in TBSV and SBMV, and about the fivefold axes in rhinovirus and poliovirus, occurs through the N-terminal arms rather than through internal loops.

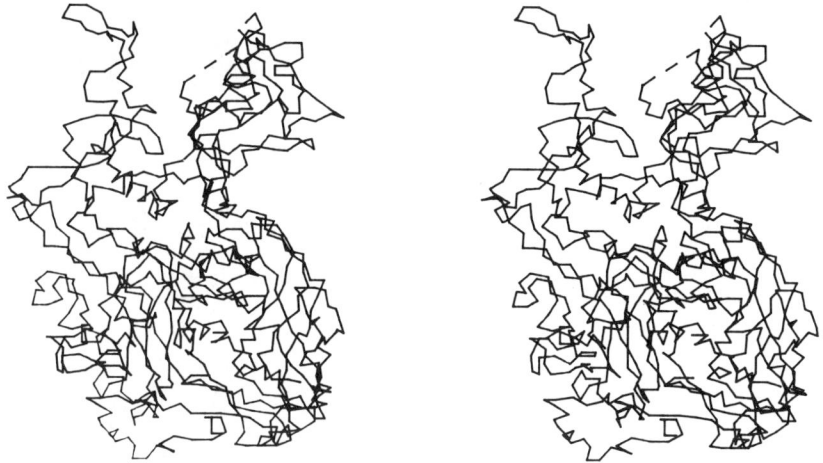

Fig. 3. A stereo view of the hexon monomer represented as a C_α plot. The line of view is from the cavity at the center of the trimer toward the corner of the base formed by P2 (see Fig. 1b). The N-terminal arm can be seen extending toward the viewer from the bottom of the base. Dashed lines indicate the chain connectivity across untraced regions within the map.

Again, such features stabilize groups of polypeptides and in turn strengthen the whole viral capsid. The complex interweaving in hexon, involving internal loops rather than termini, is probably the basis for hexon's unique requirement for another protein (adenovirus 100K) in its folding pathway (Cepko and Sharp, 1982).

One final feature of the hexon structure remains to be described. The N-terminal arm extends underneath the neighboring subunit to a disordered portion (residues 1–41) which clearly does not contribute to the overall stability of the molecule (Figs. 3 and 4). In the assembled capsid, this arm probably anchors onto an underlying protein, such as polypeptide VI (Everitt *et al.*, 1975). Similar features are found in the coat proteins of the spherical RNA viruses, where the anchor is located in the viral RNA. The N-terminal arm of ad2 hexon does not have an unusual proportion of basic residues, which reflects interactions with protein rather than with nucleic acid.

The hexon polypeptide consists of two topologically similar segments. The similarity of domains P1 and P2 was evaluated by a least-squares minimization of the C_α–C_α distances of spatially matching residue pairs. The calculation revealed that 96 such pairs, or 20% of the polypeptide, were separated by an average distance of 3.5 Å. To obtain superposition, the P2 atomic coordinates were rotated clock-

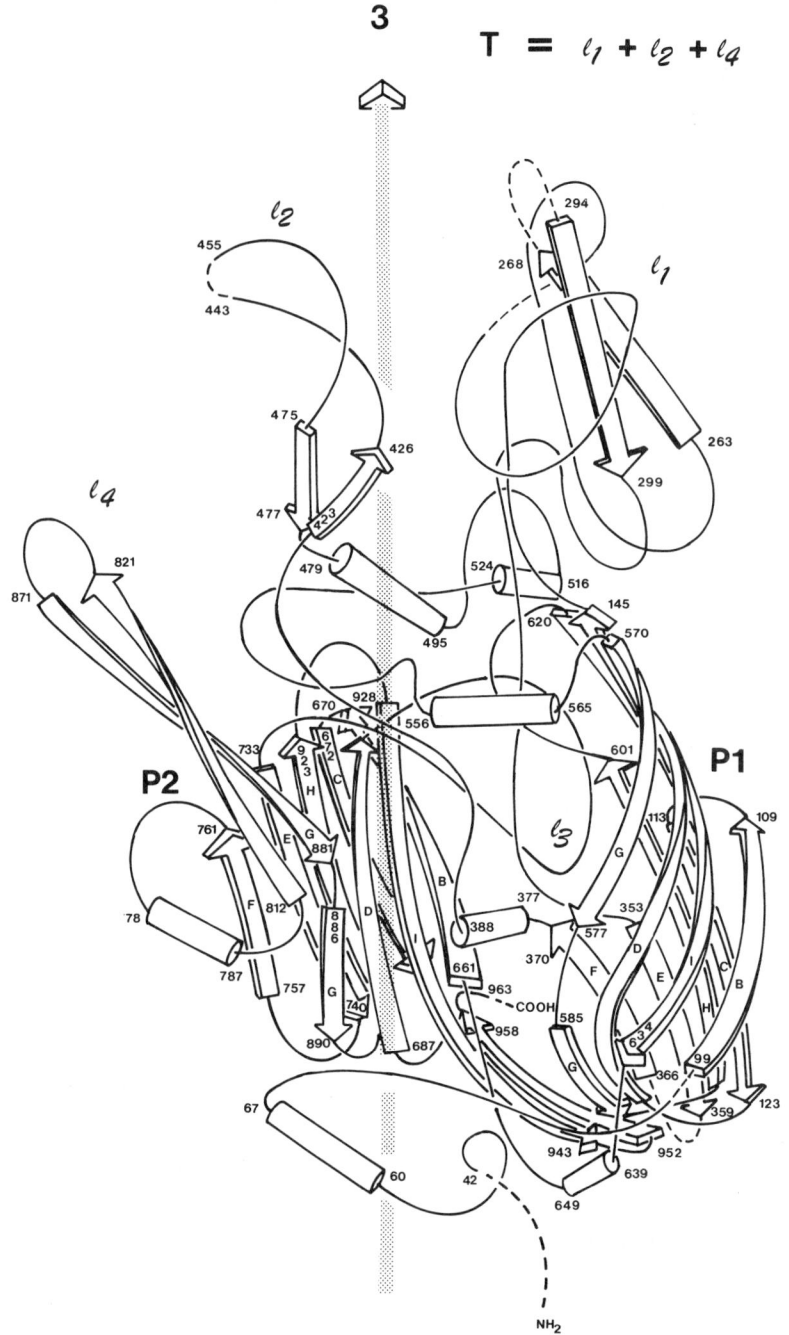

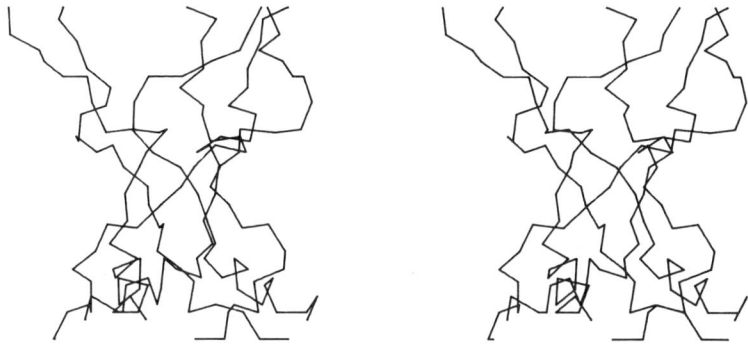

Fig. 5. A stereo pair of a C_α plot connecting residues 414–437 and 464–490 within the l_2 loops from each subunit. These define the inner core of the top region of the hexon trimer, including the β-constriction about the threefold axis. The β-constriction consists of three β-bulges arising from alternating strands of four- (423–426) and three- (475–477) residue segments to give a six-stranded antiparallel β-barrel. The view is from outside the molecule and perpendicular to the A face.

wise by 57.4° into those of P1 about an axis inclined at 25° from the vertical molecular axis. The inclination in the rotation axis results from the tilt of the β-sheets in P1 relative to their more upright arrangement in P2. The small deviation of the angle of rotation from 60° is a measure of the closeness of the pseudosixfold arrangement of the P domains to exact sixfold symmetry. Despite the similarity of the β-barrels, there is no obvious sequence homology between residues at related spatial positions. However, the overall similarity is emphasized by the occurrence of similar structural features such as the four loops l_1 to l_4, which occur between the same β-strands in both P domains. Moreover, in both l_2 and l_4 each loop is preceded by an α-helix (Fig. 4). In addition, the β-strands in both P domains exhibit a similar pattern of size distribution within the β-barrels, with the shortest strands situated at the midpoint of the jellyroll loops (Fig. 6). All

Fig. 4. A sketch of the hexon monomer backbone viewed in the same orientation as Fig. 3. The meandering of the random coil has been reduced to emphasize the secondary structure, which consists of 8% α-helix and 22% β-sheet. Dashed lines indicate the connections across invisible segments of the chain. Sequence numbers for the amino acids are indicated for each element of secondary structure. The β-strands in domains P1 and P2 are lettered with the same convention as those in the β-barrels of TBSV. A semitransparent threefold molecular axis is superimposed on the sketch to indicate the relationship of the three identical subunits. From Roberts *et al.* (1986). Copyright 1986 by the American Association for the Advancement of Science.

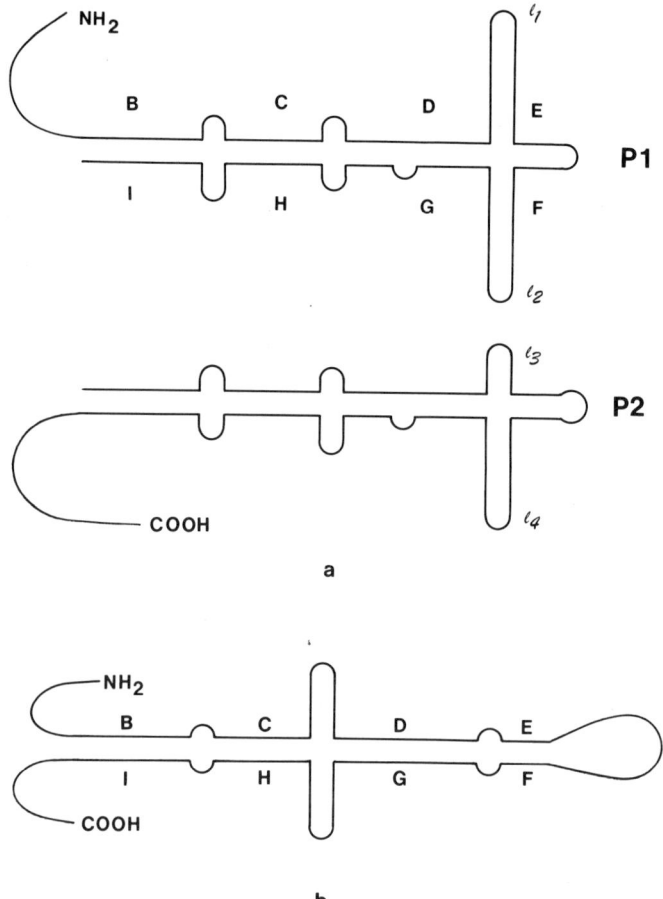

Fig. 6. A comparison of the viral β-barrels in hexon and the spherical RNA viruses. (a) The hexon P1 and P2 domains showing the situation of long intervening loops between barrel strands D : E and F : G. (b) A composite showing the positions of the intervening loops in the RNA virus coat proteins. Here, long loops occur between C : D, E : F, and G : H. The result is to place the loops at the opposite end of the β-barrel to that used in hexon.

these aspects of the structure strongly suggest that domains P1 and P2 have arisen from gene duplication and that the precursor gene itself had the same primitive ancestor as that corresponding to the related β-barrels in the small spherical RNA viruses.

The arrangement of the strands within the hexon P domains is in a progressively decreasing order of size in the sequence G-D-I-B-C-H-

E-F. The strands G-D-I-B have a right-handed twist of about 90°, which is accentuated by the presence of β-bulges on strands G, D, and I of P1 and on strand B of P2. The more exaggerated β-bulges in P1 result in a greater twist angle for its β-sheet. This enables the external residues on the lower portion of the G-D-I-B sheet to be involved in intramolecular contacts. The upper portions of strands G-D-I-B in P1 and D-I-B in P2 expose residues on the surface of the molecule that are available for intermolecular contacts. The strands C-H-E-F make up the remaining external surface of the base. The flattening in the β-barrels therefore occurs between the β-sheets G-D-I-B and C-H-E-F, positioning strands B and C to define the corners of the pseudohexagon. Although the β-barrels in hexon have the same topology and similar spatial dispositions, there are some differences in their hydrogen bonding patterns. The β-barrel in P1 is similar to that in SBMV, in that both the G and D strands contain noticeable β-bulges which are not present in the same location in P2. Instead, the top portion of strand G in P2 forms hydrogen bonds with an additional strand which is not part of the eight-stranded β-barrel. Both features require breaks of several residues in the G strands of P1 and P2. Also, the B strand in P1 forms hydrogen bonds with both I and the end strand C, whereas in P2 it bonds only to I.

The detailed differences between the hexon β-barrels are less dramatic than their similarity to each other and to those found in the spherical RNA viruses. Both hexon β-barrels have a similar distribution of strand size and the same twist of the G-D-I-B sheet as the corresponding features in the RNA viruses. In all the viruses, the C-H-E-F strands form the shorter sheet and have a less pronounced twist. All six β-barrels in the hexon trimer are separated by an α-helix, which follows strand F and is situated counterclockwise to each P domain (to the left in Fig. 4). The β-barrels in the RNA viruses are similarly flanked on the same side by an α-helix running parallel to the virion surface. In hexon, the axes of the α-helices are directed toward the internal basal cavity of the molecule.

The β-barrel supersecondary structure occurs widely in other nonviral proteins (Richardson, 1981). The different types can be grouped according to their topology, which is defined by the folding pattern of their polypeptide chain. For example, β-barrels with the same topology are found within the immunoglobulin Fab' fragment (Poljak *et al.*, 1974), copper, zinc superoxide dismutase (Tainer *et al.*, 1982), and hemocyanin (Gaykema *et al.*, 1984) despite the widely differing biological functions of these proteins (Richardson *et al.*, 1976). Concanavalin A, by contrast, contains a β-barrel with a different topology from all these examples (Reeke et al., 1975). What is particularly striking is that the β-barrels in the viral coat proteins fall in a unique topological class. The common structural motif is formed from a double-stranded antiparallel polypeptide loop with four successive pairs of β-strands (Fig. 7). The double strand is simply rolled into two complete turns of a right-handed helix. The helix then assumes the flattened form that is characteristic of most β-barrel structures.

Despite the topological similarity of the β-barrels in the P domains of hexon to those found in all the spherical RNA viruses, hexon presents an unusual case within the group. The β-barrels in hexon contain long intervening loops that protrude from strands at the oppo-

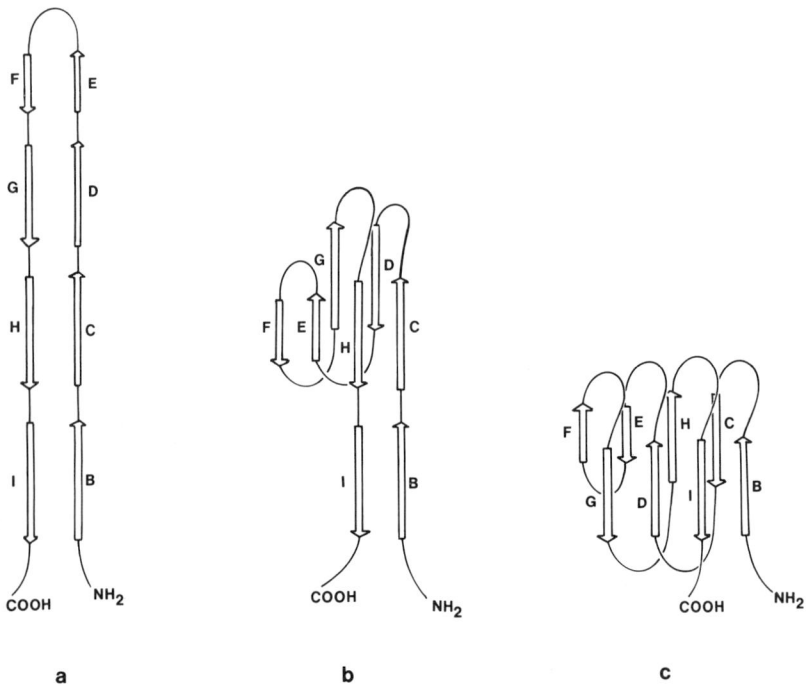

Fig. 7. A sketch showing the stages in the folding of a β-barrel into the jellyroll greek key topology of a viral protein. A loop consisting of four antiparallel pairs of β-strands (a) folds over and up in one complete turn to form the first four-stranded antiparallel β-sheet C-H-E-F shown in (b). The viral eight-stranded β-barrel (c) arises when the loop folds again to form the second β-sheet G-D-I-B. The folding pathway results in a flattened β-barrel with little hydrogen bonding across its end strand pairs, B : C and F : G.

site end of the barrel to that containing the N and C termini. The corresponding intervening loops in the RNA viral proteins are situated at the other end of the barrel (Figs. 6 and 7). These differences are related to the different architectural role played by the β-barrel in constructing the adenovirus capsid. It is at first surprising that the strands in the hexon β-barrels are perpendicular to the surface of the virion, whereas the β-strands in the RNA viruses lie parallel to the virion surface. However, it is noticeable that in both cases the strands C-H-E-F form an external surface of both structural assemblies. The role of the viral β-barrel is therefore that of a small building block of a protein shell.

If the β-barrel is regarded as the elementary unit, three different levels of symmetry relate this to the whole virion (Fig. 2). At the lowest level, the β-barrels are related by pseudosixfold symmetry to give the hexon. This acts as a complex structure unit in forming the outer shell (Burnett, 1985). At the next level, the 12 hexons on each of the $p3$ capsid facets are related by translational and threefold rotational symmetry. Point group symmetry elements of the icosahedral capsid then relate the facets to give a shell of 240 hexons. Within hexon, the C-H-E-F β-sheets will face each other, leaving the towers to form the surface of the adenovirus virion. The planar array of hexons on each facet is favored by interactions across the β-sheets that lie perpendicular to its surface.

In summary, the viral β-barrel forms the fundamental unit of the base of the hexon molecule. Since the packing of hexons within the adenovirus capsid is known from electron microscopic studies on capsid fragments (Burnett, 1985; van Oostrum and Burnett, 1985; van Oostrum *et al.*, 1987), we know the relationship of the hexon molecule to the whole biological assembly. This has allowed us to identify the role of the β-barrel in an entirely different form of virus construction from that occurring in the RNA viruses. We can now see just how versatile the viral β-barrel can be, either in directly forming the capsid of an RNA virus, or in first making a structural building block such as hexon, which then is used to construct the outer shell of adenovirus.

ACKNOWLEDGMENTS

We thank Jan van Oostrum for helpful discussions. The work has been supported by Public Health Service Grant AI 17270 from the National Institute of Allergy and Infectious Diseases and by an Irma T. Hirschl Career Scientist Award to R.M.B.

REFERENCES

Abad-Zapatero, C., Abdel-Meguid, S. S., Johnson, J. E., Leslie, A. G. W., Rayment, I., Rossmann, M. G., Suck, D., and Tsukihara, T. (1980). *Nature (London)* **286,** 33–39.
Akusjärvi, G., Alestrӧm, P., Pettersson, M., Lager, M., Jӧrnvall, H., and Pettersson, U. (1984). *J. Biol. Chem.* **259,** 13976–13979.
Blow, D. M., Janin, J., and Sweet, R. M. (1974). *Nature (London)* **249,** 54–57.
Burnett, R. M. (1984). *In* "Biological Macromolecules and Assemblies" (F. A. Jurnak and A. McPherson, eds.), Vol. 1, pp. 337–385. Wiley, New York.
Burnett, R. M. (1985). *J. Mol. Biol.* **185,** 125–143.
Burnett, R. M., Grütter, M. G., and White, J. L. (1985). *J. Mol. Biol.* **185,** 105–123.
Cepko, C. L., and Sharp, P. A. (1982). *Cell (Cambridge, Mass.)* **31,** 407–415.
Everitt, E., Lutter, L., and Philipson, L. (1975). *Virology* **67,** 197–208.
Franklin, R. M., Pettersson, U., Åkervall, K., Strandberg, B., and Philipson, L. (1971). *J. Mol. Biol.* **57,** 383–395.
Gaykema, W. P. J., Hol, W. G. J., Vereijken, J. M., Soeter, N. M., Bak, H. J., and Beintema, J. J. (1984). *Nature (London)* **309,** 23–29.
Ginsberg, H. S., ed. (1984). "The Adenoviruses." Plenum, New York.
Grütter, M., and Franklin, R. M. (1974). *J. Mol. Biol.* **89,** 163–178.
Hogle, J. M., Chow, M., and Filman, D. J. (1985). *Science* **229,** 1358–1367.
Liljas, L., Unge, T., Jones, T. A., Fridborg, K., Lövgren, S., Skoglund, U., and Strandberg, B. (1982). *J. Mol. Biol.* **159,** 93–108.
McLachlan, A. D. (1979). *J. Mol. Biol.* **133,** 557–563.
Maizel, J. V., Jr., White, D. O., and Scharff, M. D. (1968). *Virology* **36,** 126–136.
Olson, A. J., Bricogne, G., and Harrison, S. C. (1983). *J. Mol. Biol.* **171,** 61–93.
Poljak, R. J., Amzel, L. M., Chen, B. L., Phizackerley, R. P., and Saul, F. (1974). *Proc. Natl. Acad. Sci. U.S.A.* **71,** 3440–3444.
Reeke, G. N., Becker, J. W., and Edelman, G. M. (1975). *J. Biol. Chem.* **250,** 1525–1547.
Richardson, J. S. (1981). *Adv. Protein Chem.* **34,** 167–339.
Richardson, J. S., Richardson, D. C., Thomas, K. A., Silverton, E. W., and Davies, D. R. (1976). *J. Mol. Biol.* **102,** 221–235.
Roberts, M. M., White, J. L., Grütter, M. G., and Burnett, R. M. (1986). *Science* **232,** 1148–1151.
Rossmann, M. G., Arnold, E., Erickson, J. W., Frankenberger, E. A., Griffith, J. P., Hecht, H.-J., Johnson, J. E., Kamer, G., Luo, M., Mosser, A. G., Rueckert, R. R., Sherry, B., and Vriend, G. (1985). *Nature (London)* **317,** 145–153.
Sweet, R. M., Wright, H. T., Janin, J., Chothia, C. H., and Blow, D. M. (1974). *Biochemistry* **13,** 4212–4228.
Tainer, J. A., Getzoff, E. D., Beem, K. M., Richardson, J. S., and Richardson, D. C. (1982). *J. Mol. Biol.* **160,** 181–217.
van Oostrum, J., and Burnett, R. M. (1984). *Ann. N.Y. Acad. Sci.* **435,** 578–581.
van Oostrum, J., and Burnett, R. M. (1985). *J. Virol.* **56,** 439–448.
van Oostrum, J., Smith, P. R., Mohraz, M., and Burnett, R. M. (1987). Submitted.

8
The Structure of an Antineuraminidase Monoclonal Fab Fragment and Its Interaction with the Antigen

P. M. COLMAN
CSIRO
Division of Protein Chemistry
Parkville 3052, Australia

R. G. WEBSTER
St. Jude Children's Research Hospital
Memphis, Tennessee 38101

INTRODUCTION

The three-dimensional structure of a Fab fragment of an antineuraminidase monoclonal antibody, S10/1, has been determined by X-ray diffraction. The binding site of the Fab on the antigen has been identified by analyzing variants of influenza neuraminidase selected by monoclonal antibodies, including S10/1. Reactivity patterns of the S10/1 antibody to field strain viruses are consistent with the laboratory variant data, as are direct electron microscope images of the Fab–neuraminidase complex.

FAB STRUCTURE

Crystallization of the Fab fragment S10/1 (1) from ammonium sulfate has been reported (2). The space group is $P3_121$ with $a = 131.5$ Å and $c = 72.2$ Å. The structure was solved by multiple isomorphous replacement using diaminodinitroplatinum, uranyl acetate, and potassium hexachloroplatinate as heavy atom derivatives. Data were collected photographically and processed using the method and programs of Rossman (3). Merging statistics are given in Table I. Heavy atom sites were located in the usual way by difference Patterson and difference Fourier syntheses and heavy atom parameter refinement resulted in an overall figure of merit of 0.44 to 3.5 Å resolution. The sensitivity of the c axis dimension to heavy metal reagents is the most likely cause of the poor phasing statistics (Table II).

The map was interpreted using known Fab structures. The fitting was done in a 6-Å-resolution electron density map, which showed much better contrast between solvent and protein than the 3.5-Å image. However, the high-resolution image was used to determine possible starting origins for the search procedure. High features of the electron density were selected from a 4.0-Å map on the expectation that the disulfide bonds of the immunoglobulin domains might be represented among these peaks. In retrospect, two of the five disulfides are associated with peaks 3 and 4 out of a list of the 15 densest

TABLE I
Data Collection

Compound[a]	No. of data	No. of independent observations	R_{merge}[b]
Native	17,726	8,417	9.1
DANP	7,621	4,061	8.9
UO2A	3,894	2,666	14.4
UO2B	11,379	5,977	10.5
PTC6	2,833	2,021	12.8

[a] DANP, diaminodinitroplatinum (saturated solution); UO2A, uranyl acetate (10 mM, 5 days); UO2B, uranyl acetate (10 mM, 18 hr); PTC6, potassium hexachloroplatinate (10 mM, 18 hr).

[b] $R_{\text{merge}} = \sum_h (I_h - \bar{I}_h)/\sum_h \bar{I}_h$.

TABLE II
Phasing Statistics

	Resolution (Å):	18.8	11.6	8.4	6.5	5.4	4.6	4.0	3.5
DANP	RMS f_H	1.19	1.10	0.94	0.85	0.77	0.74	0.73	0.72
	RMS E	1.40	0.90	0.66	0.58	0.57	0.71	0.78	0.64
UO2B	RMS f_H	1.28	1.20	1.15	1.13	1.04	0.96	0.89	0.77
	RMS E	1.44	0.94	0.75	0.67	0.66	0.68	0.68	0.57
UO2A	RMS f_H	1.93	1.99	1.99	1.92	1.90	1.87	1.84	1.70
	RMS E	1.26	1.35	1.59	1.17	1.25	1.53	1.50	1.24
PTC6	RMS f_H	1.54	1.31	1.28	1.40	1.28	1.27	1.26	1.24
	RMS E	1.56	0.98	1.02	0.78	0.90	0.97	0.98	0.90

features in the map. In both cases the high-density features are associated with Cα positions rather than sulfur atoms, indicating that rigid parts of the structure are represented by such features at 4 Å resolution. No high density is associated with the two light chain domain disulfides or the interchain disulfide in the 4-Å electron density map.

The Cα skeleton of the Kol Fab (4) was then appropriately located at the corresponding position in the 6-Å image and a full three-dimensional orientation search was computed. At each high-density value the model was positioned with three different origins, representing the center of the S–S bridge and the two Cα positions either side of it. All four immunoglobulin domain disulfides were searched against every high-density feature in this way. $V_L V_H$ were treated as a rigid module as were $C_L C_H$. An example of the sensitivity of the orientational search to the translational placement of the model is demonstrated by the observation that around the correct solution for the C module, no signal $>3\sigma$ is found for the correct orientation on either the S–S center or one of the corresponding Cα positions. The alternate Cα position showed a peak of 5σ (cf. max peaks at the alternate origins of $<4\sigma$), which was then further explored by finer interval searches in both orientational and translational coordinates. A trace through one angular variable covering the final orientational and translational solution for the $C_L C_H$ module of the Kol protein in the S10/1 Fab 3.5-Å image is shown in Fig. 1. The significance of this peak is 9σ measured against all possible orientations for the correct translation. A similar result is found for the V module.

Subsequently, a limited search around the known solution using the mouse myeloma M603 Fab coordinates (5) resulted in the solution shown in Table III. The near identity of the orientational part of the

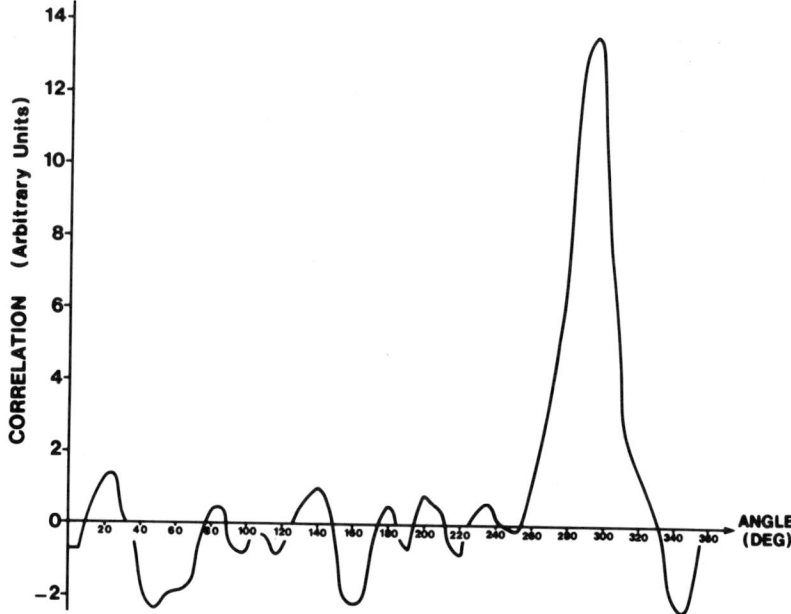

Fig. 1. Trace of correlation between Kol Fab Cα coordinates for C module rotated through one angular variable embracing the final solution for placement of the model into the 6-Å electron density of S10/1 Fab.

TABLE III
Fit of M603 Fab C^α Coordinates to 6-Å Image of S10/1 Antineuraminidase Fab[a]

	Shift 1 (Å)	Eulerian angles	Shift 2 (Å, trigonal frame of map)
V module	−(56.5,53.9,−21.8)	(50,113,97)	(70.5,85.8,9.4)
C module	−(70.5,54.7,4.6)	(47,114,96)	(79.3, 61.0,11.2)

[a] Shift 1 places heavy chain atoms 98C^α for the V module and 206C^α for the C module at the origin of the rotating frame. Shift 2 places the correctly oriented model in the map. Eulerian angle definition is θ_1 around OZ, θ_2 around new OX, and θ_3 around new OZ.

solution for the V and C modules testifies to the similarity of the elbow angle (6) in these two proteins. The possibility that heavy and light chains need to be interchanged cannot be excluded. The solution presented has the same sense as the M603 and other Fab fragments with bent elbows, namely, with longitudinal contact between V_H and C_H but not between V_L and C_L.

The crystal packing is head to tail with the complementarity-determining regions of one molecule associated with the C-terminus end of a 2_1 screw related molecule in the crystal. The amino acid sequence is, at this stage, partly determined (7). Several cycles of constrained crystallographic refinement (8) of the model against the 3-Å diffraction data have led to an R value of 0.323. The model geometry remains good with RMS deviations from bond length ideality of 0.03 Å and bond angle distance ideality of 0.06 Å. Further refinement awaits sequence data.

Although the unit cell parameters of this Fab fragment are very similar to those of the human Kol IgG molecule [a = 135.6 Å c = 82.1 Å (6)], the space group enantiomorph is opposite and the quaternary structure of the Fab fragment is grossly different. The relevance of the Fab elbow angle to antibody function remains obscure. The longitudinal contacts, always between CH1 and V_H, are few in number and involve only a small buried surface area. Details of this interface in S10/1 and other bent elbow Fab fragments must await high-resolution crystallographic structure refinement.

BINDING PROPERTIES OF S10/1 TO MONOCLONAL VARIANTS

The monoclonal antibody S10/1 was raised against A/Tokyo/3/67 influenza neuraminidase (1) and used to select a variant neuraminidase which does not bind to the antibody and which shows a single amino acid substitution of K to E at residue 368 (8). S10/1 also fails to bind five other variant neuraminidases selected with different antibodies and all showing a change of R to I at residue 344 (1,9). An R to G substitution at position 344 also abolishes binding of S10/1 to Tokyo neuraminidase but R to K does not (7). The three-dimensional structure of Tokyo neuraminidase shows residues 344 and 368 to be close together (Fig. 2), separated only by some 7 Å (10). Both residues are in upper surface loops of the structure showing substantial field strain sequence variation (11,12).

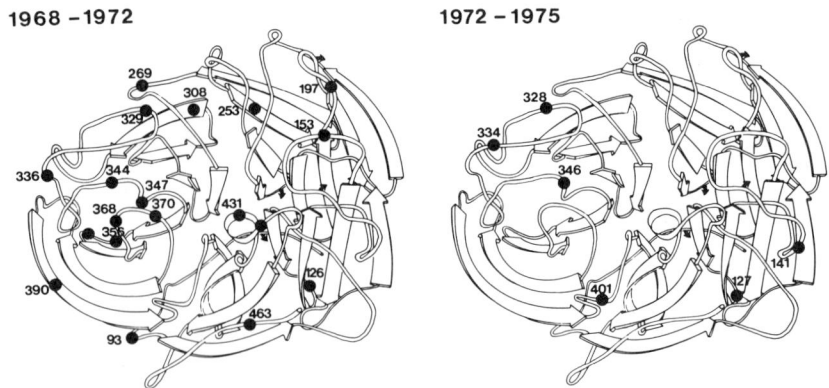

Fig. 2. Schematic of polypeptide folding in influenza neuraminidase indicating residue changes from 1968 to 1972 and 1972 to 1975.

BINDING PROPERTIES OF S10/1 TO NEURAMINIDASE FIELD STRAINS

The S10/1 antibody binds to neuraminidase from field strains of influenza isolated between 1961 and 1972 (1). From the available sequence data, the characteristics of neuraminidase which facilitate this binding must be found in sequences from A/Tokyo/3/67 through A/England/42/72, but not in A/RI/5$^+$/57 or in A/Udorn/72 and later strains. The England sequence is incomplete, being based on peptide map data (9). Five changes distinguish the England and Udorn (13) sequences: I 153 to T, R 253 to K, R 344 to K, K 368 to E, and S 390 to L (see Fig. 2). One of these, that at 368, is exactly the change in the variant selected by S10/1 and is sufficient to explain the binding data.

Twenty-five amino acid sequence changes distinguish the head regions of RI/5$^+$/57 from Tokyo/3/67 (12,14,15). In particular, changes in this period on the upper surface loops at 331 (S to R), 367 (N to S), 369 (E to D), 400 (N to S), 401 (N to D), and 403 (W to R) could affect binding of S10/1 (see Fig. 2).

On the other hand, residues 346 (T to N), 370 (Q to H), and 431 (K to E) all change during the period 1967 to 1972 without apparent effect on the binding of S10/1 to neuraminidase; S10/1 antibody shows >90% neuraminidase inhibition of all field strains between 1967 and 1972 that were tested (1). Whether this implies that residues 346, 370, and 431 are not contact residues for S10/1 binding is difficult to say. While 431 is some distance from the others, 346 and 370 are very close to 344 and 368 at which certain point changes are known to abolish binding. It may be that residues 346 and 370 are indeed contact residues, but that the particular changes observed here are adaptable to the antigen–antibody interface.

ELECTRON MICROSCOPY OF FAB–NEURAMINIDASE COMPLEXES

Crystalline complexes of S10/1 Fab and neuraminidase (2) were dissolved and examined in negative stain (2% potassium phosphotungstate, pH 7) in the electron microscope. Figure 3 shows typical

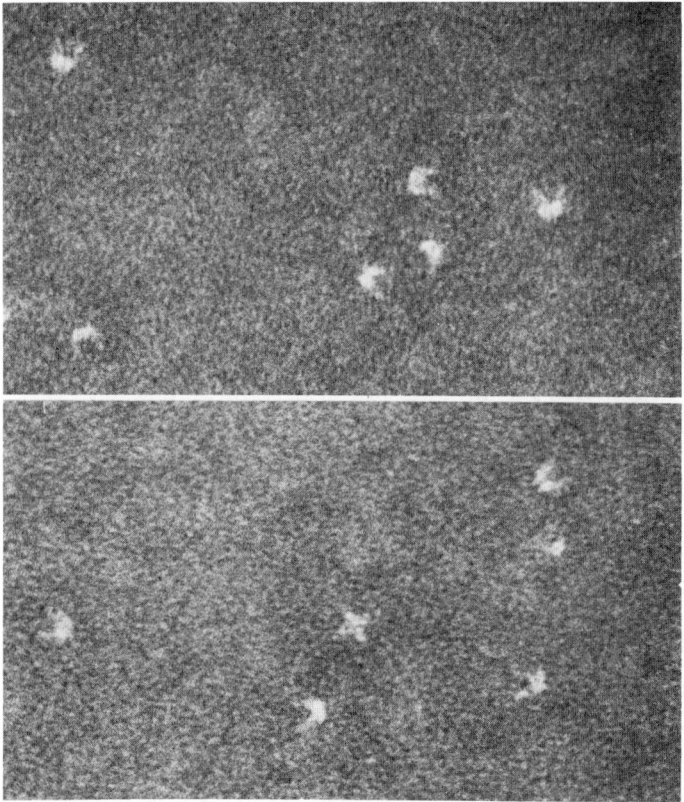

Fig. 3. Electron micrograph of S10/1 Fab–neuraminidase complexes showing predominantly side views of Fab arms attached to top of neuraminidase head.

side view images with the Fab arms subtending angles of 80° to the plane of the neuraminidase head. *En face* views are captured infrequently. The images are consistent with the sequence and binding data discussed above and suggest direct involvement of residue 368 and its near neighbors on the upper surface of the neuraminidase head in forming a complex with this Fab fragment. We cannot rigorously preclude the possibility of the antibody binding to the underside of the head and sensing amino acid sequence changes on the top side by conformational distortions transmitted through the head. However, residue 368 is oriented toward the solvent in a very exposed surface loop of the structure and replacements at that position are not likely to have major conformational consequences.

We anticipate that the complete amino acid sequence data for the Fab fragment will permit a docking study.

ACKNOWLEDGMENTS

We thank Peter Tulloch for the electron microscopy and Graeme Laver for influenza virus neuraminidase.

REFERENCES

1. Webster, R. G., Hinshaw, V. S., and Laver, W. G. (1982). *Virology* **117**, 93–104.
2. Colman, P. M., Gough, K. H., Lilley, G. G., Blagrove, R. J., Webster, R. G., and Laver, W. G. (1981). *J. Mol. Biol.* **152**, 609–614.
3. Rossmann, M. G. (1979). *J. Appl. Crystallogr.* **12**, 225–238.
4. Marquart, M., Deisenhofer, J., Huber, R., and Palm, W. (1980). *J. Mol. Biol.* **141**, 369–392.
5. Segal, D. M., Padlan, E. A., Cohen, G. H., Rudikoff, S., Potter, M., and Davies, D. R. (1974). *Proc. Natl. Acad. Sci. U.S.A.* **71**, 4298–4302.
6. Colman, P. M., Deisenhofer, J., Huber, R., and Palm, W. (1978). *J. Mol. Biol.* **100**, 257–282.
7. Air, G. M., Els, M. C., Brown, L. E., Laver, W. G., and Webster, R. G. (1985). *Virology* **145**, 237–248.
8. Hendrickson, W. A., and Konnert, J. H. (1980). *In* "Computing in Crystallography" (R. D. Diamond, S. Ramaseshan, and K. Ventatesan, eds.), pp. 13.01–13.23. Indian Acad. Sci. Int. Union Crystallogr., Bangalore.
9. Laver, W. G., Air, G. M., Webster, R. G., and Markoff, L. J. (1982). *Virology* **122**, 450–460.
10. Varghese, J. N., Laver, W. G., and Colman, P. M. (1983). *Nature (London)* **303**, 35–40.
11. Colman, P. M., Varghese, J. N., and Laver, W. G. (1983). *Nature (London)* **303**, 41–44.

12. Colman, P. M., and Ward, C. W. (1985). *Curr. Top. Microbiol. Immunol.* **114**, 177–255.
13. Markoff, L., and Lai, C. J. (1982). *Virology* **119**, 288–297.
14. Elleman, T. C., Azad, A. A., and Ward, C. W. (1982). *Nucleic Acids Res.* **10**, 7005–7015.
15. Ward, C. W., Elleman, T. C., and Azad, A. A. (1982). *Biochem. J.* **207**, 91–95.

9
Influenza Hemagglutinin

J. J. SKEHEL
National Institute for Medical Research
Mill Hill, London NW7 1AA, England

D. C. WILEY
Department of Biochemistry and Molecular Biology
Harvard University
Cambridge, Massachusetts 02138

The process of infection by influenza viruses appears to involve binding of virus particles at the cell surface, internalization by endocytosis, and fusion of virus and endosomal membranes to transfer the virus transcriptase complex into the cell. Both binding and membrane fusion processes involve the hemagglutinin and this is a brief description of our work on the molecular basis of these functions.

Hemagglutinins associate with cells by recognizing sialic acid residues of cell surface glycoconjugates. The specificity of this reaction is known primarily from observations of the receptor-destroying activity of neuraminidase (1) but it has also been known for some time that hemagglutinins differ in detailed specificity. For example, certain viruses are unable to agglutinate erythrocytes of certain species and agglutination by others can be inhibited by sialic acid-rich components of nonimmune sera (2). We have taken advantage of these latter observations to isolate mutants of X-31 virus which are resistant to such inhibition and using specifically derivatized erythrocytes have determined that unlike wild-type virus, which recognizes sialic acid in $\alpha 2:6$ linkage to galactose in carbohydrate side chains, the mutants bind to sialic acid residues in $\alpha 2:3$ linkage. Comparative nucleotide sequencing of the wild-type and mutant genes for hemagglutinin indicate that the only difference in amino acid sequence is at residue 226 of the HA_1 chain, which is leucine in wild type and glutamine in the mutant (3). We have also made the reverse selection by using derivatized erythrocytes to select $\alpha 2:6$-recognizing mutants of an avian virus which naturally recognizes $\alpha 2:3$ linkages, and in these cases also single changes from glutamine to leucine at HA_1 226 were observed (4). These observations are consistent with initial suggestions (5) that the site of receptor binding includes a number of conserved amino acid residues in the distal region of the molecule, in particular, tyrosine 98, tryptophan 153, and histidine 183 of HA_1, since residue 226 is also located at this site. We are at present analyzing the structure of the α2:3-binding hemagglutinin of the X-31 mutant in the presence of sialyllactose to obtain information on the residues involved in substrate contact and on the basis of binding specificity. In similar experiments to those above, we are also analyzing additional determinants of binding specificity following the observations (4) that the α2:6-binding avian virus mutants recognized N-acetylneuraminic acid but not N-glycolyl derivatives, whereas the α2:6-binding X-31 hemagglutinin binds to both.

Initial suggestions that the hemagglutinin is involved in membrane fusion were based on a number of properties which the molecule shares with the fusion glycoprotein of Sendai virus. These indicated that in both cases the fusion active molecules were generated by trypsinlike processing of an inactive precursor (6–8) and that the NH_2-terminal sequences of the polypeptides produced were analogous (9–11) and in influenza viruses were highly conserved (9). Additional experiments indicated, however, that unlike Sendai virus which mediates membrane fusion at neutral pH, influenza viruses, presumably because of their endosomal site of cell entry, cause fusion between pH 5 and pH 6; in the case of X-31 at pH 5.5. We have investigated the involvement of X-31 hemagglutinin in fusion by analyzing the structural consequences of incubating the molecule at this pH.

Incubation of soluble bromelain-released hemagglutinin at the pH of fusion causes aggregation; if the pH is lowered in the presence of liposomes, liposome binding is observed, or in the presence of detergents the ability to bind detergent is acquired. These changes are consistent with exposure of a hydrophobic region of the molecule which is buried at neutral pH. CD analyses indicate, however, that the structural modifications involved occur with retention of secondary structure (12), and we have attempted to obtain information on the nature of the change and the residues sensitive to low pH in two ways: by analyzing fragments produced by proteolysis of hemagglutinin in the fusion active form and by sequencing mutants which fuse membranes at different pH.

At the pH of fusion the hemagglutinin becomes extremely susceptible to proteolysis. Brief digestion with trypsin leads to the release of a soluble component from the aggregated molecule, which amino-terminal sequence analyses indicate consists of residues HA_1 28–328 (12). The release of this fragment implies that the molecular contacts between HA_1 and HA_2 in the fibrous region of the structure are broken at the pH of fusion. Furthermore, HPLC analyses indicate that the released fragment is a monomer and therefore HA_1–HA_1 interactions in the distal region of the molecule are also lost at low pH. The residual aggregate following tryptic digestion consists of HA_2 disulfide linked to peptide 1–27 of HA_1, and it appears, therefore, that this part of the molecule contains the hydrophobic self-aggregating region exposed at the pH of fusion. Further digestion of the aggregate with thermolysin causes deaggregation and amino-terminal sequence analyses of the digestion products indicate that the solubilization is accompanied by removal of the amino-terminal region of HA_2. Thus, the site of self-aggregation and presumably lipid association is the conserved hydrophobic amino-terminal region of HA_2 (13).

Mutants of X-31 which are triggered to cause membrane fusion at a pH higher than 5.5 (between 5.8 and 6.3) were isolated by selecting for resistance to amantadine hydrochloride, a reagent which has been shown to increase the pH of endosomes (14). Nucleotide sequence analyses of the genes for hemagglutinin of the mutants indicated that the amino acid substitutions which they contain fall into two groups, one that results in the destabilization of the buried location of the hydrophobic amino-terminal region of HA_2 and a second that results in the alteration of intersubunit contacts, suggesting distortion or disruption of these contacts in the fusion active conformation (15). All amino acid substitutions in the second group involve changes in charge, e.g., HA_2 81 E → G which result in the loss of salt bridges between subunits in the trimer or, e.g., HA_2 57E E → K which result in the loss of intrasubunit associations. A number of substitutions in the first group also involve changes in charge, e.g., HA_2 112 D → G, HA_1 17 H → Q, residues which are involved in associations which maintain the location of the amino-terminal region of HA_2 at neutral pH. Others which involve residues in this naturally conserved amino-terminal region are neutral substitutions, e.g., $HA_2$6 I → M, $HA_2$9 F → L.

Details of the consequences of these amino acid substitutions for the stability of the trimer are presented elsewhere (15). Our overall interpretation is that alterations in structure which lower the energy required to release the amino-terminal region of HA_2 or to rearrange the subunit interfaces lower the concentration of protons required to trigger the conformational change which allows the hemagglutinin to mediate fusion. Which amino acids or salt bridges must bind protons to cause the conformational transition is unknown.

REFERENCES

1. Gottschalk, A. (1959). *In* "The Viruses" (F. M. Burnet and W. M. Stanley, eds.), Vol. 3, pp. 51–56. Academic Press, New York.
2. Choppin, P. W., and Tamm, I. (1960). *J. Exp. Med.* **112**, 895–920.
3. Rogers, G. N., Paulson, J. C., Daniels, R. S., Skehel, J. J., Wilson, I. A., and Wiley, D. C. (1983). *Nature (London)* **304**, 76–78.
4. Rogers, G. N., Daniels, R. S., Skehel, J. J., Wiley, D. C., Wang, X.-F., Higa, H. H., and Paulson, J. C. (1985). *J. Biol. Chem.* **260**, 7362–7366.
5. Wilson, I. A., Skehel, J. J., and Wiley, D. C. (1981). *Nature (London)* **289**, 368–373.
6. Scheid, A., and Choppin, P. W. (1974). *Virology* **57**, 475–490.
7. Klenk, H-D., Rott, R., Orlich, M., and Blodorn, J. (1975). *Virology* **68**, 426–439.
8. Lazarowitz, S. G., and Choppin, P. W. (1975). *Virology* **68**, 440–454.
9. Skehel, J. J., and Waterfield, M. D. (1975). *Proc. Natl. Acad. Sci. U.S.A.* **72**, 93–97.
10. Gething, M. J., White, J. M., and Waterfield, M. D. (1978). *Proc. Natl. Acad. Sci. U.S.A.* **75**, 2737–2740.
11. Scheid, A., Graves, M., Silver, S., and Choppin, P. W. (1978). *In* "Negative Strand Viruses and the Host Cell" (B. Mahy and R. Barry, eds.), pp. 183–193. Academic Press, New York.
12. Skehel, J. J., Bayley, P. M., Brown, E. B., Martin, S. R., Waterfield, M. D., White, J. M., Wilson, I. A., and Wiley, D. C. (1982). *Proc. Natl. Acad. Sci. U.S.A.* **79**, 768–972.
13. Daniels, R. S., Douglas, A. R., Skehel, J. J., Waterfield, M. D., Wilson, I. A., and Wiley, D. C. (1983). *In* "The Origin of Pandemic Influenza Viruses" (W. G. Laver, ed.), pp. 1–7. Am. Elsevier, New York.
14. Gonazlez-Noriega, A., Grubb, J. H., Talkad, V., and Sly, W. S. (1980). *J. Cell Biol.* **85**, 839–852.
15. Daniels, R. S., Downie, J. C., Hay, A. J., Knossow, M., Skehel, J. J., Wang, M. L., and Wiley, D. C. (1985). *Cell (Cambridge, Mass.)* **40**, 431–439.

10

The Budding of Enveloped Viruses: A Paradigm for Membrane Sorting?

KAI SIMONS AND STEPHEN FULLER
European Molecular Biology Laboratory
D-6900 Heidelberg, Federal Republic of Germany

THE PROBLEM

The localization of proteins in the cell is determined by signals within their sequences (1). Many proteins in the cell are translated with amino-terminus extensions, termed signal peptides (2,3). The sequence allows recognition of the nascent polypeptide by the signal recognition particle (4), which binds in turn to the docking protein (5) located in the endoplasmic reticulum and, as a result, the polypeptide chain is transferred through the membrane of the endoplasmic reticulum. This mechanism is employed by proteins destined for different organelles in the cell, including the endoplasmic reticulum itself, the Golgi complex, the lysosomes, the secretory granules, and the cell surface (6). All these proteins are initially found in the same compartment, the endoplasmic reticulum. From there proteins destined beyond the endoplasmic reticulum are transported to the Golgi complex. Golgi proteins are left behind in the correct subcompartments of the Golgi complex, whereas the other proteins are directed to further destinations (7).

This complicated pathway of intracellular protein transport is a eukaryotic invention and is not found in prokaryotes. The mechanisms by which proteins are sorted from each other and directed to their correct locations in the cell are poorly understood. The traffic between cellular compartments is thought to be mediated by membrane vesicles which bud from one compartment to fuse with the next compartment in the pathway (8). Despite the continuous traffic of membrane components between compartments, the organelles maintain their characteristic protein and lipid compositions so that the traffic must be selective to avoid intermixing. When a membrane vesicle is formed from one organelle and transported to fuse with another, there must also be replenishment of the membrane lipids lost. This cannot in most cases be due to synthesis of new lipids, but is probably due to a compensating backward traffic of membrane. The membrane vesicles mediating traffic in both directions must both recognize and selectively fuse with their target.

With the exception of coated vesicles which endocytose surface proteins, little is known about the membrane vesicles mediating traffic between cellular compartments. There are at least four problems which must be solved to understand the selective formation of membrane vesicles (9,10). First, what generates the forces which distort the membrane of a budding vesicle? Second, how are membrane proteins which should remain in the organelle selectively excluded from the forming vesicle? Third, how are the proteins to be transported to the next organelle selectively included in the budding vesicle? Fourth, how efficient is the sorting process?

THE ENVELOPED VIRUS SOLUTION

The best-studied case of membrane sorting is the formation of enveloped viruses from the plasma membrane of infected cells. Since so little is known of intracellular sorting processes, it might be informative to consider this model for membrane sorting in more detail. The enveloped viruses are released from the cell by a budding from the host cell plasma membrane to envelop the viral nucleocapsid (11). The lipid composition of the viral envelope is almost identical to that of the host cell plasma membrane (12), but practically all of the host proteins are excluded, and instead 99% or more of the protein included in the budding virion are virally encoded. The exact mechanism of the budding process is not yet known. It is believed that transmembrane interactions between the viral spike glycoproteins and the underlying nucleocapsid provide one force for assembling the virus particles (13,14). Budding could be initiated by the nucleocapsid binding to a cluster of virus glycoproteins at the cell surface. Envelopment of the nucleocapsid by the plasma membrane would then proceed by trapping of laterally diffusing spike proteins. The nucleocapsid would act as a template for the budding process. Host membrane proteins could be excluded sterically from the developing bud. The close apposition of the spike proteins on the external side of the bilayer and the proximity of the nucleocapsid to the cytoplasmic face of the bilayer may effectively prevent the entry of plasma membrane proteins into the viral particle.

BUDDING OF ALPHAVIRUSES

Studies of alphaviruses, a family of simple and well-characterized enveloped viruses, provide a useful approach to the mechanism of membrane protein sorting (15). Alphaviruses are isometric, enveloped RNA viruses which include the well-studied Sindbis and Semliki Forest viruses. Sindbis virus contains approximately 200 copies of each of three proteins: E1 (49 kD) and E2 (47 kD), transmembrane glycoproteins forming the virus spike, and the nonglycosylated capsid protein (30 kD), which lies wholly within the viral membrane and interacts with single positive strand of viral RNA (MW = 4.2×10^6) to form the nucleocapsid (15). Semliki Forest virus is virtually identical to Sindbis but also contains equimolar amounts of a third glycoprotein, E3 (10 kD), as part of its spike. The dimensions of Sindbis virus have been established by low angle X-ray diffraction studies (16) and electron microscopy (17,18,19). The center of the virus is a ball of RNA approximately 120 Å in radius. It is surrounded by a shell of RNA and capsid protein between a radius of 140 Å and 205 Å and a lipid bilayer from radius 210 Å to 258 Å from which spike proteins project to a radius of 350 Å (16).

Both the alphavirus envelope and nucleocapsid are believed to be icosahedral, however, the assignment of a triangulation number to either has been difficult. Results of classical electron microscopy have been interpreted as supporting $T = 4$ arrangement for the spikes on the virus surface (20,17,18). This would yield a total mass significantly larger than that determined by neutron diffraction (21) or by scanning transmission electron microscopy (22). The nucleocapsid has been inferred to possess $T = 4$ symmetry from the requirement that the capsid proteins interact with the spike proteins on the virus surface (17,18). Conventional microscopy has yielded equivocal results for the symmetry of the nucleocapsid (23), and images have been interpreted as supporting $T = 3$ (19,20), $T = 4$ (17,24) and $T = 9$ (25,26).

Much of the confusion over the structure of the virus probably results from artifacts due to drying and treatment with the heavy metal negative stains used in conventional electron microscopic techniques.

Cryo-electron microscopy of virus in vitrified solution allows one to avoid these problems (27) and so obtain reliable structural information. This technique maintains the specimen in a thin layer of vitrified water so that it can be introduced to the microscope and viewed without drying (28). Work with unstained specimens is possible due to the low-background scattering of the vitrified water layer and the use of underfocus phase contrast to enhance the features in a given resolution range. The low temperature of the specimen (near −160°C) mitigates the damaging effects of the electron beam so that several low-dose images can be taken of the same specimen without loss of intermediate resolution information. Together these advantages make cryo-electron microscopy in combination with image reconstruction techniques ideal for studies of labile structures such as membrane systems (29).

A three-dimensional reconstruction of Semliki Forest virus (30) has been published in which a novel Reconstruction by Optimized Series Expansion (ROSE) method was used to combine images of virus in vitrified solution. This work confirmed the prediction that the structure of the virus would be well preserved in vitrified solution. Indeed, the departures from icosahedral symmetry of the particles appeared much smaller than observed in previous icosahedral reconstructions performed from negatively stained material. The packing of the spikes on the surface of the virion was unambiguously shown to be $T = 4$ in the ROSE map, and the spikes themselves were seen to be trimers. This was the first time that such information was obtained so clearly for a membrane virus. Although the ROSE map showed the spikes clearly, it did not reveal the shape of the membrane, although the authors inferred its position from the X-ray work of Harrison et al. (16), nor did it reveal the structure of the nucleocapsid within the virus.

Recently a three-dimensional reconstruction of Sindbis virus to a resolution of 35 Å by combining cryo-electron micrographs of vitrified particles in various orientations at four levels of defocus using a common lines procedure (31). The spikes of the virus appear as columnar trimers which are arranged on a $T = 4$ lattice in agreement with the previous negative stained work and the reconstruction of the Semliki Forest virus (30). In this structure one could see both the lipid bilayer and the nucleocapsid. Surprisingly, the lipid bilayer of the virus envelope is polyhedral rather than spherical and bends at the positions of the spikes. The nucleocapsid appears as a smooth but fenestrated

$T = 3$ structure beneath the membrane. Hence, a complete Sindbis virion contains 240 copies of each of the spike proteins and 180 copies of the capsid protein, yielding a total molecular weight of approximately 46.4×10^6. The arrangement of the spike proteins is complementary to the nucleocapsid structure. The cytoplasmic domains of the spikes near the virus fivefold axes are locked into holes formed by triplets of capsid elements. The spikes on the virus threefold axes interact very differently, if at all, with the capsid.

A reconstruction has also been performed from isolated Sindbis nucleocapsids and shows the same smooth $T = 3$ structure (32). This $T = 3$ arrangement of proteins is similar to that found for the capsids of a number of nonenveloped plant and animal viruses including several which have been solved to high resolution. The Sindbis nucleocapsid differs from that of these previously known $T = 3$ structures in being a relatively open structure. This is supported by the fenestrated structure, the RNA's sensitivity of isolated nucleocapsids (33,34), the fact that the Sindbis nucleocapsid is 30% larger than that of southern bean mosaic virus (35), which has a similar subunit molecular weight, and the observation that the isolated Semliki Forest virus nucleocapsid can contract by 60 Å upon treatment with acid pH (24,34,36). A simple interpretation of these observations is that the alphavirus nucleocapsid resembles an "expanded" structure such as those adopted by tomato bushy stunt virus or southern bean mosaic virus upon removal of divalent ions at elevated pH.

Use of a sensitive sequence alignment technique has provided further evidence for a fundamental similarity between the structures of nonenveloped $T = 3$ viruses and the nucleocapsid of Sindbis virus (32). The carboxyterminal two-thirds of the Sindbis and Semliki Forest virus capsid protein sequences align with the beta-barrel region of several picornavirus sequences. This alignment predicts that this region of the alphavirus protein will be folded into an eight stranded beta-barrel with very small loops and nonstrand regions. Such a folding is consistent with the smooth surface of the capsid observed in the reconstruction. The predicted folding also places the three residues which mutation studies have shown are involved in the proteolytic activity of capsid at one end of the barrel so that they could form an active site.

The nature of the interactions between the nucleocapsid and the envelope remain an intriguing aspect of the structure. The cytoplasmic domains of the spikes near the fivefold axes each interact with three capsid elements. If the positions of the symmetry axes are used to draw a correspondence between the nucleocapsid structure and the three conformations of the coat proteins in the tomato bushy stunt virus and southern bean mosaic virus structures (35,37), the site for interaction with the cytoplasmic domains of the spike proteins would be created by a slippage of the A subunits relative to the B and C subunits. Such a slippage occurs in the expanded structure of tomato bushy stunt virus (38). If this description is accurate, it suggests that the interaction of the nucleocapsid with the spikes may not only help organize the envelope of the virus but also stabilize the nucleocapsid.

This observed structure suggests a model for alphavirus formation in which budding alphaviruses results from a hierarchy of structural interactions. The alphavirus surface can be described in terms of three levels of organization: the trimeric spike, the hexagonal packing of trimers, and the $T = 4$ icosahedral arrangement of the virus surface. The most fundamental level of organization is the trimer. Where trimerization occurs during intracellular transport is not yet known. One possibility is that trimer formation is a late event in the biosynthetic pathway, perhaps induced by the posttranslational cleavage of the precursor p62 into E2 and E3. If this were the case, budding would be inhibited in the endoplasmic reticulum and in the Golgi stack. The next level of organization is the local packing of trimers as seen around the virus three-fold axis. von Bonsdorff and Harrison (39) provided evidence that this organization may result from the lateral interaction of the spikes when they examined detergent-fused virus envelopes in which the interaction with the virus capsid had been lost. Under these conditions the spike proteins were arranged on a hexagonal net. The formation of clusters of spikes may be the prerequisite for the budding process which must involve the interaction of spikes with the preformed icosahedral nucleocapsid. The complementary nature of the capsid structure suggests that the role of the capsid is to curve the forming hexagonal membrane lattice into a closed shell. This process involves the introduction of pentagonal elements into the net, rather than fixing each of the spikes individually into its position on the virus surface. In this view, budding is a dynamic process involving both the lateral interaction of trimers and transmembrane interaction of the spikes and the capsid proteins in the nucleocapsid, which lead to a curved $T = 4$ icosahedral surface arrangement of the virus envelope. This is a different view of virus assembly than that derived from work with nonmembranous icosahedral viruses (14,40,41). It takes into account the extra constraint that the spike proteins of a membrane virus must assemble in a plane. This hierarchy of interactions ensures that virus budding cannot occur until hexagonal clusters of spikes have been formed. The exclusion of nonviral membrane proteins would result from the lateral interactions of the spikes. It should be pointed out that not every position in the $T = 4$ surface lattice need be filled by spikes since the preformed nucleocapsid functions as a template for the budding process. If such a flexibility were allowed in virus budding, then the molecular weight measurements could be consistent with a $T = 3$ nucleocapsid enclosed by an incomplete $T = 4$ envelope.

BUDDING OF VESICULAR STOMATITIS VIRUS

The budding process is more complicated for enveloped viruses such as vesicular stomatitis virus (VSV) and influenza virus, which have an additional layer consisting of an M protein between the nucleocapsid and the lipid bilayer (11). The budding of these viruses involves interactions between the viral glycoproteins, the M proteins, and the nucleocapsid. The budding mechanism is further obscured by findings which suggest that there is no precise stoichiometry between the spike glycoproteins and the underlying M or nucleocapsid proteins (e.g., 42,43). The ratio between the G protein and internal proteins in VSV has been claimed to vary severalfold, whereas the ratio between M and the nucleocapsid proteins remains constant. (42).

There is even evidence that particles containing no G protein can be produced during infection with a VSV mutant whose G protein is not transported to the surface at the restrictive temperature. The studies which showed this variation did not consider the possibility that the apparent G to M ratios vary due to cleavage of the G protein from the membrane by protease during or after the budding process (44). This would leave membrane anchors derived from G protein in the viral envelope that would be hard to detect. Metsikkö and Simons (45) have shown that such a cleavage of G protein does in fact occur for the ts045 mutant in the Vero cell line which had been used for the previous studies. An antibody directed against the cytoplasmic tail of the G protein was used to guantitate the number of G proteins and stubs in the virion under various conditions of infection. The total remained constant although the fraction of the G cleaved did vary.

One most intriguing aspect of the membrane sorting occurring during the budding of the M protein-containing viruses is the production of mixed phenotypes during double infection of a cell (46). Under these conditions virus particles are formed which contain the nucleocapsid and M protein of one virus and the glycoproteins from both viruses used to infect the cell. VSV can include into its envelope during budding the surface glycoproteins from alphaviruses, retroviruses, bunyaviruses, arenaviruses, paramyxoviruses, myxoviruses, herpes viruses, or pox viruses. Also endogenous retrovirus spike glycoproteins can form mixed phenotypes during single infection with VSV (46). On the other hand, almost all cellular proteins are efficiently excluded. The one clearly defined exception is Thy-1 protein. Studies using monoclonal antibodies have demonstrated that this cell surface antigen can be included into viral buds (47).

The presence of homologous spike proteins is required for directing the budding process. Mixed phenotype formation does not occur in polar MDCK cells double-infected with VSV and influenza virus (48). Mixed phenotypes do not form early in double infection because VSV buds through the basolateral membrane domain and influenza virus through the apical membrane of the polarized epithelial cells. Formation of mixed viruses began only after the cytopathic effects of the infection led to a loss of surface polarity with mixing of the basolateral VSV and the apical influenza glycoproteins on the cell surface. It has also been shown that a critical amount of G protein is necessary to initiate the budding of mixed phenotypes of VSV particles in other doubly infected cells (49). Contradictory results suggesting that the presence of G protein is not essential for VSV budding have also been reported (50) but these can be explained by the cleavage of G protein described above (45).

The M protein also has an essential role in the budding process. Temperature-sensitive mutants of VSV, in which the M proteins were defective, demonstrated that no budding occurred at the nonpermissive temperature (51). In our tentative scheme for VSV budding, the M protein is placed as the adaptor between the nucleocapsid and the G protein. In this model, a central role is proposed for the G protein: the budding proceeds first when a critical concentration of G protein has accumulated at the cell surface. M protein binds to the cytoplasmic domains of a G protein cluster and this interaction is stabilized by the binding of the nucleoprotein precursor. The nucleocapsid is assembled at the budding site (52). Interactions between the G and the M proteins force the membrane to wrap around the forming nucleocapsid. Foreign viral spike proteins are included in the budding virion either by direct transmembrane interaction with the underlying M protein layer or by lateral interactions with G proteins bound to the M proteins. More G proteins and foreign spike proteins are included into budding vesicles by diffusion-mediated trapping. The cellular

Thy-1 protein which is also included into VSV is not a transmembrane protein (53). It is attached to the outer aspect of the plasma membrane by a covalently bound phospholipid tail. In this case, interactions with the VSV proteins would be possible only on the extracytoplasmic side. Copatching experiments suggested that Thy-1 antigens did not associate with the viral spike proteins outside budding virions (47). However, these findings do not rule out weak lateral interactions between the Thy-1 and the G proteins that could have been disturbed by the added antibodies. In our opinion the most likely mechanism for phenotypic mixing is based on weak lateral interactions between the extracytoplasmic parts of the spike proteins. These interactions may be enhanced by a cooperative effect due to the filling of the surface lattice by the spike proteins forming around the nucleocapsid during budding. The foreign spike proteins could be included in the lattice in positions which are analogous to those occupied by the spikes on the threefold axes of the $T = 4$ Sindbis envelope which have no direct interaction with the $T = 3$ nucleocapsid. The curving of the membrane which is responsible for the budding of the mixed virion would then result from interactions between the G protein and the underlying M protein which in turn interacts with the nucleocapsid. The budding process may be restricted to the cell membrane for several reasons. Budding could depend on a critical concentration of spike proteins that is reached only in the plasma membrane. Alternatively, lateral interactions between the spike proteins may be favored by the conditions prevailing in the extracellular milieu (pH, ionic concentrations, etc.) which would differ from those in the intracellular compartments, through which the G proteins are transported. It is also possible that a posttranslational modification could take place late in the pathway that would favor budding.

VIRUS BUDDING AS A PARADIGM FOR INTRACELLULAR MEMBRANE SORTING

To form the basis of a model for intracellular membrane sorting, this hypothetical scheme for VSV budding has to be modified by reversing the orientation of the membrane vesicles (10). In the virus particle the nucleocapsid scaffold is inside the vesicle toward the cytoplasmic aspect of the lipid bilayer, whereas in intracellular membrane carrier vesicles (e.g., clathrin-coated vesicles) the cytoplasmic face of the membrane faces outward and is enclosed into the polyhedral clathrin scaffold (54). The model proposes that the proteins to be included in the membrane vesicles are analogous to the foreign viral spike proteins in phenotypic mixing. These bind either directly to an underlying adaptor protein or to a transmembrane sorting protein, the function of which would be analogous to the proposed role of G protein in the inclusion of foreign viral glycoproteins. The lateral interactions responsible for inclusion of proteins into the budding vesicle might be promoted by the conditions in the donor compartment (pH, ionic gradients, etc.) to be reversibly broken by the environment prevailing in the acceptor compartment. The proposed adaptor protein, equivalent to the VSV M proteins, would interact with both the transmembrane sorting proteins and the scaffold proteins responsible for generating the curvature which leads to the closure of the membrane vesicle. Exclusion of proteins from transport would be due to their inability to fit into the lattices formed on either side of the membrane. The stringency of exclusion could vary depending on the sorting event. For instance, in the sorting of apical proteins in epithelial cells about 10% of an apical viral protein has been found to be missorted to the basolateral cell surface domain. In contrast, a much smaller proportion of a viral basolateral protein was found on the wrong surface domain (55).

One interesting feature of the viral budding model for membrane vesicle formation is the flexibility in the stoichiometry of proteins that are selectively included. A VSV particle can be formed which consists of G proteins alone or varying proportions of phenotypically included foreign spike proteins. The same could hold for intracellular membrane sorting. Stoichiometry of the proteins included into the vesicle membrane would depend on concentration of the proteins in the membrane of the donor compartment.

The heavy reliance on the normal functions of the host cell by viruses has been the key to the use of viruses as tools to study the molecular mechanisms of processes in the animal cell. Enveloped virus budding may seem at first sight to be a purely virus-related phenomenon since it results in the formation of a particle which escapes from the host cell and moves on to infect others. However, the underlying mechanism may be a general one. Since both enveloped virus formation and intracellular sorting events have similar spatial constraints, they may utilize analogous mechanisms. It seems likely that the first virus which invented the budding mechanism modified host cellular proteins for its own purposes. Seen from this viewpoint, an understanding of viral budding should lead to a better understanding of intracellular membrane sorting.

ACKNOWLEDGMENTS

We are grateful to Jacques Dubochet, Henrik Garoff, Barry Gumbiner, and Kalervo Metsikkö for stimulating discussions and for a critical reading of this manuscript. We would also like to thank Hilkka Virta for excellent technical assistance in the experimental work behind the ideas expressed here, and Annie Steiner for typing the manuscript.

REFERENCES

1. Blobel, G. (1980). *Proc. Natl. Acad. Sci. U.S.A.* **77**, 1496–1500.
2. Milstein, C., Browlee, G. G., Harrison, T. M., and Mathews, M. B. (1972). *Nature (London), New Biol.* **239**, 117–120.
3. Blobel, G., and Dobberstein, B. (1975). *J. Cell Biol.* **67**, 835–862.
4. Walter, P., and Blobel, G. (1980). *Proc. Natl. Acad. Sci. U.S.A.* **77**, 7112–7116.
5. Meyer, D. I., Krause, E., and Dobberstein, B. (1982). *Nature (London)* **297**, 647–650.
6. Sabatini, D. D., Kreibich, G., Morimoto, T., and Adesnik, M. (1982). *J. Cell Biol.* **92**, 1–22.
7. Farquhar, M. G., and Palade, G. (1981). *J. Cell Biol.* **91**, 77s–103s.
8. Palade, G. (1975). *Science* **189**, 347–358.
9. McCloskey, M., and Poo, M. (1984). *Int. Rev. Cytol.* **87**, 19–81.
10. Simons, K., and Fuller, S. D. (1985). *Annu. Rev. Cell Biol.* **1**, 245–288.
11. Simons, K., and Garoff, H. (1980). *J. Gen. Virol.* **50**, 1–21.
12. Patzer, E. J., Wagner, R. R., and Dubovi, E. J. (1979). *CRC Crit. Rev. Biochem.* **6**, 165–217.
13. Garoff, H., and Simons, K. (1974). *Proc. Natl. Acad. Sci. U.S.A.* **71**, 3988–3992.
14. Harrison, S. C. (1983). *Adv. Virus Res.* **28**, 175–240.
15. Simons, K., and Warren, G. (1984). *Adv. Protein Chem.* **36**, 79–132.
16. Harrison, S. C., David, A., Jumblatt, J., and Parnell, J. E. (1971). *J. Mol. Biol.* **60**, 523–528.
17. von Bonsdorff, C.-H. (1973). *Commentat. Biol.* **74**, 1–53.
18. von Bonsdorff, C.-H., and Harrison, S. C. (1975). *J. Virol.* **16**, 141–145.
19. Enzmann, P. J., and Weiland, F. (1979). *Virology* **95**, 501–510.
20. Horzinek, M., and Mussgay, M. (1969). *J. Virol.* **4**, 514–520.
21. Jacrot, B., Cruiell, M., and Söderlund, H. (1987). In preparation.
22. Freeman, R., and Leonard, K. (1981). *J. Microsc. (Oxford)* **122**, 275–286.
23. Murphy, F. A. (1980). In "The Togaviruses" (R. W. Schlesinger, ed.), pp. 241–310. Academic Press, New York.
24. Söderlund, H., Kääriänen, L., and von Bonsdorff, C.-H. (1975). *Med. Biol.* **53**, 412–417.
25. Brown, D. T., and Gliedman, J. B. (1973). *J. Virol.* **12**, 1534–1539.
26. Brown, D. T., Waite, M. F., and Pfefferkorn, E. R. (1972). *J. Virol.* **10**, 524–536.
27. Adrian, M., Dubochet, J., Lepault, J., and McDowall, A. (1984) *Nature (London)* **308**, 32–36.
28. Dubochet, J., Lepault, J., Freeman, R., Berriman, J. A., and Homo, J.-Cl. (1982). *J. Microsc.* **128**, 219–237.

29. Dubochet, J., Adrian, M., Lepault, J., and McDowall, A. W. (1985). *Trends in Biochem. Sci. (Pers. Ed.)* **10**, 143–146.
30. Vogel, R. H., Provencher, S. W., von Bonsdorff, C.-H., Adrian, M., and Dubochet, J. (1986). *Nature (London)* **320**, 533–535.
31. Fuller, S. D. (1987). *Cell* (in press).
32. Fuller, S. D., and Argos, P. (1987). *EMBO J.* (in press).
33. Kääriïanen, L., and Söderlund, H. (1971). *Virology* **43**, 291–299.
34. Söderlund, H., von Bonsdorff, C.-H., and Ulmanen, I. (1979). *J. Gen. Virol.* **45**, 15–26.
35. Abad-Zapetero, C., Abel-Meguid, S. S., Johnson, J. E., Leslie, A. G. W., Rayment, I., Rossman, M. G., Suck, D., and Tsukihara, T. (1980). *Nature (London)* **286**, 33–39.
36. Söderlund, H., Kääriäinen, L., von Bonsdorff, C.-H., and Weckström, P. (1972). *J. Virol.* **47**, 753–760.
37. Harrison, S. C., Olson, A. J., Schutt, C. E., Winkler, F. K., and Bricogne, G. (1978). *Nature (London)* **276**, 368–373.
38. Robinson, I. K., and Harrison, S. C. (1982). *Nature (London)* **297**, 563–568.
39. von Bonsdorff, C.-H., and Harrison, S. C. (1978). *J. Virol.* **28**, 578–583.
40. Caspar, D. L., and Klug, A. (1962). *Cold Spring Harbor Symp. Quant. Biol.* **27**, 1–24.
41. Rossmann, M. G. (1984). *Virology* **134**, 1–11.
42. Lodish, H. F., and Porter, M. (1980). *J. Virol.* **33**, 52–58.
43. Ruta, M., Murray, M. J., Webb, M. C., and Kabat, D. (1979). *Cell* **16**, 77–88.
44. Little, S. P., and Huang, A. S. (1978). *J. Virol.* **27**, 330–339.
45. Metsikkö, K., and Simons, K. (1986). *EMBO J.* **5**, 1913–1920.
46. Zavada, J. (1982). *J. Gen. Virol.* **63**, 15–24.
47. Calafat, J., Janssen, H., Demant, P., Hilgers, J., and Zavada, J. (1983). *J. Gen. Virol.* **64**, 1241–1253.
48. Roth, M. G., and Compans, R. W. (1981). *J. Virol.* **40**, 848–860.
49. Witte, O. N., and Baltimore, D. (1977). *Cell* **11**, 505–511.
50. Weiss, R. A., and Bonnett, P. L. P. (1980). *Virology* **100**, 252–274.
51. Knipe, D. M., Baltimore, D., and Lodish, H. F. (1977). *J. Virol.* **21**, 1149–1158.
52. Dubois-Dalcq, M., Holmes, K., and Rentier, B. (1984). *In* "Assembly of Enveloped RNA Viruses" pp. 21–43. Springer, New York.
53. Low, M. G., and Kincade, P. W. (1985). *Nature (London)* **318**, 62–64.
54. Crowther, R. A., Finch, J. T., and Pearse, B. M. F. (1976). *J. Mol. Biol.* **103**, 785–798.
55. Pfeiffer, S., Fuller, S. D., and Simons, K. (1985). *J. Cell Biol.* **101**, 470–476.

PART III

ANTIBODY AND PROTEIN–PROTEIN INTERACTIONS

11
Intramolecular Localization of Antigenic Determinants by Molecular Immunoelectron Microscopy

J. LAMY
Laboratory of Biochemistry
School of Pharmacy
University François Rabelais
Tours, France

INTRODUCTION

The nature of antigenic determinants has in the past decade led to controversy. In an initial theory, the determinants were considered to be either of the sequential or of the conformational type, depending on whether they were composed of residues contiguous or distant in the sequence. In the 1970s, it was generally accepted that the presence of a few sequential determinants explained most of the protein antigenicity. However, in the past five years the wide use of monoclonal antibodies has led to a reexamination of this view, and now the antigenic determinants are rather considered as small hydrophilic and flexible surface domains. It is widely accepted that a large part of the protein surface (80% or more) is composed of overlapping antigenic surface domains which are mainly, if not all, completely or partially composed of amino acid residues not contiguous in the sequence (Todd et al., 1).

The methods employed to localize antigenic determinants are based mainly on chemical modifications of amino acid residues, on inhibition of immunological monoclonal recognition by synthetic peptides and by proteolytic peptides, and on the comparison of antibody affinities for evolutionary related proteins. Of course, all these methods require the knowledge of the three-dimensional structure and the amino acid sequences of the antigenic molecules. Therefore, a small number of antigens, such as myoglobin, hemoglobin, lysozyme, cytochrome c, insulin, and the coat protein of the tobacco mosaic virus, have been studied in detail. In the 1970s, our laboratory was working on the quaternary structure of arthropod hemocyanin, a respiratory pigment of invertebrates, which is probably the best immunogen substance in existence. The methodological approach used was molecular immunoelectron microscopy (MIEM), a low-resolution method to localize antigenic determinants in high molecular weight proteins. At that time, the resolution of MIEM was limited because of the necessity of using polyclonal antibodies and because of the poor resolution of the negative stain. At the end of the 1970s and in the early 1980s, the development of monoclonal antibodies and image processing methods allowed one to overcome these limitations so that, if the direct localization of antigenic determinants is still a challenge, it is not perhaps any longer a chimera.

The following sections describe the two main steps of our MIEM work: (1) the precise intramolecular localization of the 24 subunits which compose the hemocyanin of the scorpion *Androctonus australis* and of the 48 subunits of the hemocyanin of the horseshoe crab *Limulus polyphemus*; (2) the results of the first attempts of intrasubunit localization of antigenic epitopes in our favorite material, arthropod hemocyanins.

STRUCTURE OF ARTHROPOD HEMOCYANINS: THE SUBJECT OF THIS STUDY

Hemocyanins are the blue, copper-containing respiratory pigments of arthropods and mollusks. Arthropodan hemocyanins are composed of ≃75-kDa polypeptide chains called monomeric subunits. In the oligomeric native forms, the monomeric subunits are grouped by multiple of six subunits, producing 1 × 6-mers or hexamers (450 kDa), 2 × 6-mers or dodecamers (900 kDa), 4 × 6-mers (1.8 MDa), and 8 × 6-mers (3.6 MDa). In addition to these well-known forms, recent results of Mangum et al. (2) indicate that a 6 × 6-meric structure occurs in centipedes. The aggregation level of a hemocyanin molecule is strongly dependent on its phylogenic origin. For example, most of the Crustacea hemocyanins belong to the 1 × 6-meric or 2 × 6-meric types, whereas tetrahedral 4 × 6-mers are exclusively found in the thalassinid shrimps. In the Chelicerata subphylum, 8 × 6-mers are found in the horseshoe crabs, and flat 4 × 6-mers and 2 × 6-mers occur in the arachnid class. For this work, we selected the two most complex oligomers: the 4 × 6-meric hemocyanin of *A. australis* and the 8 × 6-meric hemocyanin of *L. polyphemus*. For more detail on the structural and phylogenic aspects of arthropod hemocyanins, the reader is referred to van Holde and Miller (3) and Ellerton et al. (4).

ARCHITECTURE OF *ANDROCTONUS AUSTRALIS* AND *LIMULUS POLYPHEMUS* HEMOCYANINS

Until recently no three-dimensional structure for arthropod hemocyanin had been determined by X-ray crystallography, so that hemocyanins were known mainly from electron microscopy (EM) and image processing of the EM views. In 1984, Gaykema et al. (5) determined the crystallographic structure of the hemocyanin of the spiny lobster *Panulirus interruptus* at 3.4 Å resolution. Their results demonstrated that most of the assumptions that had been made to interpret the EM views were correct so that only minor adjustments and refinements had to be made (Lamy et al., 6). Now that we have a perfect understanding of the protein architectures required for interpreting the immunolabeling experiments, we can describe the models of *A. australis* and *L. polyphemus* hemocyanins.

The monomeric building block of all arthropod hemocyanins is the 75-kDa subunit. There is general agreement that the foldings of all arthropod hemocyanin subunits are similar to the folding of the *P. interruptus* subunit. The X-ray diffraction patterns of isolated subunits of *A. australis* (Fearon et al., 7) and *L. polyphemus* (Magnus and Love, 8) hemocyanins are in good agreement, and the strong preservation of amino acid sequences of hemocyanin subunits from Crustacea and Chelicerata (Gaykema et al., 5; Jollès et al., 9; Moore and Riggs, 10; Nemoto and Takagi, 11; Schartau et al., 12; Schneider et al., 13; Lamy et al., 14; Yokota and Riggs, 15) suggests that the folding of the *P. interruptus* subunit is general in arthropodan hemocyanins, with only minor differences. Figure 1a shows the common structural features of the hemocyanin subunits of which the various oligomer models are composed (Lamy et al., 16).

These features may be summarized as follows. First, the overall shape of the subunit looks like a bean or a kidney, with approximate

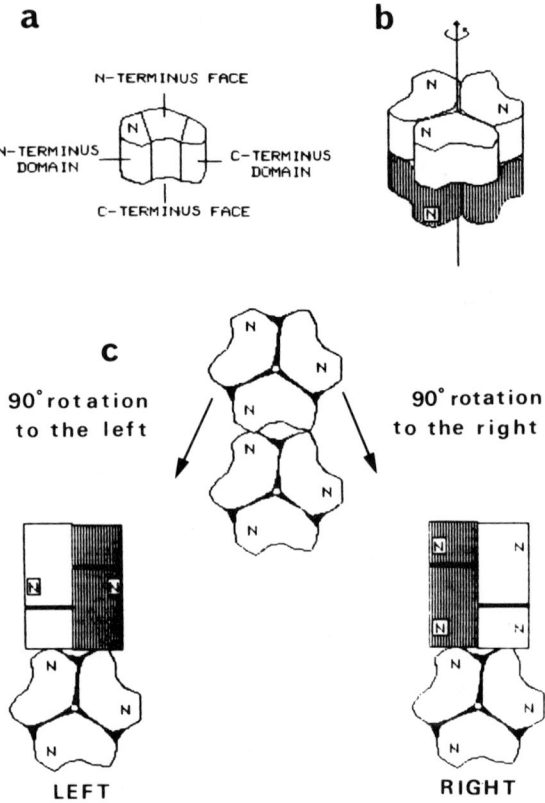

Fig. 1. Schematic representation of the structural components of highly aggregated arthropod hemocyanins. (a) The 75-kDa subunit; N designates the N-terminus in the N-terminus face. (b) The 450-kDa hexamer; the arrow indicates the pseudocrystallographic threefold axis. (c) The 900-kDa 2 × 6-mers of the right and left isomeric types. According to Lamy *et al.* (16), the right form results from a 90° rotation of the upper hexamer to the right.

dimensions of the *P. interruptus* subunit, 48 × 55 × 80 Å (Gaykema *et al.*, 5). Second, the external surface of the subunit is composed of two flat faces, one convex and one concave. To distinguish the flat faces, advantage is taken of the fact that in *P. interruptus* hemocyanin, the face containing the N-terminus amino acid is accessible from the outside, whereas the face containing the C-terminus amino acid is not accessible from the outside because of intersubunit contacts. Third, the subunit is composed of three domains designated as domain 1 (or N-terminus domain), domain 2 (or central), and domain 3 (C-terminus). Domain 1 is smaller (177 residues in *P. interruptus* hemocyanin) than domain 3 (263 residues) and, when viewed from the top, its contour line is rather rectangular whereas the outline of domain 3 is more triangular. Furthermore, domain 1 is more prominent than domain 3. The central domain (233 residues) contains the binuclear oxygen binding site.

The hexameric building block of all arthropod hemocyanins is supposed to be similar to the native hemocyanin of *P. interruptus*. A schematic representation of this symmetric structure is shown in Fig. 1b. Notice that the subunits of the two trimeric layers have their C-terminus faces in contact, which indicates that only their N-terminus faces and the sides of the concave and convex faces are accessible from the outside. Furthermore, the twofold axes rotating threefold layers are located such that the N-terminus domains in the two layers are superimposed. The same pattern is of course observed with the C-terminus domains. This configuration results in an alternation of superposed rectangular domain 1 and of superposed triangular domain 3 when the molecule is viewed from its threefold pseudocrystallographic axis.

The basic dodecameric structure of arthropod hemocyanins is more complex because of the so-called enantiomorphic convention. Let us suppose that, as indicated in Fig. 1c, two hexamers are brought into contact in such a way that the lower hexamer derives from the upper one by translation. If the upper hexamer is rotated to the left or to the right, then a pair of isomeric dodecamers appears. By convention (Sizaret *et al.*, 17), the right isomer is the structure resulting from a 90° rotation of the *upper* hexamer to the right. In our previous reports, the subunits were supposed to have a twofold axis passing through the center of the concave and convex faces. Therefore, the rotation of the upper hexamer to the left or to the right produced a pair of enantiomorphs. The discrimination of the N- and C-terminus domains introduces a new element of asymmetry in the model so that the left and right isomers can no longer be considered enantiomers. In *A. australis* and *L. polyphemus* hemocyanins, the dodecameric units are of the *right* isomeric type. This choice, which was a matter of controversy, has been recently reexamined (Lamy *et al.*, 6). It is supported by three experimental sets of data. First, in the 4 × 6-mer molecule, the rocking axis described by van Heel and Frank (18) passes through the lower right and the upper left hexamers of the top view when the longitudinal cleft is oriented from top to bottom. Second, in the 4 × 6-mer 45° view, described by van Heel *et al.* (19), the molecule stands on its left dodecamer when the longitudinal cleft is oriented from top to bottom and when the hexamers with hexagonal contours are located on the top of the figure. Third, the EM views of the dodecamers resulting from the dissociation of L. polyphemus hemocyanin are classified by the image processing method of van Heel and Frank (18) into two populations corresponding to the two faces of the molecule. In the average images, the alternation of the rectangular and triangular vertices and the position of the intersubunit clefts are in perfect agreement with the choice of the right isomeric type (Lamy et al., 6).

Bringing side by side two copies of the right dodecamer in such a way that their 1/1 edges come into contact produces a 4 × 6-mer. According to Lamy et al. (16), in the 1/1 edge, the two subunits located away from the middle of the edge have their concave faces accessible from the outside. Two additional features discovered by van Heel and Frank (18) have still to be introduced in the model, the rocking effect and the FLIP/FLOP effect. The rocking effect describes the fact that the 4 × 6-mer molecule, in its top view, does not stand on the support plane on four hexamers but on only three. The result is a shaky molecule which "rocks" around an axis, joining the lower right and the upper left hexamers when the longitudinal cleft between the dodecamers is oriented from top to bottom in the picture. The FLIP/FLOP effect results from another observation of van Heel and Frank (18) that the molecule is not perfectly square because of the slight shift of one dodecamer along the longitudinal cleft direction. Thus, the projection of the 4 × 6-mer in its top view is no longer a rectangle but a parallelogram. This FLIP/FLOP effect allowed van Heel and Frank to separate by the CORAN (correspondence analysis) method two different new views that they called FLIP and FLOP views. In fact, these views reveal the existence of different faces that Sizaret et al. (17) called FLIP and FLOP faces because of the origin of their discovery. According to these authors, in the top view, the molecule is standing on its FLIP *face* when its contact points with the support are the two hexamers of the short parallelogram diagonal and one hexamer of the long diagonal. Conversely, the molecule stands on its FLOP *face* when the contact points with the support are the two hexamers of the long diagonal and one hexamer of the short diagonal. As shown below, the definitions of the FLIP and FLOP faces are critical elements of the definition of an absolute quaternary configuration. An architectural model integrating all the experimental data presently available is shown in Fig. 2. The agreement between the three main views of the model, called top view, side view, and 45° view, and the corresponding negatively stained EM views is pretty good (Lamy et al., 20). Notice that in Fig. 2a the long diagonal is oriented perpendicularly to the rocking axis, therefore the molecule is standing on its FLIP face.

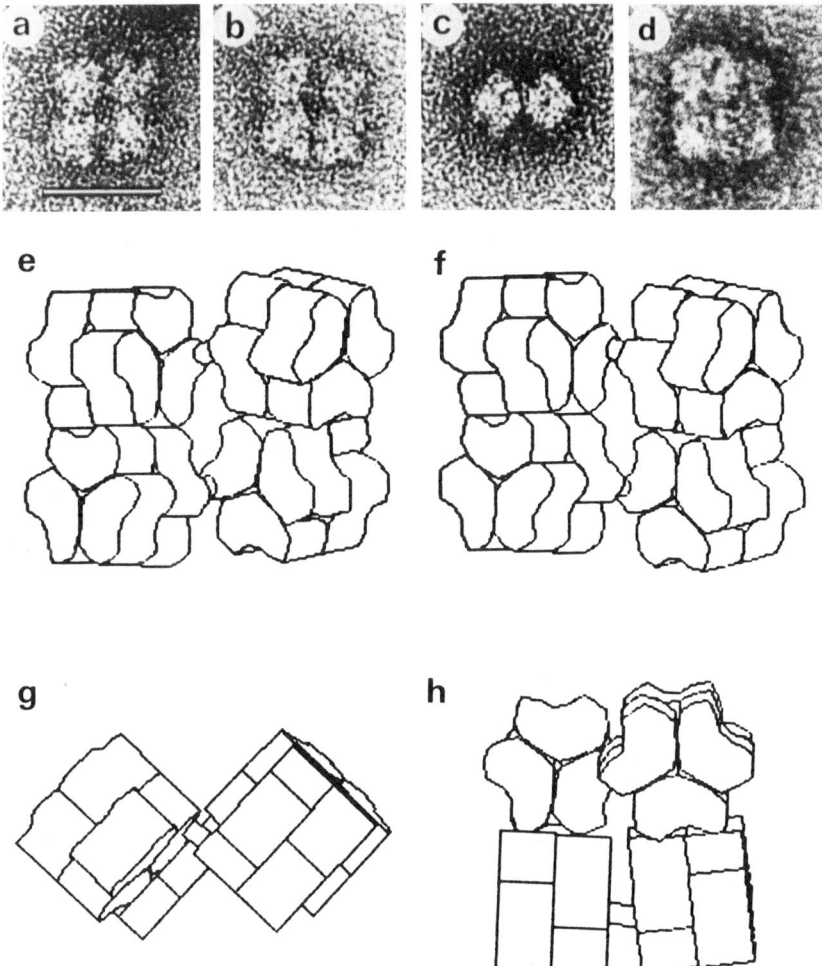

Fig. 2. Model of architecture and electron microscopic views of 4× 6-meric chelicerate hemocyanin. (a,e) Top view standing on its FLIP face; (b,f) top view standing on its FLOP face; (c,g) side view; (d,h) 45° view. The length of the bar is 25 nm.

In Fig. 2b, the molecule is standing on its FLOP face, the long diagonal being oriented to the upper left.

Under mild conditions, the 8 × 6-mer molecule dissociates into 4 × 6-mers, then into dodecamers. Therefore, the clarification of its architecture only requires an understanding of how the two 4 × 6-mer copies are assembled to produce the five main EM views. These views, cataloged by Lamy et al. (16), are designated as cross view, bow tie view, ring view, asymmetric pentagon, and symmetric pentagon. Figure 3 shows a model built by assembling two copies of the 4 × 6-mer model of Fig. 2, as described below. First, one copy of the 4 × 6-mer is positioned on a support so that it presents its top view. On this is placed another copy of the 4 × 6-mer so that the longitudinal cleft between the dodecamers is superposed. Then, the upper molecule is rotated in order to align its rocking axis with the cleft of the lower molecule. Finally, the upper 4 × 6-mer is translated down and is slightly adjusted to give the best fit. The five views of this model, shown in Fig. 3, are in almost perfect agreement with the corresponding EM views.

With the exception of the dodecameric hemocyanin of the spider *Cupiennius salei* (Markl, 21), no covalent bond is involved in the cohesion of the hemocyanin architecture at either the hexameric level or higher levels.

SUBUNIT HETEROGENEITY AND SUBUNIT COMPOSITION OF *ANDROCTONUS AUSTRALIS* AND *LIMULUS POLYPHEMUS* HEMOCYANINS

The subunit heterogeneities of *A. australis* and *L. polyphemus* hemocyanins have been and are still a matter of controversy. Indeed, the higher the resolving power of the separation method, the higher is the number of the different "subunits" reported by the authors. The degree of heterogeneity of the chelicerate hemocyanin subunits is based on their chromatographic, electrophoretic, immunological and functional properties, on their amino acid sequences, and on their role in reassembly experiments.

Removing calcium ions by EDTA or by extensive dialysis, then dialyzing the native molecule at neutral pH in the presence of 1 M urea or at alkaline pH, produces a mixture of monomeric and dimeric subunits called dissociated hemocyanin. Then the subunits are separated by various techniques such as ion–exchange chromatography (Sullivan *et al.*, 23), polyacrylamide gel electrophoresis (Lamy *et al.*, 24), or crossed immunoelectrophoresis (Lamy *et al.*, 22). Figure 4 shows a crossed immunoelectrophoresis of dissociated hemocyanin versus its homologous antiserum. The nine peaks correspond to eight antigenically distinct monomeric subunits termed *Aa* 2, 3A, 3B, 3C, 4, 5A, 5B, and 6 and to a stable heterodimer called subunit *Aa* 3C-5B (or dimeric fraction *Aa* 1) (Lamy *et al.*, 22). All the monomeric subunits were shown to have different primary structures (Jollès *et al.*, 9). In *L. polyphemus*, hemocyanin crossed immunoelectrophoresis also en-

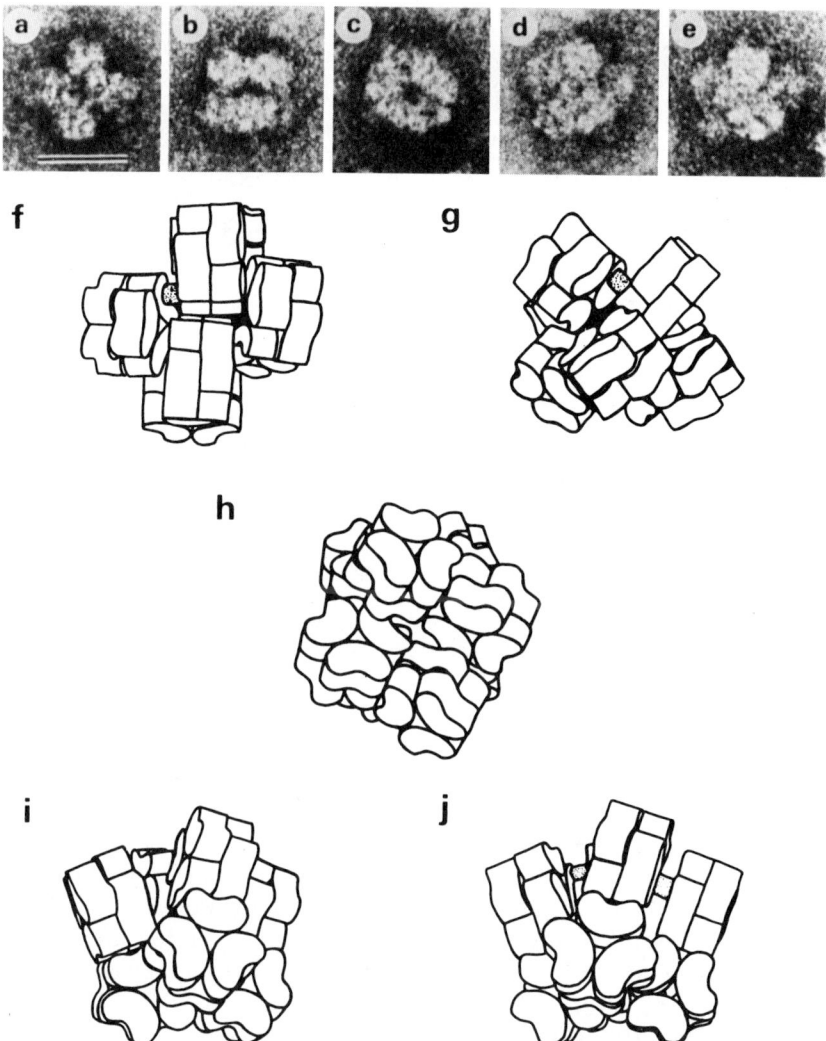

Fig. 3. Model of architecture and electron microscopic views of 8 × 6-meric horseshoe crab hemocyanin. (a,f) Cross view; (b,g) bow tie view; (c,h) ring view; (d,i) asymmetric pentagon; (e,j) symmetric pentagon. The length of the bar is 25 nm.

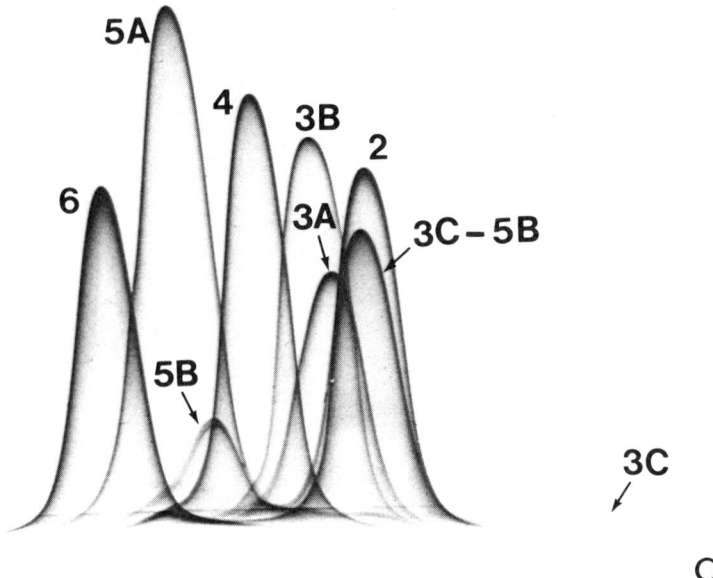

Fig. 4. Crossed immunoelectrophoresis of dissociated hemocyanin of the scorpion *Androctonus australis* against its homologous antiserum. O = deposit well. Subunits are termed according to Lamy et al. (22).

abled Lamy et al. (25) to separate eight subunits which they called I, II, IIA, IIIA, IIIB, IV, V, and VI. Subunits V and VI are the components of a heterodimer similar to the dimeric subunit *Aa* 3C-5B of *A. australis* hemocyanin. However, whereas in the *Androctonus* system there was perfect agreement between the subunit heterogeneities, determined by immunoelectrophoresis and by polyacrylamide gel electrophoresis, a strong discrepancy occurred in the *Limulus* system. Brenowitz et al. (26) found no less than 16 electrophoretic species, whereas Lamy et al. (25) had observed 8 immunologically different species. This difference implies that some of the electrophoretically different subunits are immunologically identical. For example, the electrophoretic subunits designated as *Lp* I, I', and I" by Brenowitz et al. had the same N-terminus sequences, the same intramolecular locations, the same roles in reassembly experiments, and the same functional properties (Lamy et al., 14; Brenowitz et al., 27). These three electrophoretic fractions could be distinguished only by their electric charges in polyacrylamide gel electrophoresis and in ion–exchange chromatography. Therefore, it is now generally accepted that the antigenically distinct subunits are the authentic subunits, but because their amino acid sequences are slightly different they may exhibit microheterogeneity. The reason for such heterogeneity is not clear. Several complete amino acid sequences or important segments of sequences have been elucidated in subunits from merostome and arachnid hemocyanins, but no one reported to have observed any microheterogeneity in the sequenced subunits (Gaykema et al., 5; Nemoto and Takagi, 11; Schartau et al., 12; Schneider et al., 13; Yokota and Riggs, 15).

The determination of the subunit number of copies in the whole molecule is usually performed in two steps. The first step is the dissociation of the native molecule into subunits and the second is an assay of each subunit in the mixture of free subunits. Although the dissociation step does not raise any problems, the assay step is more difficult. Two methods were used: planimetric integration of electrophoretic and chromatographic elution profiles and immunoassays. If densitometric methods are rather imprecise in cases of overlapping and trailing fractions, immunoassays are perturbed by immunological cross-reactivities. For example, Markl et al. (28) and Brenowitz et al. (29) proposed a planimetric integration of the overlapping peaks produced by dissociated hemocyanin in crossed immunoelectrophoresis against anti-dissociated hemocyanin antisera. Because of the intersubunit cross-reactivities (Lamy et al., 30), the results produced by this method are certainly over- or underestimated to an unknown extent. For *A. australis* and *L. polyphemus* hemocyanins, two complementary methods were used: a densitometric integration of the chromatographic and electrophoretic elution profiles and a quantitative immunoassay based on rocket immunoelectrophoresis. For this purpose, a series of eight antibody preparations was made perfectly specific for the eight subunits by immunoadsorption to the seven cross-reacting subunits present in dissociated hemocyanin. Then each subunit was assayed by rocket immunoelectrophoresis, using a calibration curve prepared with an immunologically pure subunit. The subunit composition of *A. australis* and *L. polyphemus* hemocyanins determined by this method are shown in Table I (Lamy et al., 31). To make the results easier to understand, the number of copies has been rounded to the nearest integer. As shown in Table I, *A. australis* subunits are all present as multiples of two subunits, a result in perfect agreement with the hypothesis that the whole molecule is composed of identical dodecameric halves. Conversely, the relative amounts of several subunits of *L. polyphemus* hemocyanin are incompatible with this hypothesis. Thus, subunit *Lp* IIA is present as one copy per 4×6-mer, and subunit *Lp* I as three copies, demonstrating that the 4×6-meric

TABLE 1
Number of Copies per 4×6-mer (Rounded to the Nearest Integer) of the Subunits of *Androctonus australis* (*Aa*) and *Limulus polyphemus* (*Lp*) Hemocyanins

Aa^a		Lp^b	
Subunit	Number of copies	Subunit	Number of copies
2	4	I	3
3A	2	II	4
3B	2	IIA	1
3C	2	IIIA	4
4	4	IIIB	4
5A	4	IV	4
5B	2	V	2
6	4	VI	2

[a] After Lamy et al. (31).
[b] After Lamy et al. (14).

half-molecule cannot be composed of identical dodecamers. This means either that two (or more) types of dodecameric quarters of molecule exist or that some subunits, such as *Lp* IIa and/or *Lp* I, are randomly distributed in the 4 × 6-mer.

INTRAMOLECULAR LOCALIZATION OF SUBUNITS WITH POLYCLONAL ANTIBODIES

THE QUATERNARY STRUCTURE OF THE 4 × 6-MERIC HEMOCYANIN OF *ANDROCTONUS AUSTRALIS*

The intramolecular localization of the subunits within a protein obviously requires the knowledge of the protein architecture. However, the knowledge of the three-dimensional structure of the protein does not necessarily allow discrimination between the different polypeptide chains. For example, the five subunits which compose *P. interruptus* hemocyanin (van Eerd and Folkerts, 32; Folkerts and van Eerd, 33) could not be distinguished within the hexamer by X-ray crystallography (Gaykema et al., 5). In *A. australis* hemocyanin, the problem is still more complex for three reasons. First, the hemocyanin has not yet been crystallized, second, the number of different subunits is 8 instead of 5, and, third, the native molecule is composed of 24 subunits instead of 6. Therefore, another method had to be devised to solve the quaternary structure of *A. australis* hemocyanin.

The principle of this method is simple. A subunit-specific label is bound to the native molecule, then the position of the label is observed in the electron microscope. If the label is an antibody or an antibody fragment, the method is called molecular immunoelectron microscopy. A review of the technical features of MIEM applied to the *Escherichia coli* ribosome has been published (Stöffler and Stöffler-Meilicke, 34). Practically, the antigenic particle is incubated with an antibody preparation directed against one or several antigenic determinants, then the immunocomplexes are freed from the antibody excess and the purified material is examined with the electron microscope.

In preliminary experiments, we used rabbit antibodies raised against purified subunits. This method had several inconveniences. First, since hemocyanin is an excellent immunogen, the corresponding polyclonal immune sera produced abundant precipitates with their homologous antigen. This effect was further amplified because native hemocyanin contained several (two or four) copies of each subunit. Second, the immune serum was directed toward all the epitopes of the subunit so that it cross-reacted more or less strongly with other subunits. Third, the purification step was difficult because most of the immunocomplexes were insoluble and remained on the gel filtration column. Moreover, the few hemocyanin molecules appearing in the effluent were not labeled or had important deformations, suggesting that subunits had been extracted from the oligomer possibly as a result of antibody binding. These disappointing preliminary results suggested that a large part of the problem came from the divalency of the immunoglobulin molecule and that if we could prevent the immunoprecipitation, soluble immunocomplexes would be easier to observe. We attempted various methods, such as the use of very high or very low antigen to antibody ratios, but the results were still not satisfactory. After months of frustrating work, we reexamined the problem of the reduction of the effectiveness of our antibodies and found that the solution consisted in a splitting of the divalent IgG's into monovalent Fab fragments.

When applied to Fab fragments, the MIEM method produced good soluble immunocomplexes with Fab fragments, pointing out the contour line of the hemocyanin molecule. Figure 5 shows selected EM views of some of these complexes with their interpretation in the architectural model of Fig. 2. These results indicated that we were on

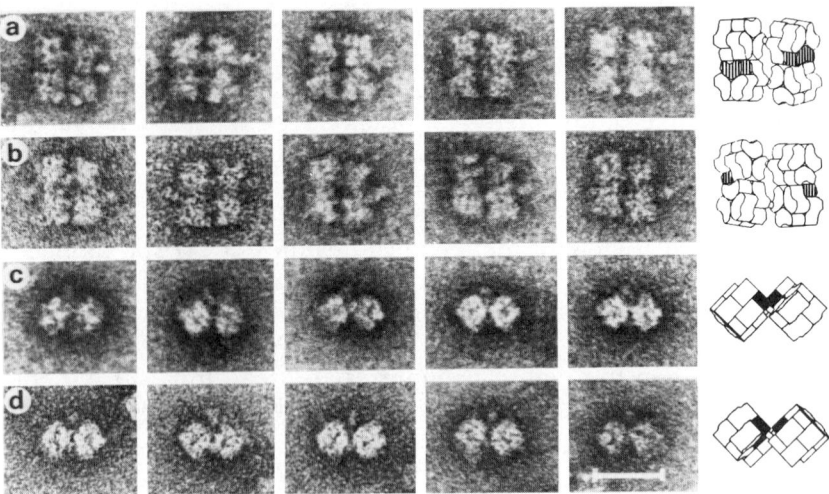

Fig. 5. Simple immunolabeling of *Androctonus australis* hemocyanin with Fab fragments specific for (a) subunit *Aa* 3A; (b) subunit *Aa* 3B; (c) subunit *Aa* 3C; and (d) subunit *Aa* 5B. The models show the corresponding intramolecular locations of the subunits in the model of the quaternary structure described in the section on Intramolecular Localization of Subunits with Polyclonal Antibodies. The length of the bar is 25 nm.

the right track to solve the quaternary structure of *A. australis* hemocyanin, but there was still a long way to go. Actually, as shown in Figs. 5a and 5b, antibody preparations made perfectly specific for subunit *Aa* 3A and *Aa* 3B produced exactly the same type of immunocomplexes. Similarly, antibodies against subunits *Aa* 3C and *Aa* 5B produced the same kind of immunocomplexes (Figs. 5c and 5d). In fact, there was a small difference between the location of subunit *Aa* 3A and *Aa* 3B, but at that time we did not recognize it. The difference is that anti-*Aa* 3A Fab fragments are bound to hexamers of the short diagonal whereas anti-*Aa* 3B Fab fragments bind to hexamers of the long diagonal. This observation was the key to the determination of the absolute quaternary structure. Let us now return to Fab fragments specific for subunits *Aa* 3C and *Aa* 5B, the two components of the stable heterodimer *Aa* 3C-5B. Figures 5c and 5d show that in both cases the subunit is located in the longitudinal cleft between the dodecamers. Lamy *et al.* (35), van Bruggen *et al.* (36), and Lamy *et al.* (37) had shown that subunit *Aa* 3C-5B is required for the reassembly of 4 ×6-mers from free subunits. Thus, the intramolecular location and the structural role of subunit *Aa* 3C-5B were in good agreement, suggesting that the dimeric subunit was involved in the interdodecamer bridges. Even though the EM patterns of Fig. 5 revealed considerable information with respect to the location of subunits Aa 3A, 3B, 3C, and 5B, they also presented new problems such as "Where are the two copies of the subunit localized in the 4 × 6-mer?" and "Are Aa 3A and Aa 3C or Aa 3A and Aa 5B located in the same hexamer?" Both problems were solved by a series of double immunolabeling experiments, using the six possible binary mixtures of the four subunit-specific Fab fragments.

Figure 6a shows that when the labeling agent is a mixture of anti-3A and anti-3B Fab fragments, the 4 × 6-mer molecule bears two Fab fragments slightly away from the middle of the lateral edge. A comparison of these results with those of Figs. 5a and 5b demonstrates that subunits Aa 3A and Aa 3B occupy symmetrical positions in the dodecamer with respect to the middle of the lateral edge, meaning that they belong to different hexamers. Similarly, as shown in Fig. 6b, a mixture of anti-Aa 3C and anti-Aa 5B Fab fragments produces twice-labeled 4 × 6-mers. These images and those of Figs. 5c and 5d are similar, except that the Fab fragments protrude out of the two opposing clefts in Fig. 6b. Therefore, it can be concluded that subunits Aa 3C and Aa 5B are similarly and symmetrically located in the interdodecamer bridge area. Moreover, the labeled epitopes of the two copies of subunit Aa 3C are located on one face of the molecule, whereas those of the two copies of subunit Aa 5B are on the opposite face.

The second problem was to determine whether subunits Aa 3A and Aa 3C are components of the same or of different hexamers. Again, the problem was solved by double immunolabeling experiments, the results of which definitely establish that subunits Aa 3A and Aa 3C are on the same hexamer whereas Aa 3B and Aa 5B are on different hexamers (see Figs. 6c–6f). In these experiments, prior to the immunolabeling step, the 4 × 6-mers were partially dissociated into 2 × 6-mers under mild conditions. Figures 6d and 6e show twice-labeled dodecamers, with one Fab fragment bound to each hexameric quarter of molecule. In the confirmation experiments of Figs. 6c and 6f, the Fab fragments are obviously bound to the same hexamer. These images, together with the fact that subunit Aa 3B is a constituent of the long diagonal hexamers, allowed us to definitely localize subunits Aa 3A, 3B, 3C, and 5B with respect to the FLIP and FLOP faces. Thus, subunits Aa 3B and Aa 3C have their flat accessible (N-terminus) face oriented toward the FLOP face, whereas subunits Aa 3A and Aa 5B have their flat face oriented toward the FLIP face of the 4 × 6-mer. The location of subunits Aa 3A, 3B, 3C, and 5B is summarized in the model of quaternary structure shown in Fig. 7.

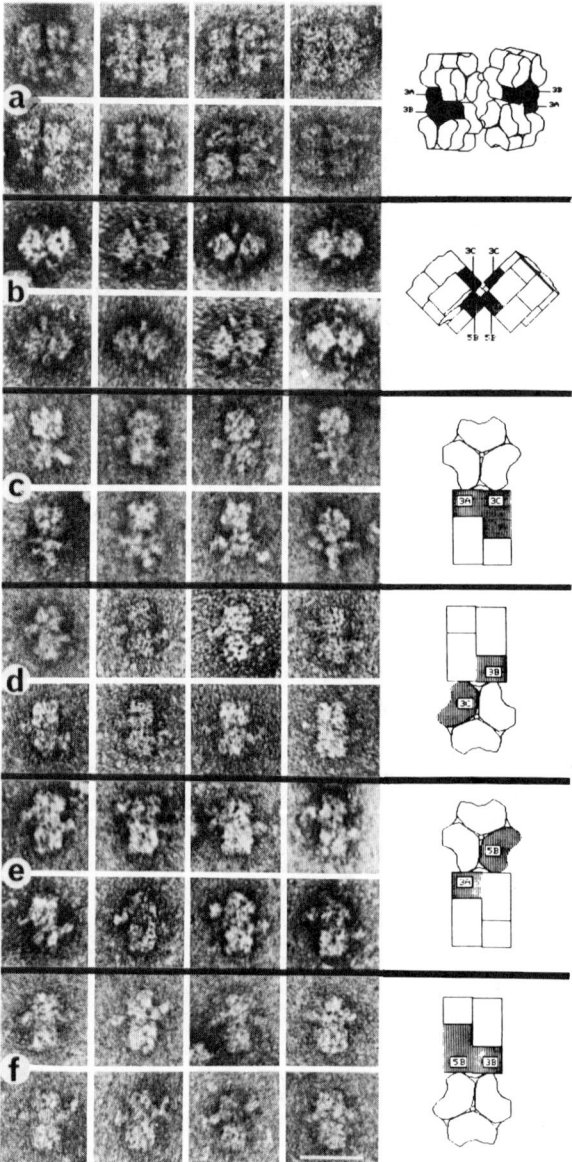

Fig. 6. Double immunolabeling of *Androctonus australis* 4 × 6-mer (a,b) and 2 × 6-mer (c–f) with binary mixture of Fab fragments specific for (a) subunit *Aa* 3A + *Aa* 3B; (b) subunit *Aa* 3C + *Aa* 5B; (c) subunit *Aa* 3A + *Aa* 3C; (d) subunit *Aa* 3B + *Aa* 3C; (e) subunit *Aa* 3A + *Aa* 5B; (f) subunit *Aa* 3B + *Aa* 5B. The models show the corresponding intramolecular locations of the subunits in the model of quaternary structure described in the section on Intramolecular Localization of Subunits with Polyclonal Antibodies. For comparison of the model to the EM views, note that the stain exclusion is produced mainly by the part of the molecule in contact with the support film, which is not directly visible on the model. The length of the bar is 25 nm.

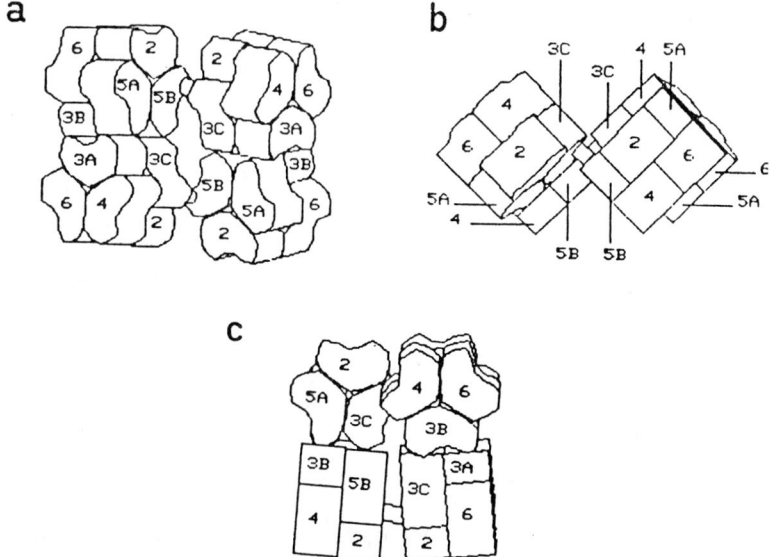

Fig. 7. Model of quaternary structure of *Androctonus australis* hemocyanin. (a) Top view (the FLIP face is oriented toward the reader); (b) side view; (c) 45° view.

The other four subunits present in four copies each per whole molecule (one copy per hexamer) were localized by simple immunolabeling (Lamy et al., 38). The model shown in Fig. 7 perfectly agrees with all the presently available data concerning the structure, function, and evolution of arthropod hemocyanin. For example, as stated above, certain subunits (*Aa* 3C-5B, *Aa* 2, *Aa* 3B, and *Aa* 5A) are required for the reassembly of nativelike 4 × 6-mer structures. This requirement suggests an important structural role that the model explains. Actually, all the required subunits, except subunit *Aa* 2, are involved in interhexamer contacts. In addition, the dimeric subunit *Aa* 3C-5B is obviously involved in interdodecamer junctions. The symmetrical positions of subunits *Aa* 3A and *Aa* 3B explain that subunit *Aa* 3B could replace *Aa* 3A, which is not required for 4 × 6-mer autoassembly (Lamy et al., 37). However, the reason why subunit *Aa* 3A did not replace subunit *Aa* 3B is not yet understood. That subunit *Aa* 2 can replace subunits *Aa* 4 and *Aa* 6, as demonstrated by hexamer reassembly from subunits *Aa* 2 + *Aa* 4 or *Aa* 2 + *Aa* 6 or *Aa* 4 + *Aa* 6 or *Aa* 2 + *Aa* 4 + *Aa* 6 (Lamy et al., 39), is also correctly explained by the model. Indeed, these subunits are the only three subunits of *A. australis* hemocyanin not involved in interhexamer contacts.

AN APPROACH TO THE QUATERNARY STRUCTURE OF THE 8 × 6-MERIC HEMOCYANIN OF *LIMULUS POLYPHEMUS*

As described above, the structure of *L. polyphemus* hemocyanin is twice as complex as the structure of *A. australis* hemocyanin. Fortunately, the problem can be simplified because, under mild conditions (overnight dialysis against an $I = 0.1$ Tris–HCl buffer, pH 7.0 containing 10 mM EDTA), the native 8 × 6-mers dissociate into 4 × 6-meric halves (Brenowitz et al., 40). Therefore, we expected that the determination of the quaternary structure of *L. polyphemus* hemocyanin would be a simple repetition of the previous work on *A. australis* hemocyanin. Unfortunately, upon the binding of Fab fragments, the 4 × 6-mers underwent structural changes, producing, in some cases, a complete dissociation. To prevent this antibody-induced dissociation, the 4 × 6-mers were cross-linked with dimethyl suberimidate, a bifunctional agent, which bridges amino groups exposed on the external surface of the protein. The 4 × 6-mers frozen in their native state by this treatment retained, at least partially, their ability to bind Fab fragments (Lamy et al., 14).

With the cross-linking modification, the technique developed for *A. australis* hemocyanin became applicable to *L. polyphemus* half-molecules. Figure 8 shows selected views of some immunocomplexes obtained with the various subunit-specific antisera, and Table II summarizes the topographical locations of the subunits. Table II also indicates, for comparison, the subunits occupying equivalent posi-

TABLE II
Topology of the Various Subunits in *Limulus polyphemus* 4 × 6-mer Hemocyanin[a]

Subunit	Location	Subunit located in the same position in *A. australis* hemocyanin
Lp I	Corners in the top view	*Aa* 6
Lp II	Middle of the lateral edges in the top view	*Aa* 3A + *Aa* 3B
Lp IIA	Corner and/or end of 2 × 6-mers in the top view	*Aa* 6
Lp IIIA	Middle of the end of the 2 × 6-mers in the top view	*Aa* 4
Lp IIIB	End of the 2 × 6-mer in the top view in the immediate neighborhood of the cleft	*Aa* 2
Lp IV	Near the top/bottom edges in the side view	*Aa* 5A
Lp V + VI	Interdodecamer bridge area	*Aa* 3C + *Aa* 5B

[a] After Lamy et al. (14).

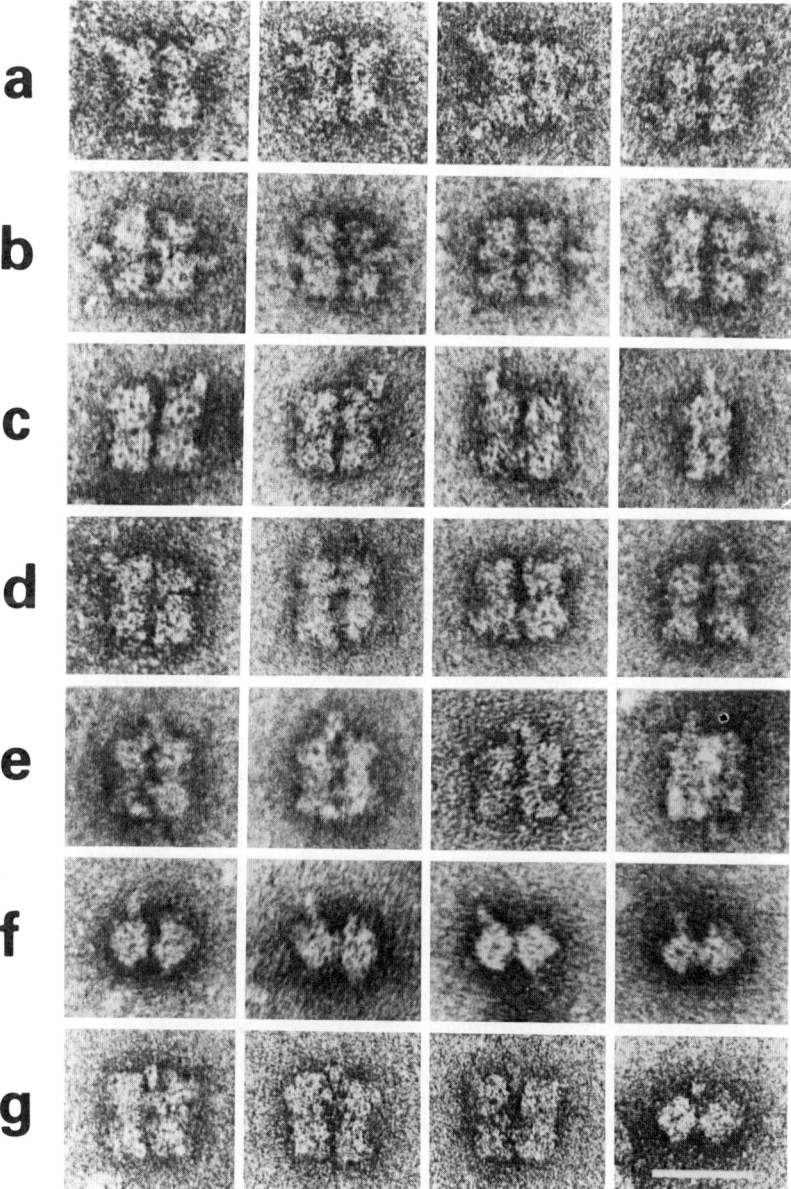

Fig. 8. Selected EM views of soluble immunocomplexes between cross-linked 4 × 6-meric half-molecule of *Limulus polyphemus* hemocyanin and Fab fragments specific for (a) subunit *Lp* I; (b) subunit *Lp* II; (c) subunit *Lp* IIA; (d) subunit *Lp* IIIA; (e) subunit *Lp* IIIB; (f) subunit *Lp* IV; (g) dimeric subunit *Lp* V-VI. The length of the bar is 25 nm.

tions in *A. australis* hemocyanin. The precision of the quaternary structure of *L. polyphemus* 4 × 6-mers is not so good as that of *A. australis* hemocyanin for several reasons. First, as shown above, the subunit stoichiometry demonstrates that the molecule cannot be composed of identical dodecamers. Second, the high antigenic relatedness of subunits *Lp* V and *Lp* VI has prevented their separation from each other so that their positions with respect to the FLIP and FLOP faces could not be determined. Third, the two positions, which in *A. australis* hemocyanin are filled by subunits *Aa* 3A and *Aa* 3B, are occupied in the *Limulus* system by two copies of subunit *Lp* II. This disposition contributes to the difficulty of assigning positions to subunits *Lp* V and VI with respect to the FLIP and FLOP faces. Four, the binding positions of anti-*Lp* IIA Fab fragments to the hemocyanin contour line resemble those of subunits *Lp* I and *Lp* IIIA. Finally, we assigned the same locations to subunits *Lp* I and *Lp* IIA essentially because the sum of their number of copies reaches 4, but this choice may have to be reexamined.

A large part of the uncertainties in the above-described results comes from specific aspects of the negative stain. Actually, the three perpendicular views which would be necessary to localize with precision a label in a three-dimensional model are usually not available, probably because of specific interactions between the support and the protein which favor the frequency of certain views. This obstacle may be overcome in the future by using electron microscopy at low temperature in vitreous water. However, in this work, since the equipment was not available, we tried to improve the localization of the Fab fragments with respect to the contour line of the 4 × 6-mer molecule. For this purpose, we subjected a sample set of 4 × 6-mers labeled with anti-*Lp* II Fab fragments to the CORAN method of image processing. We expected that this method, which had already allowed the observation of the rocking and of the FLIP and FLOP effects, would select subsets of images with the Fab fragments exactly in the same orientation. In the first attempt, anti-*Lp* II fragments were chosen because only one Fab fragment is usually visible on the lateral edge, suggesting that despite the polyclonal character of the immune serum a single epitope of subunit *Lp* II would be accessible to the antibody from the outside. The results, shown in Fig. 9, were exactly those expected (Frank *et al.*, 41). Figure 9a shows an anti-*Lp* II Fab fragment appearing as two gaps in the stain layer on both lateral edges of the 4 × 6-mer. Figure 9c is a difference image between labeled (Fig. 9a) and unlabeled (Fig. 9b) particles. The attachment point of the Fab fragment appears as a sharp peak of stain exclusion (in white on the

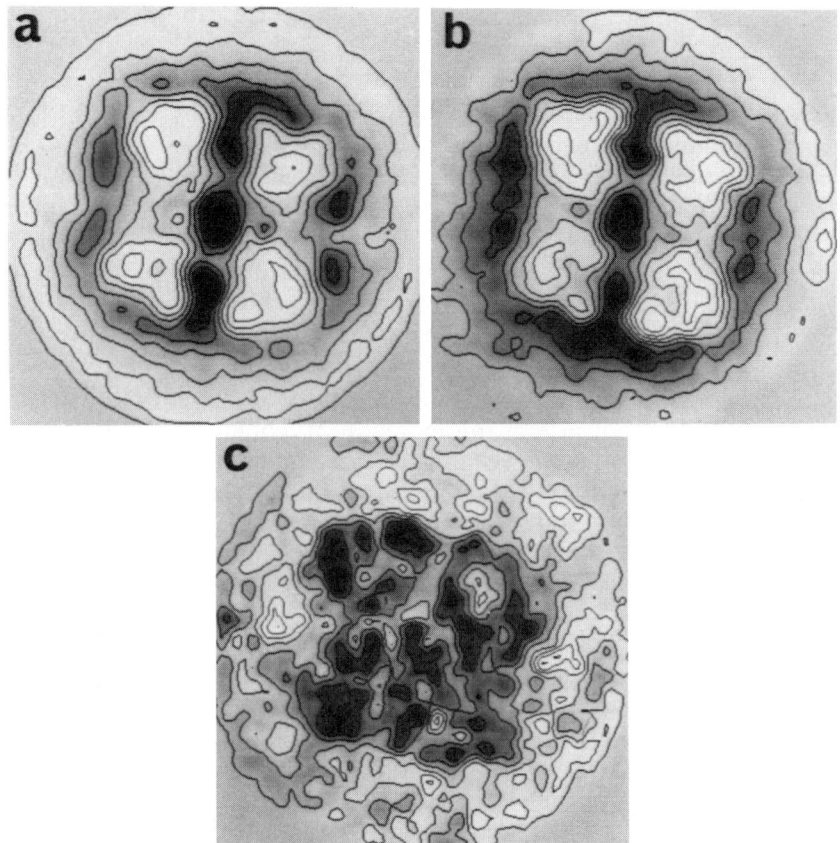

Fig. 9. Average (a,b) and difference images (c) of *Limulus polyphemus* half-molecules labeled (a) and unlabeled (b) with Fab fragments specific for subunit *Lp* II. Prior to averaging, the subset of images was selected by the correspondence analysis method of van Heel and Frank (18).

figure). Figure 9c also shows that because of the flexibility of the Fab fragment around its attachment point, the averaging resulted in only a small unvariable region among the molecule images analyzed. The center of the binding point of the Fab fragment is determined with a precision of about 1.3 nm. Though this result did not change our knowledge of the quaternary structure, it demonstrated the high resolving power of the method, and it suggested that substituting monoclonal for polyclonal antibodies would still increase the precision of the epitope localization. A later section shows that this is actually the case.

Despite some uncertainties, the topological locations of the subunits in the 4 × 6-mer were used to build the model of quaternary structure shown in Fig. 10a. Then two copies of this model were assembled to produce the model of the quaternary structure of the whole molecule shown in Figs. 10b through 10f. To evaluate the fit between the quaternary structure of the whole molecule and the model, immunolabeling with subunit-specific Fab fragments was repeated on cross-linked native 8 × 6-mers. To compare the characteristic selected views of the immunocomplexes shown in Fig. 11 with the model of Fig. 10, the mechanism of stain exclusion must be taken into account. Indeed, the EM image is formed by the electron-dense stain layer surrounding the electron-transparent protein. Thus, the EM view is comparable to a partial molding of the protein, the molded part being the lower portion of the molecule in contact with the support film. This effect is further complicated by the meniscus phenomenon which tends to make visible parts of the protein emerging from the stain layer. Therefore, the part of the molecule which produces the highest degree of stain exclusion corresponds to the part of the model which is not directly visible to the observer. That the molecule can stand on different faces producing similar contour lines but different

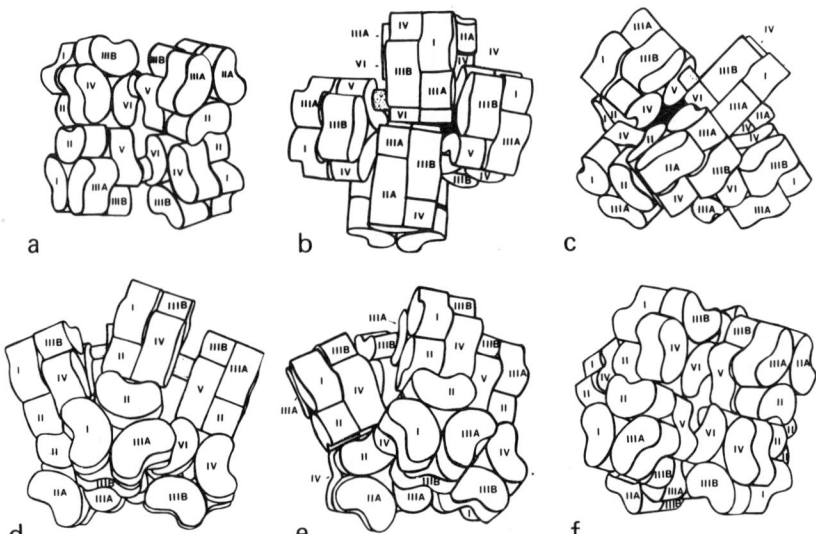

Fig. 10. Model of quaternary structure of *Limulus polyphemus* hemocyanin. (a) 4 × 6-mer: top view; (b) 8 × 6-mer: cross view; (c) 8 × 6-mer: bow tie view; (d) 8 × 6-mer: symmetric pentagon; (e) 8 × 6-mer: asymmetric pentagon; (f) 8 × 6-mer: ring view.

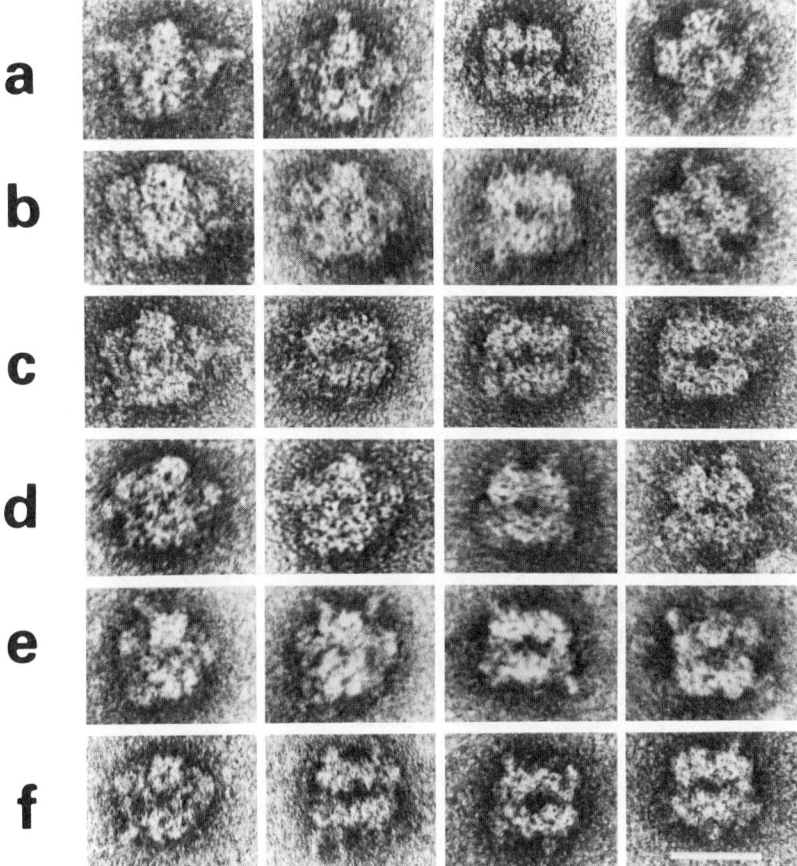

Fig. 11. Selected EM views of soluble immunocomplexes between cross-linked 8 × 6-mers of *Limulus polyphemus* hemocyanin and Fab fragments specific for (a) subunit Lp I; (b) subunit Lp II; (c) subunit Lp IIA; (d) subunit Lp IIIA; (e) subunit Lp IIIB; (f) subunit Lp IV. The length of the bar is 25 nm.

subunit topographies must also be taken into account. For example, in its pentagonal view, the molecule stands on the support on one of its four dodecamers. These four orientations may produce immunocomplexes with Fab fragments differently bound for subunits located in nonequivalent positions in the four dodecamers (i.e., subunits Lp I and Lp IIA). Taking into account these possible causes of misinterpretation, the immunocomplexes of Fig. 11 are in excellent agreement with the model of Fig. 10. Thus, subunit Lp I (Fig. 11a) is located in the micrograph in the upper left, upper right, upper central, and lower right corners, in positions exactly equivalent to those of subunit Lp I in the model. Subunit Lp II (Fig. 11b) is located in the middle of the left and right sides of the pentagons. Subunit Lp IIA (Fig. 11c) has the same location as subunit Lp I, especially in the upper right corner of the pentagonal view. In the case of subunit Lp IIIA (Fig. 11d), the agreement between the model and the EM views is quite good. Actually, in the bow tie view, the Fab fragments are bound to the upper two vertices corresponding to the top/bottom edges of the 4 × 6-mer. Fab fragments specific for subunit Lp IIIB also produce characteristic images. The second view from the left of row e (Fig. 11) shows two Fab fragments approximately bound to the middle of the left and right upper sides of the pentagon. This view exactly fits the model as is shown in Fig. 10d. Finally, Fig. 11f shows anti-Lp IV Fab fragments bound to the lower left corner of the pentagon and to the upper left and right corners of the bow tie view. Again the agreement with the locations, shown in Figs. 10c and 10d, is quite good. No anti-Lp V or Lp V+VI Fab fragments could be observed in any cross-linked 8 × 6-mer view. This is also partially explained by the model. Indeed, subunits V and VI are located in the center of the molecule so that if Fab fragments would bind to these subunits in most cases they would not appear on the outline of the 8 × 6-mer. The reason why immunocomplexes were not observed in the cross view or in the bow tie view while they are visible in equivalent positions of the 4 × 6-mer is not clear.

AUTOASSEMBLY AND MIEM OF OLIGOMERS COMPOSED OF HEMOCYANIN SUBUNITS FROM DIFFERENT SPECIES

The above subsections show that subunits Lp II and Aa 3A,3B occupy the same positions in the 4 × 6-mer architecture of the two systems. Such intramolecular locations in specific positions suggest that these subunits play special structural and/or functional roles. On the other hand, reassembly experiments from free subunits in the presence of calcium also demonstrate specific structural roles for subunits. As described above, subunit Aa 3B is required for the reassembly of *A. australis* 4 × 6-mers. In the *Limulus* system subunit Lp II is also needed for 4 × 6-mer and 8 × 6-mer reassembly. This similar behavior of subunits occupying the same intramolecular location suggested that they may be interchanged. The fact that subunit Lp II really takes the place of subunits Aa 3A and Aa 3B is demonstrated by the following experiment. The 4 × 6-mers were reassembled, at neutral pH in the presence of 10 mM CaCl$_2$, from a complete mixture of A.

australis subunits in which subunit *Lp* II had been substituted for subunits *Aa* 3A and 3B. The 4 × 6-mers were cross-linked by dimethyl suberimidate, then incubated with anti-*Lp* II Fab fragments and purified by gel filtration. Figure 12 shows that the Fab fragments bind to the middle of the lateral edge exactly where subunit *Lp* II is known to be located in the *Limulus* system. This experiment was the first *direct* demonstration that in reassembled oligomers a subunit actually replaces the subunit(s), which it is expected to do (Lamy et al., 14).

IMPROVING THE SPECIFICITY OF MIEM BY USING HETEROLOGOUS POLYCLONAL ANTIBODIES

Given an oligomeric hemocyanin, i.e., the 4 × 6-meric hemocyanin of *A. australis*, and a polyclonal antibody preparation, raised against one subunit of this hemocyanin, it is clear that MIEM allows *at most* the localization of the subunit within the oligomer. Specifically, labeling a portion of the subunit external surface corresponding to a domain or to a particular face is impossible. To attempt to overcome this limitation, we tried to select a small number of the antigenic determinants for a subunit, taking advantage of the evolutionary modification of the protein. In this hypothesis, an antibody preparation raised against a hemocyanin possessing a single antigenic determinant in common with *A. australis* hemocyanin would bind to a single determinant in *A. australis* hemocyanin. In 1980–1981, we investigated by this method the cross-reactivities between hemocyanin subunits of about 100 species, and we demonstrated that all the arthropod hemocyanins cross-react, producing soluble immunocomplexes or immunoprecipitates. As expected, evolutionary related species produce immunoprecipitates, indicating that more than one common antigenic determinant is present in their hemocyanins. Conversely, hemocyanins of evolutionary distant species weakly cross-reacted producing only soluble immunocomplexes. For example, antisera raised against crustacean hemocyanins usually did not precipitate chelicerate hemocyanin subunits. However, in a few cases, a crossed immunoprecipitation occurred between a native chelicerate hemocyanin and an antiserum raised against a crustacean hemocyanin. In all cases, these precipitation reactions were abolished on the dissociation of the chelicerate oligomer.

Figure 13 shows an example of soluble immunocomplexes produced by an antiserum raised against the hemocyanin of the fiddler crab *Uca pugnax* and subunit *Aa* 6 of *A. australis* hemocyanin. Clearly, some of these soluble immunocomplexes were composed of

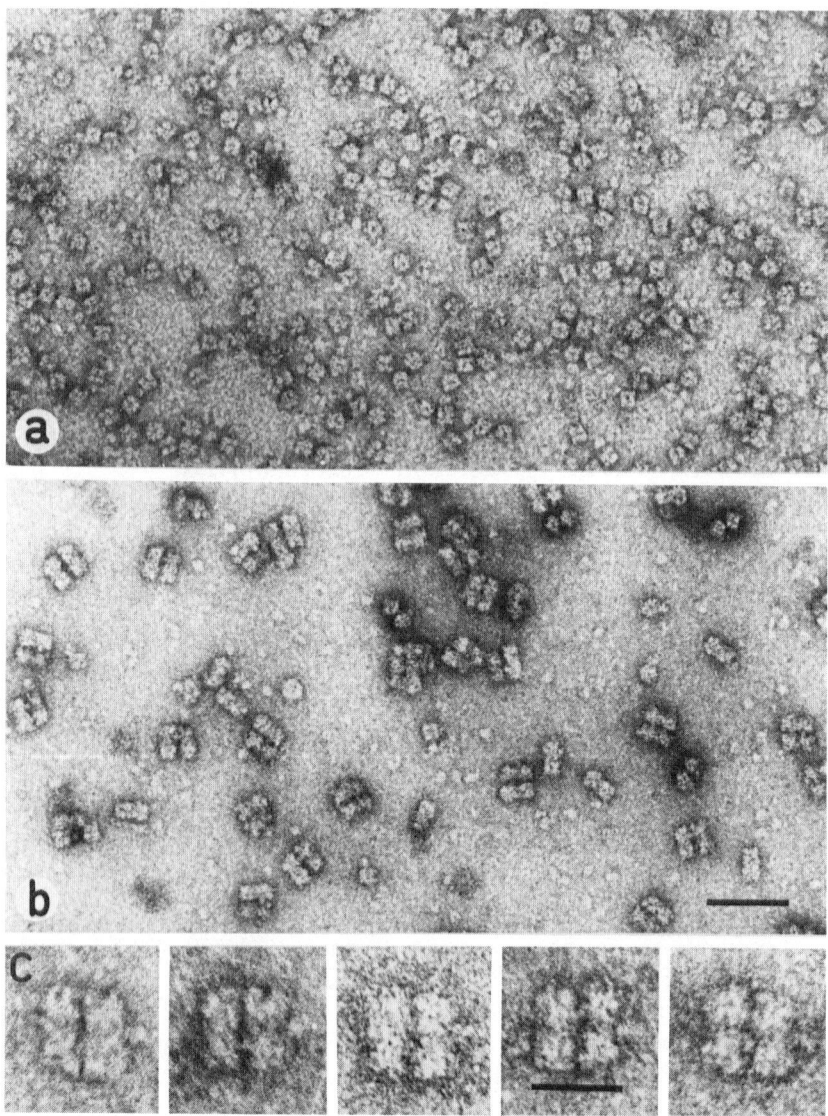

Fig. 12. Demonstration of the role of subunit Lp II in the reassembly of hybrid oligomers of *Androctonus australis* and *Limulus polyphemus* subunits. (a) Complete mixture without Aa 3A and Aa 3B; (b) same mixture + Lp II; (c) selected views from the same mixture as in (b) cross-linked and labeled with anti-Lp II-specific Fab fragments. The length of the bar is 50 nm in (a) and (b) and 25 nm in (c). From Lamy et al. (14).

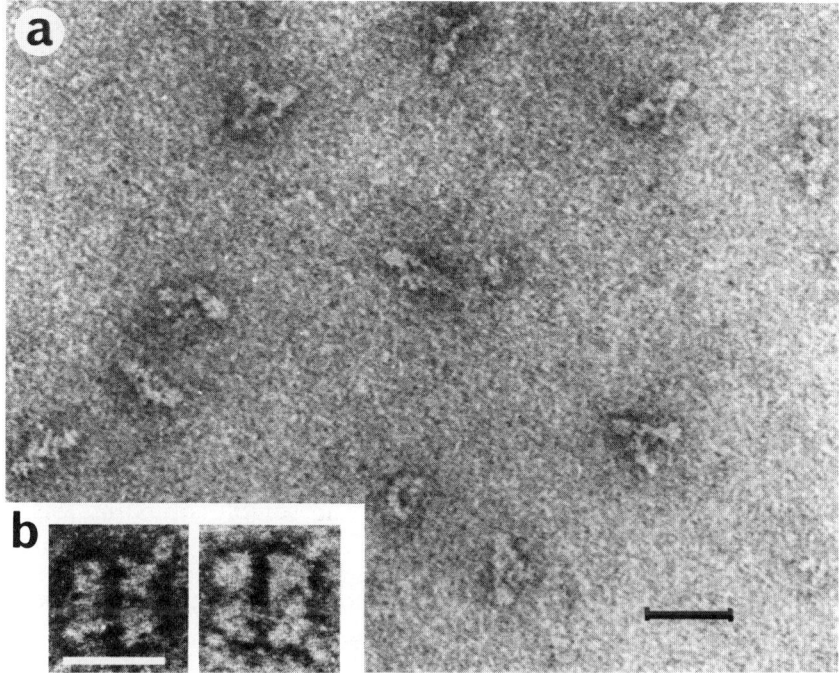

Fig. 13. Demonstration of the cross-reactivity between the hemocyanin of the scorpion *Androctonus australis* and the hemocyanin of the fiddler crab *Uca pugnax* by molecular immunoelectron microscopy. (a) Immunocomplexes composed of isolated subunit *Aa* 6 and anti-*U. pugnax* rabbit IgG; (b) immunocomplexes composed of native *A. australis* hemocyanin and anti-*U. pugnax* Fab fragments. The length of the bar is 25 nm.

two immunoglobulin molecules and two copies of subunit *Aa* 6, demonstrating that subunit *Aa* 6 and *Uca pugnax* hemocyanin have two antigenic determinants in common. When the anti-*U. pugnax* hemocyanin antiserum was used with native *A. australis* 4 × 6-mers, an immunoprecipitation reaction occurred which prevented the observation of the complexes (Lamy et al., 42). The Fab fragments were then prepared and immunolabeling was carried out. As expected, Fab fragments were visible in the corners of the 4 × 6-mers, explaining why an immunoprecipitate was formed with divalent immunoglobulins.

This experiment demonstrates how using cross-reacting antisera may increase the resolution of MIEM in binding only a few Fab fragments per subunit. This work opened a new line of research that we could designate as submolecular IEM. Unfortunately, it was clear that the method was highly hazardous and that we may have to screen thousands of species to find a given favorable cross-reactivity limited to a single determinant between certain pairs of subunits. At that time, the preparation of monoclonal antibodies became widely practiced in France, and we stopped using the polyclonal cross-reacting antisera for monoclonal antibodies.

INTRAMOLECULAR LOCALIZATION OF EPITOPES WITH MONOCLONAL ANTIBODIES

Labeling an epitope within a subunit with a monoclonal antibody is exactly equivalent to labeling a subunit with an oligomer using a polyclonal antibody preparation. Theoretically, if the antigenic particle contains more than one copy of the epitope, a precipitate could occur as in the case of *A. australis* 4 × 6-mers incubated with immune serum directed toward *U. pugnax* hemocyanin. However, as shown below, high molecular weight ordered and soluble structures easily visible in the microscope were often obtained. Therefore, the additional work of preparation, selection, and characterization of monoclonal antibodies was partially compensated for by the possibility of using the antibodies as such without preparing and purifying the Fab fragments.

BINDING MONOCLONAL ANTIBODIES TO FREE SUBUNITS

The incubation of a monoclonal antibody with a protein produces soluble immunocomplexes provided that the epitope is present in a single copy per antigen molecule. This condition is always filled in arthropod hemocyanin subunits. Figure 14 shows characteristic immunocomplexes composed of two copies of subunit *Aa* 3B and one copy of mAbL4, a monoclonal antibody raised against subunit *Aa* 3B. These images do not precisely indicate the attachment point of the immunoglobulin molecule to the subunit, but they demonstrate that the antibody molecule recognizes an epitope present on the free subunit surface. In the specific case of mAbL4, despite repeated attempts, no soluble immunocomplex could be obtained with the native 4 × 6-mers. This apparent nonaccessibility of the epitope in the native oligomer may be due to masking or steric hindrance by other subunits. Another possible explanation is that aggregation-linked conformational changes could induce structural modifications, leading to a dramatic reduction of the epitope affinity for mAbL4. In Fig. 14, the two copies of subunit *Aa* 3B are located on the left side of the Fab arms, in

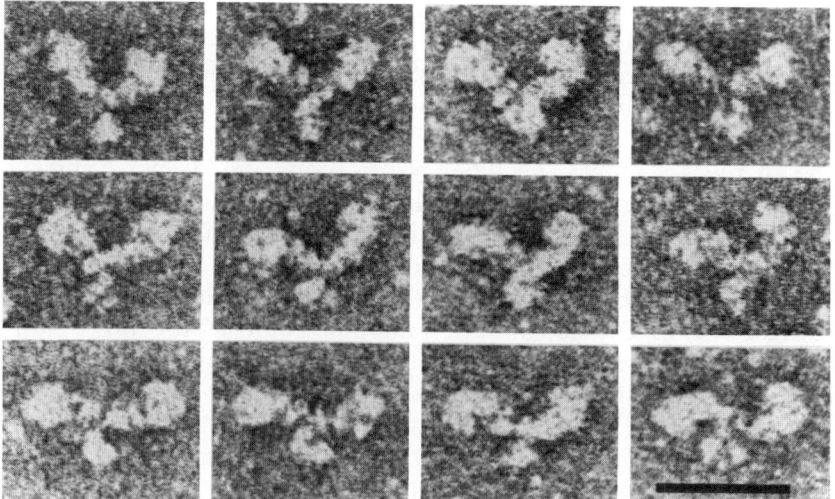

Fig. 14. Soluble immunocomplexes composed of one copy of monoclonal antibody mAbL4 and two copies of free pure subunit Aa 3B. The length of the bar is 25 nm.

positions suggesting an important rotational flexibility around the long axis of the Fab arm. Other examples of this phenomenon are given below.

BINDING MONOCLONAL ANTIBODIES TO 4 × 6-MERIC HEMOCYANINS

In Fig. 14, the attachment point of the Fab fragment to the subunit is clearly visible, but the lack of subunit asymmetry makes it difficult to determine the subunit face to which the Fab is bound. Image processing might allow precise identification of the attachment point, but the experiment has not yet been done. In another approach to this problem, advantage was taken of the fact that in the 4 × 6-mer the orientations of the subunits are precisely known (Figs. 2 and 7). Therefore, localizing the binding point of the Fab arm with respect to the contour line of the 4 × 6-mer allows one to decide on which surface domain of the subunit the epitope is located. The following results demonstrate the performance of the method.

The first example reports the localization of an epitope within subunit Aa 2 in A. australis 4 × 6-mers (Lamy et al., 6). The monoclonal antibody, designated as mAb6302 in the nomenclature of the laboratory, was demonstrated to have a high affinity for the isolated subunit Aa 2 and for the 4 × 6-mer molecule. In addition to its high affinity, mAb6302 was perfectly specific for subunit Aa 2, meaning that it did not bind to any other subunit of *A. australis* hemocyanin either free or incorporated in the 4 × 6-mer. Figure 15 shows *A. australis* 4 × 6-mer labeled with mAb6302. The immunocomplexes appear as long strings composed of whole hemocyanin molecules mostly in the side view and in the 45° view. The hemocyanin molecules do not come into contact and extra material is clearly visible between them. Figure 15 also shows enlargements of smaller immunocomplexes obtained at lower antibody to antigen ratios. These structures, denoted as "parachute" and "butterfly," are obviously the building blocks of the strings. The appearance of these immunocomplexes suggests that the mAb6302 epitope is located near the top/bottom edges of the 4 × 6-mer. These linearly ordered immunocomplexes also confirm that the four copies of the epitope are distributed in the two faces of the molecule as expected from previous intramolecular location of subunit Aa 2 (Fig. 7). However, the resolution of the negative stain was not good enough to precisely localize the binding point of the monoclonal antibody.

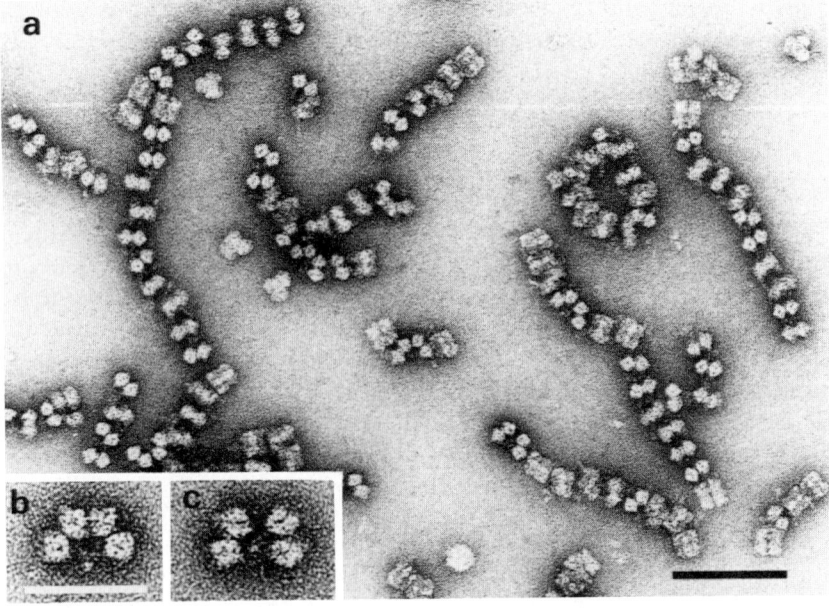

Fig. 15. Stringlike structures (a), parachute form (b), and butterfly form (c) resulting from the incubation of monoclonal antibody 6302 with native *Androctonus australis* hemocyanin. The length of the bar is 100 nm in (a) and 50 nm in (b) and (c).

To increase the precision of the method, the immunolabeling experiment was repeated with monoclonal Fab fragments. Of course, as shown in Fig. 16a, the stringlike structures were no longer visible, and the immunocomplexes were composed of 4 × 6-mers in the side view bearing one or two Fab fragments on the *top edge of the left dodecamer* and/or on the *bottom edge of the right dodecamer*. The reason why only two Fab fragments were simultaneously seen on a given molecule, while the 4 × 6-mer molecule contains four copies of subunit Aa 2, is explained by the mechanism of the negative stain (Fig. 16b). Actually, the Fab fragments bound to the two subunits facing the carbon grid produce a complete stain exclusion. Conversely, the Fab fragments bound to the two copies of subunit Aa 2 located on the top of the molecule are transparent to the electron beam because they emerge from the stain layer. To precisely locate the Fab fragment attachment points, the EM views were subjected to the CORAN method of image processing. Subsets of the molecule images selected by the correspondence analysis procedure were averaged. The average image of Fig. 16c clearly shows that the Fab fragment is bound to the top of the upper left edge and is oriented almost perpendicularly to the left plane of the longitudinal cleft. The model of Fig. 16d shows that the binding point is located in the area of the subunit corresponding to the junction of the concave, convex, and flat accessible faces. The three-dimensional structure of subunit Aa 2 is not yet known at high resolution, but the intrasubunit location of the mAb6302 epitope can be compared with the corresponding area of the P. interruptus hemocyanin model (Gaykema et al., 5). If the foldings of P. interruptus and Aa 2 subunits are not very different, the epitope of mAb6302 should be located on the top of the β-barrel (domain 3) in an area containing approximately 25 residues. The model of Fig. 16d shows the location of the epitopes of mAb6302 in the model of quaternary structure.

The second example shows the results obtained with mAb5701, a monoclonal antibody directed against an epitope of subunit Aa 6 and possessing a high affinity for subunit Aa 6 (Lamy et al., 6,20). At the high magnification of Figs. 17a and 17b, the binding point of the IgG molecule is obviously in the corners of the hemocyanin molecule, exactly in the area in which subunit Aa 6 was shown to be located (Lamy et al., 38). As recently reported, the soluble immunocomplexes can be classified according to the angle between the two Fab arms and according to the rotation angle around the long axis of the Fab fragments (Lamy et al., 6). The values of the angle between the Fab arms can be approximately deduced from the orientation of the 4 × 6-meric

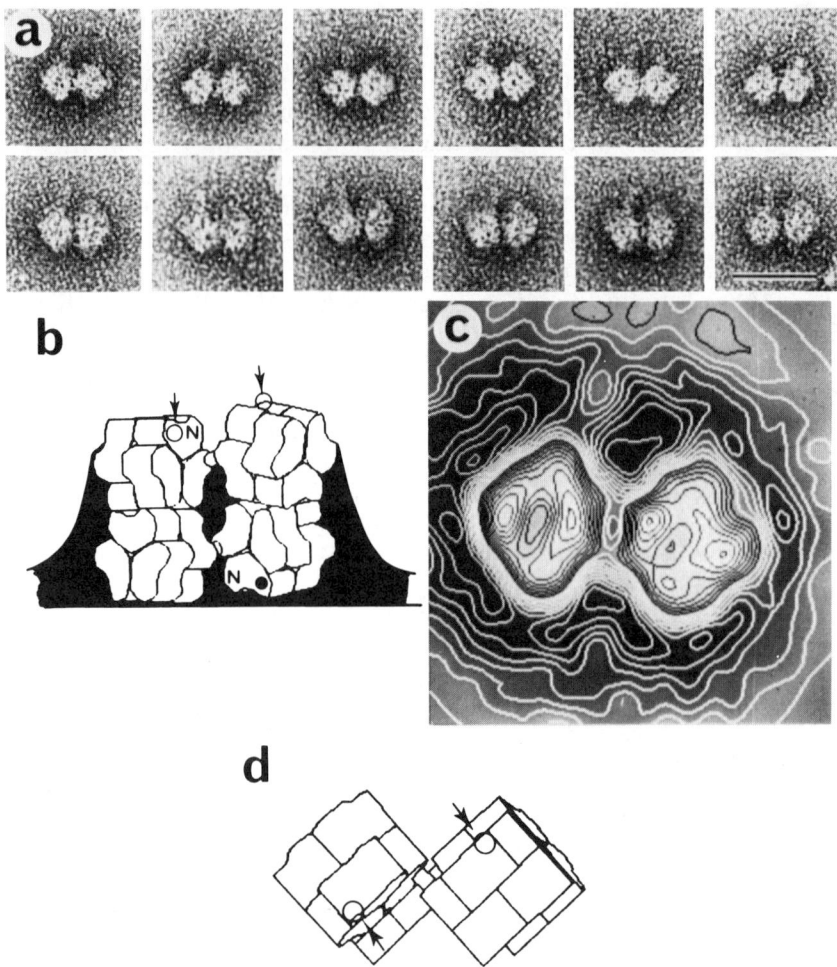

Fig. 16. Localization of the epitope of monoclonal antibody 6302 in *Androctonus australis* hemocyanin. (a) Selected views of immunocomplexes composed of one 4 × 6-mer molecule and one or two Fab fragments from mAb6302. The length of the bar is 25 nm. (b) Schematic representation of the negative stain of the side view. The solid circle shows the position of one of the epitopes embedded in the stain. The other epitope located in the left dodecamer is not visible in this orientation. The arrows indicate the epitope positions (open circle) emerging from the stain layer, not visible on the micrographs. (c) Average image of a subset of EM views similar to those of (a) selected by correspondence analysis image processing. The nucleus of stain exclusion on the upper left corner indicates the attachment point of the Fab fragment. (d) Location of the epitope of mAb6302 in the model of quaternary structure. Circles and arrows have the same meaning as in (b). Notice that because of the rocking effect the epitope of the lower right dodecamer, visible in (b), is no longer visible and that the epitope hidden in (b) is visible. N = N-terminus domain.

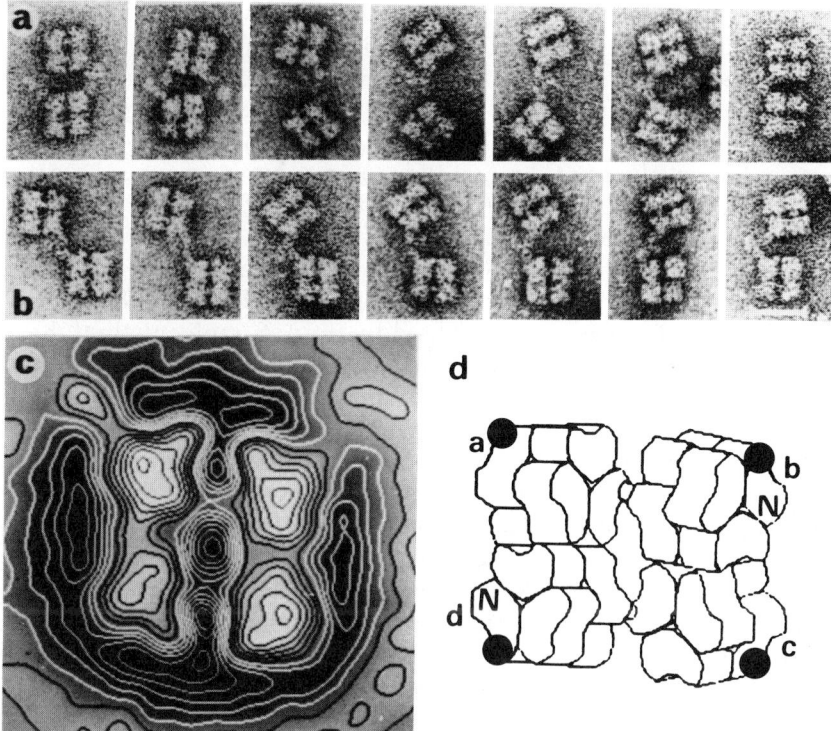

Fig. 17. Intramolecular localization in *Androctonus australis* hemocyanin of the epitope of mAb5701. (a,b) EM views of the first and second type of immunocomplexes, respectively. The length of the bar is 25 nm. (c) Average image of a subset of EM views of the first and second type, selected by correspondence analysis, showing the attachment point of the monoclonal antibody to its epitope. (d) Model of quaternary structure showing the intrasubunit location of the mAb5701 epitope. Notice that the flat accessible faces of *Aa* 6 subunit are facing backward in (a) and (c) and toward the reader in (b) and (d). N = N-terminus domain.

longitudinal clefts and from the contact point between the Fab fragments and the hemocyanin molecules. The first type of immunocomplex is shown in Fig. 17a. EM views on the left of row a are clearly composed of two copies of mAb5701 bridging two 4 × 6-mer molecules. The longitudinal clefts of the hemocyanin molecules are almost perfectly aligned. In the last view, on the right of row a, the clefts are parallel. The other pictures between these two extreme views show a continuous transition. All the immunocomplexes of Fig. 17a can be explained by the segmental flexibility of the IgG molecule in the hinge area. It is remarkable that in this type of complex the Fab arms are bound to subunits with different orientations in the hemocyanin. This is illustrated in Fig. 17d, where the four copies of subunit Aa 6 are labeled a, b, c, and d. Subunits Aa 6_a and Aa 6_c have their accessible flat faces oriented toward the support plane, whereas the accessible flat faces of subunits Aa 6_b and Aa 6_d are facing toward the observer. In complexes of the first type, an a subunit (or a c subunit) is always linked to a d subunit (or to a b subunit) of the other 4 × 6-mer molecule. In the second type of complex, shown in row b, the transition between the various views is also easily explained by the assumption of a segmental flexibility in the hinge area. However, the accessible flat faces of the subunits bridged by the IgG molecule both have the same orientation with respect to the support plan (i.e., a–c or b–d linkage). The transition between the first and the second type is more complex to explain. A high degree of rotational flexibility around the Fab axis must be postulated either between the variable and constant domains or in the hinge area. This phenomenon is in excellent agreement with similar observations of Wrigley *et al.* (43,44) and Roux (45). However, it must be noted that Wrigley, Roux, and our team have employed the same MIEM method, using a negative staining of the immunocomplexes. Therefore, one can wonder to what extent the rotation may have been forced by interactions between the protein and the support film. Figure 17 also shows an average image obtained by the CORAN method on a population of immunocomplexes similar to those shown in Figs. 17a and 17b. Clearly, the attachment point of the Fab to the 4 × 6-mer molecule is the corner corresponding to the domain of subunit Aa 6 equivalent to domain 3 (C-terminus) of *P. interruptus* hemocyanin.

The last example of immunolabeling with monoclonal antibodies also involves subunit Aa 6. Figure 18 shows an EM field containing many immunocomplexes of subunits Aa 6 and mAbL8. Here again, the immunocomplexes belong to different types related by an important rotational flexibility of the Fab arm. There are so many complexes in Fig. 18 that some 4 × 6-mer molecules have their four corners labeled. This is an *a posteriori* demonstration of the correctness of our determination of the number of copies of subunit Aa 6 in the 4 × 6-mer molecule. Indeed, on the one hand, rocket immunoelectrophoresis established that the 4 × 6-mer molecule contains an average of four copies of subunit Aa 6. On the other hand, Fig. 18 shows that individual molecules actually possess four copies of subunit Aa 6. The immunocomplex enlarged in Fig. 18 also shows that both types of linkage occur in the same molecule. The complex is composed of a string of four 4 × 6-mer molecules labeled 1 through 4 in Fig. 18b. In mole-

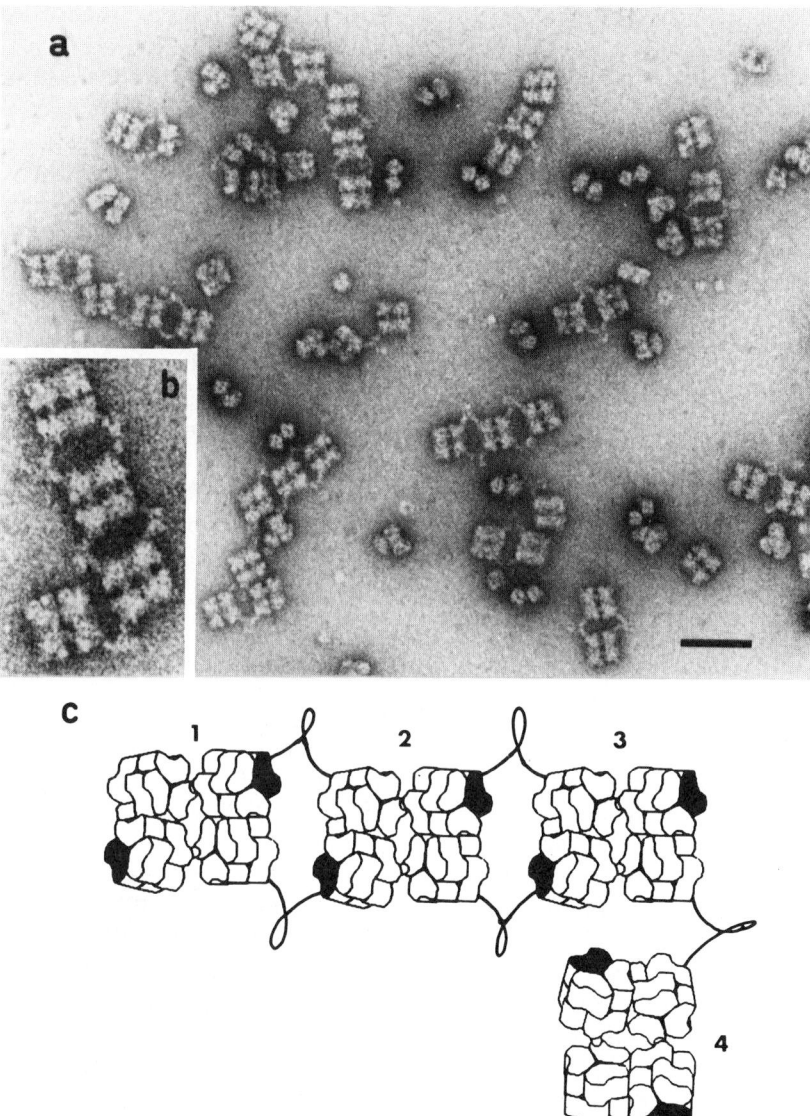

Fig. 18. Intramolecular localization in *Androctonus australis* hemocyanin of the epitope of mAbL8. (a,b) EM views of immunocomplexes; (c) interpretation of the enlarged immunocomplexes using the quaternary structure of Fig. 7. The length of the bar is 50 nm.

cule 2, the four copies of subunit Aa 6 are exclusively linked with subunits of molecules 1 and 3. In the four cases, an IgG molecule links two subunits having their flat accessible faces oppositely oriented. Conversely, in molecule 3, the subunit located in the lower right corner is linked to a subunit of molecule 4 with the same orientation, whereas the other two subunits of molecule 3 are linked with two subunits of molecule 2 with different orientations. This disposition suggests that the type of linkage does not depend on the 4 × 6-mer molecule, but only on the immunoglobulin molecule. This observation reinforces the hypothesis of the high rotational flexibility of the Fab arm described above. Besides information on the rotational flexibility of monoclonal antibodies, Fig. 18 also shows that the epitope of mAbL8 is located in the domain of subunit Aa 6 equivalent to domain 3 of *P. interruptus* hemocyanin. It is not yet clear whether the location of antigenic determinants in domain 3 is significant, but the first three epitopes with clear-cut locations were found in different subunits in the area of domain 3, which is known to be the top of a β-barrel. This location may be related to the fact that antigenic determinants have been reported to be more frequent in areas with high flexibilities (Westhof *et al.*, 46; Tainer *et al.*, 47).

PERSPECTIVES AND CONCLUSIONS

When our laboratory began to use MIEM with Fab fragments, the first results were encouraging, and we immediately had the impression that with some improvement the method would allow the direct localization of determinants in complex proteins, provided that the architecture were known. Six years later it appears that the realization of this idea is not so simple. However, things are proceeding smoothly, and image processing of molecules labeled with monoclonal antibodies now allows us to circumscribe in *one direct step* (in favorable cases) an antigenic determinant in an element of the protein surface composed of about 25 residues. If we remember that the complete protein (4 × 6-mer) contains about 16,000 residues, it is clear that a large part of the work has already been done. However, there is reason to believe that the last step will be the most difficult. Nevertheless, we can reasonably hope that technical progress, such as electron microscopy in amorphous ice, will free the investigator of the negative stain and its artifacts. In this method, the molecules are suspended in ice so that the immunoglobulin label becomes visible at all possible angles. Furthermore, the observation of projections of the randomly oriented immunocomplexes' angular relation may allow three-dimensional models to be reconstructed (van Heel, 48; Verschoor et al., 49). An additional, but soluble problem comes from the fact that the precise identification of an epitope requires that the amino acid sequence and the three-dimensional structure of the antigen be known. This is not yet the case for hemocyanin, but the fact that the first X-ray crystallographic structure and complete sequence data of hemocyanins have recently been determined leads us to be reasonably optimistic.

ACKNOWLEDGMENTS

The work presented in this paper has been carried out in the past 10 years. It would not have been possible without many fruitful collaborations that are a pleasure for me to acknowledge here. I am particularly indebted to my wife Josette Lamy and to my collaborators Michèle Leclerc, Solange Compin, and Philippe Billald for preparing, purifying, and characterizing antigens and antibodies, to Pierre-Yves Sizaret for taking all the electron micrographs and preparing the iconography, to Professor Geneviève Motta (University of Orléans) for preparing monoclonal antibodies, to Dr. Richard Feldmann (NIH) and Jean Christophe Taveau (University of Tours) for elaborating the three-dimensional models, to Drs. Joe and Celia Bonaventura (Duke University) for cooperative research on *Limulus* hemocyanin, and to Drs. Pierre and Jacqueline Jollès (CNRS, Paris) for N-terminus sequences of purified subunits.

I would also like to especially thank Dr. Joachim Frank for a very stimulating and enjoyable collaboration in the past 5 years. All the image processing work was done by Pierre-Yves Sizaret, Nicolas Boisset, and Guy Cavé (University of Tours) under his direction (especially the unpublished material of Figs. 16c and 17c of this paper) in his laboratory.

This work has been financially supported mainly by the French CNRS (RCP No. 080816) and by the University of Tours.

REFERENCES

1. Todd, P. E. E., East, I. J., and Leach, S. J. (1982). *Trends Biochem. Sci.* **7**, 212–216.
2. Mangum, C. P., Scott, J. L., Black, R. E. L., Miller, K. I., and van Holde, K. E. (1984). In "Proceedings of the First Congress of Comparative Physiology and Biochemistry, Section of IUBS," p. A152. ESPCB Publ., Liege.
3. van Holde, K. E., and Miller, K. I. (1982). *Q. Rev. Biophys.* **15**, 1–129.
4. Ellerton, H. D., Ellerton, N. F., and Robinson, H. A. (1983). *Prog. Biophys. Mol. Biol.* **41**, 143–248.
5. Gaykema, W. P. J., Hol, W. G. J., Vereijken, J. M., Soeter, N. M., Bak, H. J., and Beintema, J. J. (1984). *Nature (London)* **309**, 23–29.
6. Lamy, J., Lamy, J., Billiald, P., Sizaret, P.-Y., Cave, G., Frank, J., and Motta, G. (1985). *Biochemistry* **24**, 5532–5542.
7. Fearon, E. R., Love, W. E., Magnus, K. A., Lamy, J., and Lamy, J. (1983). *Life Chem. Rep., Suppl.* **1**, 65–68.

8. Magnus, K. A., and Love, W. E. (1983). *Life Chem. Rep., Suppl.* **1**, 61–64.
9. Jollès, J., Jollès, P., Lamy, J., and Lamy, J. (1979). *FEBS Lett.* **106**, 289–291.
10. Moore, M. D., and Riggs, A. F. (1983). *Life Chem. Rep., Suppl.* **1**, 93–97.
11. Nemoto, T., and Takagi, T. (1983). *Life Chem. Rep., Suppl.* **1**, 89–92.
12. Schartau, W., Eyerle, F., Reisinger, P., Geisert, H., Storz, H., and Linzen, B. (1983). *Hoppe-Seyler's Z. Physiol. Chem.* **364**, 1383–1409.
13. Schneider, H. J., Drexel, R., Feldmaier, G., Linzen, B., Lottspeich, F., and Henschen, A. (1983). *Hoppe-Seyler's Z. Physiol. Chem.* **364**, 1357–1381.
14. Lamy, J., Lamy, J., Sizaret, P.-Y., Billiald, P., Jollès, P., Jollès, J., Feldmann, R. J., and Bonaventura, J. (1983). *Biochemistry* **22**, 5573–5583.
15. Yokota, E., and Riggs, A. F. (1984). *J. Biol. Chem.* **259**, 4739–4749.
16. Lamy, J., Sizaret, P.-Y., Frank, J., Verschoor, A., Feldmann, R., and Bonaventura, J. (1982). *Biochemistry* **21**, 6825–6833.
17. Sizaret, P.-Y., Frank, J., Lamy, J., Weill, J., and Lamy, J. N. (1982). *Eur. J. Biochem.* **127**, 501–506.
18. van Heel, M., and Frank, J. (1981). *Ultramicroscopy* **6**, 187–194.
19. van Heel, M., Keegstra, W., Schutter, W., and van Bruggen, E. F. J. (1983). *Life Chem. Rep., Suppl.* **1**, 69–73.
20. Lamy, J., Lamy, J., Sizaret, P.-Y., Billiald, P., and Motta, G. (1985). *In* "Respiratory Pigments in Animals: Relation Structure–Function" (J. Lamy, J. P. Truchot, and R. Gilles, eds.), pp. 73–86. Springer-Verlag, Berlin and New York.
21. Markl, J. (1980). *J. Comp. Physiol. B* **140**, 199–207.
22. Lamy, J., Lamy, J., and Weill, J. (1979). *Arch. Biochem. Biophys.* **193**, 140–149.
23. Sullivan, B., Bonaventura, J., and Bonaventura, C. (1974). *Proc. Natl. Acad. Sci. U.S.A.* **71**, 2558–2562.
24. Lamy, J., Richard, M., and Goyffon, M. (1970). *C.R. Hebd. Seances Acad. Sci., Ser. D* **270**, 1627–1630.
25. Lamy, J., Lamy, J., Weill, J., Bonaventura, J., Bonaventura, C., and Brenowitz, M. (1979). *Arch. Biochem. Biophys.* **196**, 324–339.
26. Brenowitz, M., Bonaventura, C., Bonaventura, J., and Gianazza, E. (1981). *Arch. Biochem. Biophys.* **210**, 748–761.
27. Brenowitz, M., Bonaventura, C., and Bonaventura, J. (1983). *Biochemistry* **22**, 4707–4713.
28. Markl, J., Savel, A., and Linzen, B. (1981). *Hoppe-Seyler's Z. Physiol. Chem.* **362**, 1255–1262.
29. Brenowitz, M., Bonaventura, C., and Bonaventura, J. (1984). *Arch. Biochem. Biophys.* **230**, 238–249.
30. Lamy, J., Compin, S., and Lamy, J. (1983). *Arch. Biochem. Biophys.* **223**, 584–603.
31. Lamy, J., Lamy, J., Sizaret, P.-Y., and Weill, J. (1981). *In* "Invertebrate Oxygen Binding Proteins: Structure, Active Site, and Function" (J. Lamy and J. Lamy, eds.), pp. 425–443. Dekker, New York.
32. van Eerd, J. P., and Folkerts, A. (1981). *In* "Invertebrate Oxygen Binding Proteins: Structure, Active Site, and Function" (J. Lamy and J. Lamy, eds.), pp. 139–149. Dekker, New York.
33. Folkerts, A., and van Eerd, J. P. (1981). *In* "Invertebrate Oxygen Binding Proteins: Structure, Active Site, and Function" (J. Lamy and J. Lamy, eds.), pp. 215–225. Dekker, New York.
34. Stöffler, G., and Stöffler-Meilicke, M. (1984). *Annu. Rev. Biophys. Bioeng.* **13**, 303–330.

35. Lamy, J., Lamy, J., Baglin, M.-C., and Weill, J. (1977). *In* "Structure and Function of Haemocyanin" (J. V. Bannister, ed.), pp. 37–49. Springer-Verlag, Berlin and New York.
36. van Bruggen, E. F. J., Bijlholt, M. M. C., Schutter, W. G., Wichertjes, T., Bonaventura, J., Bonaventura, C., Lamy, J., Lamy, J., Leclerc, M., Schneider, H.-J., Markl, J., and Linzen, B. (1980). *FEBS Lett.* **116**, 207–210.
37. Lamy, J., Lamy, J., Bonaventura, J., and Bonaventura, C. (1980). *Biochemistry* **19**, 3033–3039.
38. Lamy, J., Bijlholt, M. M. C., Sizaret, P. Y., Lamy, J., and van Bruggen, E. F. J. (1981). *Biochemistry* **20**, 1849–1856.
39. Lamy, J., Lamy, J., Sizaret, P.-Y., Maillet, M., and Weill, J. (1977). *J. Mol. Biol.* **118**, 869–875.
40. Brenowitz, M., Bonaventura, C., and Bonaventura, J. (1984). *Biochemistry* **23**, 879–888.
41. Frank, J., Sizaret, P.-Y., Verschoor, A., and Lamy, J. (1983). *In* "Proceedings of the 41st Annual Meeting of EMSA" (G. W. Bailey, ed.), pp. 282–283. San Francisco Press, San Francisco, California.
42. Lamy, J., Lamy, J., Leclerc, M., Compin, S., Sizaret, P.-Y., and Weill, J. (1981). *In* "Invertebrate Oxygen Binding Proteins: Structure, Active Site, and Function" (J. Lamy and J. Lamy, eds.), pp. 181–195. Dekker, New York.
43. Wrigley, N. G., Brown, E. B., and Skehel, J. J. (1983). *J. Mol. Biol.* **169**, 771–774.
44. Wrigley, N. G., Brown, E. B., Daniels, R. S., Douglas, A. R., Skehel, J. J., and Wiley, D. C. (1983). *Virology* **131**, 308–314.
45. Roux, K. H. (1984). *Eur. J. Immunol.* **14**, 459–464.
46. Westhof, E., Altschuh, D., Moras, D., Bloomer, A. C., Mondragon, A., Klug, A., and van Reggenmortel, M. H. V. (1984). *Nature (London)* **311**, 123–126.
47. Tainer, J. A., Getzoff, E. D., Alexander, H., Houghten, R. A., Olson, A. J., Lerner, R. A., and Hendrickson, W. A. (1984). *Nature (London)* **312**, 127–134.
48. van Heel, M. (1984). *Ultramicroscopy* **13**, 165–184.
49. Verschoor, A., Frank, J., Radermacher, M., Wagenknecht, T., and Boublik, M. (1984). *J. Mol. Biol.* **178**, 677–698.

12
Studies of the Tertiary and Quaternary Structure of Antibody Constant Domains[1]

EDUARDO A. PADLAN, GERSON H. COHEN,
AND DAVID R. DAVIES
Laboratory of Molecular Biology
NIDDK, NIH
Bethesda, Maryland 20892

INTRODUCTION

Our current knowledge of the structure of antibody molecules is based on the crystal structure analysis of a relatively small number of molecules, mostly fragments of intact antibodies (reviewed by Davies and Metzger, 1). Nevertheless, sufficient data have emerged to demonstrate that the antibody molecule makes extended use of a common structural motif, namely, the antibody domain structure. There is also evidence that this structure is frequently used elsewhere in the immune system, e.g., in Class 1 and Class 2 antigens of the major histocompatibility complex, in β_2-microglobulin, in T-cell receptors, etc. For the antibody molecule the functional unit is usually a pair of like or unlike domains which are frequently noncovalently linked. The two variable domains interact to form a combining site ingeniously organized to express the diversity of the underlying gene structure. Similarly, in the constant regions of the molecule, pairs of domains are involved with various effector functions such as activation of the complement system and binding to cell surface Fc receptors.

---

[1] The work described in this article forms part of two papers that have been submitted for publication elsewhere.

The association of antibody domains in pairs raises questions about the mechanism of assembly of the complete antibody within the plasma cell. For the variable domains, most of the V_H and V_L have to be able to pair tightly, regardless of their sequences, and a recent analysis has shown that this is achieved by utilizing twelve invariant or highly conserved residues that interact in the interface between V_H and V_L (Chothia et al., 2). For C_H1 and C_L this must also be true regardless of class or subgroup. That is, C(kappa) and C(lambda) both must be able to associate with any C_H1, whether it be alpha, gamma, delta, epsilon, or mu. To understand the mechanism of interaction of C_H1 and C_L, we located the interface residues between pairs of domains by examining known structures and then examined their significance from the extent of variation among these residues in the different classes.

Within the Fc part of the antibody molecule the interacting domains occur in identical pairs. In the structure of the human Fc that has been analyzed crystallographically (Deisenhofer, 3), these domains belong to the gamma class, and the two heavy chains each contribute two domains. In the epsilon and mu classes, however, there are three domains in each chain of the Fc. The Fc(epsilon) binds to receptors on mast cells (reviewed by Metzger et al., 4), and when these antibodies are linked by antigen the cells are stimulated to release vasoactive amines and other pharmacologically active substances (Ishizaka and Ishizaka, 5). While the site of attachment of the Fc(epsilon) receptor to the Fc is not known directly, attempts have been made to identify it through the use of peptides that activate the mast cells (Hamburger, 6; Stanworth et al., 7). We have attempted, unsuccessfully so far, to crystallize the Fc(epsilon). Here we describe a model of the Fc(epsilon) based on the sequence homology between the domains of this Fc and those constant domains of known three-dimensional structure.

AN ANALYSIS OF THE INTERFACE RESIDUES BETWEEN C_H1 AND C_L

The atomic coordinates of three Fabs are available from the Protein Data Bank (8): NEW (9), McPC603 (10), and KOL (11). The structure of the Fab of J539, a mouse IgA, kappa, antibody, has also been refined at 2.6 Å resolution (Suh et al., 12) and its coordinates were used in this analysis. The amino acid sequences for these proteins are available in Kabat et al. (13), whose numbering scheme is employed here throughout. The $C_L : C_H1$ domains were structurally aligned by superposition using the program ALIGN (G. H. Cohen, unpublished). The amino acid sequences were aligned by first superposing the isolated domains of McPC603 and KOL based on three-dimensional structure and then using this structural alignment to align the sequences. The other domain sequences were aligned with those of McPC603 and KOL using the program of Murata et al. (14).

In Table I we see the C_H1 domain interface sequences from human and mouse heavy chains in the different classes. The numbers on either side of the first and fourth columns of sequence show the number of pair contacts made by the atoms of these residues with residues of the opposing domain, the numbers being from KOL, NEW, McPC603, and J539, respectively. A similar listing is given in Table II for the human and mouse kappa and lambda chains. These interactions are indicated schematically in Figs. 1 and 2, where it can readily be seen that interacting residues occur on each of the four interface segments of each domain, that is, they are distributed generally over the four-strand surface of the domain.

Examination of these tables shows that there are a few highly interacting residues that are also highly conserved. These include Tyr 122, Pro 123, Leu 124, Leu 143, Phe 174, and Pro 175 of the heavy chain, together with Phe 118, Ser 121, Glu 123, Glu 124, Thr 131, Val 133, and Thr 162 of the light chain. However, there are also other highly interacting residues that are not conserved. These include Thr 127H, Lys 145H, Val 177H, Leu 135L, Glu 160L, and Tyr 178L. The area of the interface excluded from contact with solvent is approximately constant for each domain with a value of about 500 $Å^2$ per domain, considerably smaller than the value 900 to 1000 $Å^2$ found by Chothia et al. (2) for the $V_H:V_L$ contact, but closer to the values found by Chothia and Janin (15) for the insulin dimer and for the trypsin–trypsin inhibitor interaction.

Further examination of the interface between C_H1 and C_L revealed the existence of a cavity, which in the case of McPC603 has a volume of 143 $Å^3$, and is lined by residues V133, L160, N161, S162, S176, and T178 of the light chain and L143, F174, P175, A177, T186, M187, and S188 of the heavy chain. Similar cavities occur in J539, NEW, and KOL with volumes of 146, 70, and 49 $Å^3$, respectively. Although we have not observed any solvent inside this cavity in the case of McPC603 or J539, this is presumably because of the low resolution of the data used in the structure determination, and we assume that the cavity is normally occupied by solvent. The reason for this cavity is not known, but a possible role may be to provide more flexibility in the interaction between the domains as a result of the introduction of

TABLE I
Amino Acid Sequence of the C_H1 Domain from Human and Mouse Heavy Chains[a]

		ALA	ALA	ALA	GLU	ALA	GLY	ALA	ALA	—	GLY	GLU
		SER	LYS	SER	SER	PRO	ASP	SER	SER	SER	SER	SER
		THR	THR	PRO	ALA	THR	LYS	THR	ILE	ILE	ALA	GLN
		LYS	THR	THR	ARG	LYS	LYS	GLN	ARG	ARG	SER	SER
		GLY	PRO	SER	ASN	ALA	GLU	SER	TRP	TRP	ALA	PHE
		PRO	PRO	PRO	PRO	PRO	PRO	PRO	PRO	PRO	PRO	PRO
120		SER	SER	LYS	THR	ASP	ASP	SER	GLU	GLU	THR	ASN
		VAL	VAL	VAL	ILE	VAL	MET	VAL	LEU	LEU	LEU	VAL
	35	PHE	1 TYR	PHE	39 TYR 40	PHE	PHE	PHE	TYR	TYR	PHE	PHE
	8	PRO	37 PRO	PRO	11 PRO 13	LEU	PRO	PRO	PRO	PRO	PRO	PRO
	25	LEU	8 LEU	LEU	24 LEU 21	ILE	LEU	LEU	LEU	LEU	LEU	LEU
	3	ALA	24 ALA	SER	11 THR 11	ILE	SER	THR	LYS	LYS	VAL	VAL
		PRO	3 PRO	LEU	17 LEU 27	SER	GLU	ARG	PRO	PRO	SER	SER
		SER	GLY	CYS	41 PRO 9	GLY	CYS	CYS	CYS	CYS	CYS	CYS
		SER	SER	SER	6 PRO	CYS	LYS	CYS	—	—	GLU	GLU
129	-9	LYS	5 ALA	THR	ALA 1	ARG	ALA	LYS	LYS	LYS	ASN	SER
		—	—	—	LEU 3	HIS	—	—	—	—	—	—
130		SER	ALA	GLN	SER	PRO	PRO	ASN	GLY	GLY	SER	PRO
133		THR	GLN	PRO	SER	LYS	GLU	ILE	THR	THR	ASN	LEU
		—	—	—	—	—	—	PRO	—	—	—	—
134		SER	THR	ASP	—	ASP	GLU	SER	ALA	ALA	PRO	SER
		GLY	ASN	GLY	—	ASN	ASN	ASN	SER	SER	SER	ASP
	2	GLY	SER	ASN	ASP	SER	GLU	ALA	MET	MET	SER	LYS
137	26	THR	MET	VAL	PRO	PRO	LYS	THR	THR	THR	THR	ASN
		—	—	—	—	—	—	—	—	—	—	LEU
138		ALA	VAL	VAL	VAL	VAL	ILE	SER	—	—	VAL	VAL
139	6	ALA	8 THR	—	12 ILE 15	VAL	ASN	VAL	—	—	ALA	ALA
		—	—	—	—	—	—	THR	—	—	—	—

Pos													
140	4 LEU	5	LEU	ILE		ILE	LEU	LEU	LEU	LEU	LEU	VAL	MET
	4 GLY	1	GLY	ALA	3	GLY	ALA	GLY	GLY	GLY	GLY	GLY	GLY
	CYS		CYS	CYS	2	CYS	CYS	CYS	CYS	CYS	CYS	CYS	CYS
	6 LEU	9	LEU	LEU	1	LEU	LEU	LEU	LEU	LEU	LEU	LEU	LEU
	VAL		VAL	VAL	7	ILE	VAL	ALA	VAL	ALA	VAL	ALA	ALA
	18 LYS	12	LYS	GLN		THR	ILE	THR	LYS	THR	ASP	LYS	ARG
	3 ASP		GLY	PHE		GLY	–	GLY	LYS	GLY	TYR	GLN	ASP
	TYR		TYR	PHE		TYR	–	TYR	TYR	TYR	PHE	ASP	PHE
	PHE		PHE	PRO		HIS	SER	PHE	PHE	PHE	PHE	PHE	PHE
	PRO		PRO	PRO		PRO	GLN	PRO	PRO	PRO	PRO	PRO	PRO
	GLU		GLU	GLN		THR	GLN	GLU	ASN	GLU	ASN	ASP	SER
150	–		–	GLN		–	–	–	–	–	–	–	–
151	PRO		PRO	PRO		SER	PRO	PRO	PRO	PRO	PRO	SER	THR
	VAL		VAL	LEU		VAL	LEU	VAL	VAL	VAL	VAL	ILE	ILE
	THR		THR	SER		THR	LYS	MET	THR	THR	THR	THR	SER
154	VAL		VAL	VAL		VAL	ILE	VAL	VAL	VAL	VAL	PHE	SER
156	SER		THR	THR		THR	SER	THR	THR	THR	THR	SER	PHE
157	TRP		TRP	TRP		TRP	TRP	TRP	TRP	TRP	TRP	TRP	THR
162	ASN		ASN	SER		GLY	TYR	GLU	ASP	TYR	LYS	ASN	TRP
	–		–	–		–	–	–	–	–	TYR	GLN	ASN
163	SER		SER	GLU		LYS	MET	PRO	THR	SER	–	LYS	TYR
	GLY		GLY	SER		SER	GLY	LYS	GLY	ASP	SER	ASP	GLN
165	ALA		SER	GLY		GLY	THR	LYS	SER	SER	LEU	ASN	LYS
	–		–	–		–	–	–	–	–	ASN	ASP	THR
166	LEU		LEU	GLN		LYS	SER	LEU	LEU	SER	–	–	GLN
167	THR		SER	GLY		ASP	SER	ASN	ASN	ASP	ASN	ILE	SER
168	–		–	–		–	–	–	–	–	MET	–	–
	SER		SER	VAL		ILE	ILE	GLY	MET	GLY	SER	MET	GLN
169	GLY		GLY	THR		THR	VAL	THR	THR	THR	SER	SER	SER

(*continued*)

TABLE I (Continued)

171		VAL	VAL	ALA		THR	GLN	GLU	THR	THR	THR	THR	ILE
		HIS	HIS	ARG	6	VAL	ARG	HIS	MET	VAL	ARG	ARG	ARG
	11	THR	THR	ASN	1	ASN	THR	VAL	THR	ASN	GLY	GLY	THR
	43	PHE	PHE	PHE	51	PHE	45	PHE	PHE	LEU	PHE	PHE	PHE	PHE
	3	PRO	PRO	PRO	18	PRO	10	PRO	PRO	PRO	PRO	PRO	PRO	PRO
	3	ALA	ALA	PRO	11	ALA	6	GLU	SER	ALA	ALA	SER	SER	THR
177	17	VAL	VAL	SER		ALA		ILE	GLU	THR	ALA	VAL	VAL	LEU
		—	—	—		—		—	—	THR	—	—	—	—
178		LEU	LEU	GLN	5	LEU	5	GLN	MET	LEU	LEU	LEU	LEU	ARG
	7	GLN	GLN	ASN	4	ALA		ARG	ARG	THR	THR	GLY	ARG	THR
180	6	SER	SER	ALA		SER		ARG	ASN	LEU	LEU	SER	GLY	GLY
182		SER	ASP	SER		GLY		ASP	GLY	SER	SER	GLU	GLY	GLY
183		GLY	—	GLY		GLY		SER	ASN	GLY	GLU	—	LYS	LYS
		—	—	ASN		—		—	—	—	—	—	—	—
184		LEU	LEU	LEU		ARG		TYR	—	HIS	LEU	LEU	TYR	TYR
		TYR	TYR	TYR		TYR		TYR	TYR	TYR	LYS	TYR	TYR	LEU
	1	SER	THR	THR	2	THR		MET	THR	ALA	VAL	ALA	ALA	ALA
	3	LEU	LEU	THR		MET		THR	MET	THR	THR	ALA	ALA	THR
	14	SER	SER	SER	5	SER		THR	VAL	ILE	THR	THR	THR	SER
		SER	SER	SER		ASN		SER	LEU	SER	SER	SER	SER	GLN
190	11	VAL	SER	GLN	23	GLN	30	GLN	GLN	LEU	GLU	GLN	GLN	VAL
	6	VAL	VAL	LEU		LEU		LEU	VAL	LEU	VAL	VAL	VAL	LEU
		THR	THR	THR		THR		SER	THR	LEU	THR	LEU	LEU	LEU
		VAL	VAL	LEU		LEU		THR	VAL	THR	THR	LEU	LEU	LEU
		PRO	PRO	PRO		PRO		PRO	LEU	VAL	—	LEU	LEU	SER
		SER	SER	ALA		ALA		LEU	ALA	SER	SER	PRO	PRO	PRO
		SER	SER	THR		VAL		GLN	SER	GLY	TRP	SER	LYS	LYS
											GLY	GLY	LYS	SER

198

Pos	1	2	3	4	5	6	7	8	9	10	11	12	13
197	SER	—	PRO	GLN	GLU	—	GLN	—	GLU	ALA	LYS	ASP	ILE
198	—	LEU	—	CYS	—	—	—	TRP	—	TRP	SER	VAL	—
199	GLY	—	PRO	ALA	—	PRO	ARG	—	ASN	ALA	ALA	MET	LEU
200	THR	—	SER	GLY	—	GLU	GLN	—	LEU	LYS	LYS	GLN	GLU
202	—	—	—	—	GLY	—	—	—	—	—	—	—	—
203	GLN	GLU	LYS	—	GLU	GLY	GLY	ASN	ASN	GLN	GLU	ASN	ASP
205	THR	THR	SER	SER	GLU	—	GLY	HIS	HIS	MET	GLY	GLU	GLU
206	TYR	VAL	VAL	VAL	VAL	TYR	TYR	—	—	PHE	HIS	TYR	TYR
207	ILE	THR	THR	LYS	—	LYS	LYS	THR	THR	THR	THR	VAL	LEU
	CYS	CYS	CYS	CYS	LYS	CYS	CYS	CYS	CYS	CYS	CYS	VAL	VAL
	ASN	ASN	HIS	SER	CYS	VAL	CYS	THR	ARG	HIS	HIS	CYS	CYS
210	VAL	VAL	VAL	VAL	SER	VAL	THR	ARG	VAL	HIS	LYS	LYS	LYS
	ASN	ALA	LYS	GLN	VAL	VAL	ILE	VAL	ALA	VAL	THR	VAL	ILE
	HIS	HIS	HIS	HIS	GLN	GLN	ASN	ALA	THR	ALA	GLN	HIS	HIS
	LYS	PRO	TYR	ASP	HIS	THR	LYS	LYS	HIS	THR	HIS	HIS	TYR
	PRO	ALA	THR	—	ASP	ALA	LYS	PRO	THR	PRO	PRO	PRO	GLY
	SER	SER	—	SER	—	SER	ARG	LYS	PRO	SER	PRO	ASN	GLY
	ASN	SER	ASN	ASN	—	LYS	LYS	ARG	SER	SER	ASN	GLY	LYS
	THR	THR	PRO	PRO	PRO	SER	LYS	LYS	PHE	ASN	GLY	ASN	ASN
218	LYS	LYS	—	VAL	VAL	LYS	GLU	LYS	GLU	ASP	LYS	LYS	ARG
	—	—	—	—	—	—	TRP	—	—	TRP	GLU	—	ASP
219	VAL	VAL	GLN	GLN	GLN	PRO	VAL	VAL	VAL	VAL	—	LYS	LEU

(*continued*)

TABLE I (Continued)

220	ASP		ASP	ASP	GLU	GLU	PHE	ASP	SER	ASP	HIS
	LYS	7	LYS	VAL	LEU	ILE	LYS	ASN	ARG	VAL	VAL
	ARG		LYS	THR	ASP	PHE	PHE	LYS	THR	PRO	PRO
223	VAL		ILE	VAL	VAL	—	PRO	THR	ILE	LEU	ILE
226	GLU		—	—	ASN	—	—	PHE	LEU	PRO	PRO
	PRO		—	—	CYS	—	—	SER	—	—	—
	4 LYS	10	—	—	—	—	—	—	—	—	—
	11 SER	5	—	—	—	—	—	—	—	—	—
230	10 CYS	15	—	—	—	—	—	—	—	—	—

[a] The sequences are (in order) from human IgG$_1$ [Sequence No. 1 in Kabat et al. (13), pp. 173–177], mouse IgG$_1$ (No. 45), human IgA (No. 34), mouse IgA (No. 70), human IgM (No. 38), and mouse IgM (No. 65). The numbering scheme (in the leftmost column) is that of Kabat et al. (13). Listed on the left and on the right of some residues in the human IgG$_1$ sequence are the total number of interactions between pairs of atoms involving these residues in the C$_H$1:C$_L$ pair of domains in NEW and KOL, respectively; alongside the mouse IgA sequence are the contacts found in the C$_H$1:C$_L$ pairs of J539 and McPC603, respectively.

TABLE II
Amino Acid Sequence of the Light Chain
Constant Domains of Human and Mouse Lambda
and Kappa Chains[a]

		GLN		GLN	ARG		ARG	
		PRO		PRO	THR		ALA	
110		LYS		LYS	VAL		ASP	
		ALA		SER	ALA		ALA	
		ALA		SER	ALA		ALA	
		PRO		PRO	PRO		PRO	
	8	SER		SER	SER		THR	
	4	VAL		VAL	VAL		VAL	
	15	THR	1	THR	PHE	5	SER	1
		LEU		LEU	ILE	9	ILE	4
	46	PHE	42	PHE	PHE	67	PHE	60
	3	PRO		PRO	PRO	4	PRO	6
120		PRO		PRO	PRO		PRO	
	12	SER	9	SER	SER	5	SER	7
	1	SER	2	SER	ASP		SER	
	6	GLU	22	GLU	GLU	16	GLU	13
	32	GLU	25	GLU	GLN	32	GLB	31
		LEU		LEU	LEU		LEU	
		GLN		GLU	LYS		THR	
		ALA		THR	SER	2	SER	2
		ASN		ASN	GLY		GLY	
	10	LYS	4	LYS	THR		GLY	
130		ALA		ALA	ALA		ALA	
	7	THR	6	THR	SER	2	SER	3
		LEU		LEU	VAL		VAL	
	6	VAL	6	VAL	VAL	8	VAL	6
		CYS		CYS	CYS		CYS	
	33	LEU	29	THR	LEU	31	PHE	40
	3	ILE	5	ILE	LEU		LEU	
	4	SER	7	THR	ASN	6	ASN	8
		ASP		ASP	ASN		ASN	
		PHE		PHE	PHE		PHE	
140		TYR		TYR	TYR		TYR	
		PRO		PRO	PRO		PRO	
		GLY		GLY	ARG		LYS	
		ALA		VAL	GLU		ASP	
		VAL		VAL	ALA		ILE	
		THR		THR	LYS		ASN	
		VAL		VAL	VAL		VAL	
		ALA		ASP	GLN		LYS	
		TRP		TRP	TRP		TRP	
		LYS		LYS	LYS		LYS	

(continued)

TABLE II (Continued)

150		ALA		VAL		VAL		ILE	
		ASP		ASP		ASP		ASP	
		SER		GLY		ASN		GLY	
		SER		THR		ALA		SER	
		PRO		PRO		LEU		GLU	
		VAL		VAL		GLN		ARG	
		LYS		THR		SER		GLN	
		ALA		GLN		GLY		ASN	
		GLY		GLY		ASN		GLY	
		VAL		MET		SER		VAL	
160	16	GLU	24	GLN		GLN	17	LEU	12
		THR		THR		GLU	3	ASN	
	10	THR	7	THR		SER	15	SER	7
		THR		GLN		VAL	5	TRP	4
		PRO		PRO		THR	16	THR	7
	2	SER	2	SER		GLU	1	ASP	
		LYS		LYS		GLN		GLN	
		GLN	6	GLN		ASP		ASP	
		SER		SER		SER		SER	
		—		—		LYS		LYS	
170		ASN		ASN		ASP		ASP	
		ASN		ASN		SER		SER	
		LYS		LYS		THR		THR	
		TYR		TYR		TYR		TYR	
	3	ALA	5	MET		SER	14	SER	7
	5	ALA	6	ALA		LEU	10	MET	10
	1	SER	4	SER		SER	9	SER	19
		SER		SER		SER		SER	
	21	TYR	27	TYR		THR		THR	
		LEU		LEU		LEU		LEU	
180		SER		THR		THR	2	THR	
		LEU		LEU		LEU		LEU	
		THR		THR		SER		THR	
		PRO		ALA		LYS		LYS	
		GLU		ARG		ALA		ASP	
		GLN		ALA		ASP		GLU	
		TRP		TRP		TYR		TYR	
		LYS		GLU		GLU		GLU	
		SER		ARG		LYS		ARG	
		HIS		HIS		HIS		HIS	
190		LYS		SER		LYS		ASN	
		SER		SER		VAL		SER	
		TYR		TYR		TYR		TYR	
		SER		SER		ALA		THR	
		CYS		CYS		CYS		CYS	
		GLN		GLN		GLU		GLU	

(continued)

TABLE II (Continued)

		VAL			VAL	VAL	ALA	
		THR			THR	THR	THR	
		HIS			HIS	HIS	HIS	
		GLU			GLU	GLN	LYS	
200		GLY			GLY	GLY	THR	
		—			—	LEU	SER	
		—			—	SER	THR	
		SER			HIS	SER	SER	
		THR			THR	PRO	PRO	
		VAL			VAL	VAL	ILE	
		GLU			GLU	THR	VAL	
	1	LYS			LYS	LYS	LYS	
		THR		5	SER	SER	SER	
		VAL			LEU	PHE	PHE	
210		ALA			SER	ASN	ASN	
		PRO			ARG	ARG	ARG	8
		THR			ALA	GLY	1 ASN	
	11	GLU	12		ASP	GLU	16 GLU	
	22	CYS	16		CYS	CYS	CYS	
		SER			SER	—	—	

^a The sequences are (in order) from human lambda [Sequence No. 16 in Kabat *et al.* (13), pp. 167–169], mouse lambda (No. 31), human kappa (No. 1), and mouse kappa (No. 23). The numbering scheme is that of Kabat *et al.* (13). The total number of atomic contacts in the $C_H1 : C_L$ pairs of NEW and KOL, respectively, are listed on the left and on the right of some residues in the human lambda sequence; alongside the mouse kappa sequence are the contacts found in the J539 and McPC603 pairs, respectively.

different amino acid side chains. Alber *et al.* (16) have observed a similar cavity in T4 bacteriophage lysozyme, in which a mutation that changes the alanine residue at position 146 to threonine propagates the movement of several side chains culminating in the movement of Met 106, which lies in the vicinity of the cavity, causing it to move into the cavity.

A MODEL OF IgE Fc

The model of the IgE Fc described here is based on the assumption that the domains of this structure would be closely related to the constant domains of other classes, and in particular to the constant domains of the human gamma Fc, whose three-dimensional structure

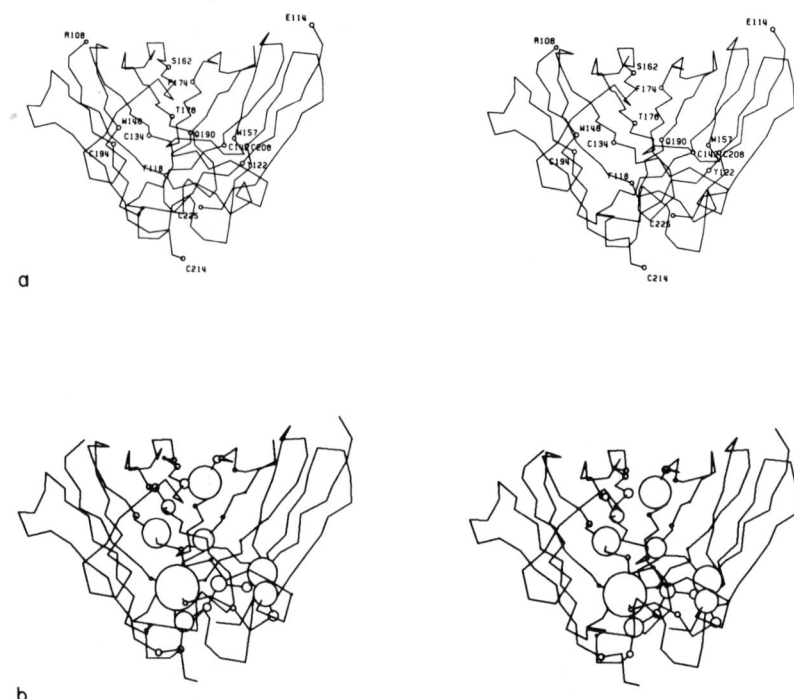

Fig. 1. (a) Stereograph of the alpha-carbon skeleton of the $C_H1:C_L$ domains of McPC603. The C_L domain is on the left. Circled residues are labeled to serve as reference points. (b) The same model illustrating the interacting residues in the interdomain interface. The radius of each circle is proportional to the extent of the interaction of that residue with the opposite domain as measured by the number of pair interactions that occur between atoms of this residue and atoms of the opposite chain.

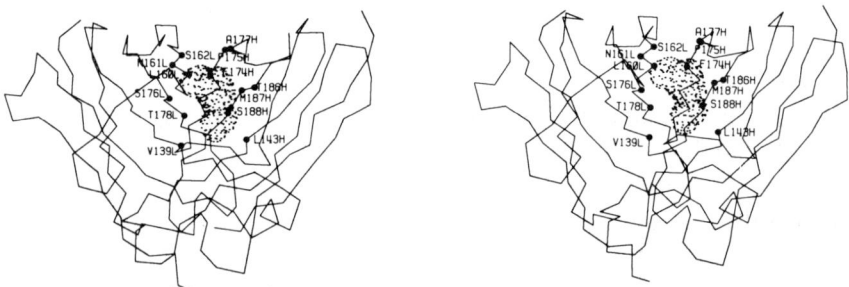

Fig. 2. The location of the interface cavity between the $C_H1:C_L$ domains of McPC603. The dotted surface shows the cavity, with the solid circles indicating the residues lining the cavity. The orientation is the same as in Fig. 1.

is known (Deisenhofer, 3). Examination of the sequence homologies between these epsilon and gamma domains shows a reasonable fit between CE3 and CG2 (32% identity) and between CE4 and CG3 (32% identity). The agreement between CE2 and CG2 (16%) and CG3 (20%) is not as strong, with CG3 slightly favored.

The model was constructed using the interactive modeling program BILDER (Diamond, 17), assuming that the backbone structures of the homologous regions of the epsilon and gamma chains were identical except where there were insertions and deletions. These were confined to the loops joining the strands of the β-sheets of the bilayer domain structure. The sequence alignment is shown in Table III (Dorrington and Bennich, 17; Edelman et al., 18; Dayhoff, 19). Every effort was made to retain the starting peptide dihedral angles as well as the relative positions of the side groups. The peptide dihedral angles at insertions and deletions were kept stereochemically proper. Energetically unfavorable backbone dihedral angles which occasionally resulted from the replacement of a glycine residue by another bulkier group were modeled to conform to more acceptable values. Other dihedral angles, which in the IgG Fc structure were energetically unfavorable, were retained in this model. Pro 284 and Pro 531 were retained as cis in the IgE model.

The quaternary interactions between laterally paired domains were as follows: CE2 and CE4 as in CG3, and CE3 as in CG2. Longitudinally, CE3 and CE4 were arranged as in CG2 and CG3. The CE2 domains were positioned by eye over the CE3 domains, while maintaining an approximate dyad axis. They are located along the dyad by 67 Å from the CE4 pair and rotated by 22.5° in a clockwise direction when looking down the dyad axis.

The inter-epsilon chain disulfide bridges were found to be more easily formed by linking Cys 261 of one chain to Cys 357 of the other. More rebuilding would be necessary if the identical disulfides were to be paired. We have assumed that the carbohydrate present in CG2 (Deisenhofer, 3) is also present in the same configuration in CE3. No attempt was made to model the two other carbohydrate moieties known to be present on Asn 288 and Asn 403.

The resulting model was regularized using the program PROLSQ of Hendrickson (18). The final coordinates have been deposited in the Protein Data Bank (Bernstein et al., 8).

Figure 3 shows the final model. Most of the model could be constructed easily and without any unexpected results. The most notable was the alignment of the disulfide bridges in CE2, which definitely seemed to favor not joining like with like across the dyad axis, al-

TABLE III
Alignment of Human IgE and IgG Sequences[a]

	IgE		IgG	
	227	R	341	G
		D		Q
		F		P
	230	T		R
250		P		E
		P		E
		P		P
		T		Q
		V		V
		K		Y
255		I	350	T
257		L		L
		Q		P
		S		P
260	240	S		S
		C		R
		D		E
		G	357	E
		G		—
		G		—
		H	358	M
		F		T
		P	360	K
		P		N
270	250	T		Q
		I		V
		Q		S
	253	L		L
	253a	L		T
	254	C		C
		L		L
		V		V
		S	370	K
		G		G
		Y		F
280	260	T		Y
284		P		P
		G		S
		T		D
		I		I
		N-[CHO]		A
		I		V
290		T	380	E
		W		W
292		L		E
295	270	E		S
296		D		N

(*continued*)

TABLE III (*Continued*)

		IgE		IgG
299		G		G
300		Q		Q
		V		P
302		M		E
304		D		N
		V	390	N
		D		Y
		L		K
	280	S		T
		T		T
310		A		P
		S		P
		T		V
		T		L
		Q		D
		E	400	S
		G		D
		E		G
	290	L		S
		A		F
320		S		F
		T		L
		Q		Y
		S		S
		E		K
		L	410	L
		T		T
		L		V
	300	S		D
		Q		K
330		K		S
		H		R
		W		W
		L		Q
		S		Q
		D	420	G
		R		N
		T		V
	310	Y		F
		T		S
340		C		C
		Q		S
		V		V
		T		M
		Y		H
		Q	430	E
		G		A
		H		L

(*continued*)

TABLE III (*Continued*).

	IgE			IgG	
	320		T		H
			F		N
350			E		H
352			D		Y
			S		T
			T		Q
			K		K
			K	440	S
			C		L
358			A		S
361	330		D		L
			S		S
			N	445	P
			P	235	L
			R		G
			G		G
			V		P
			S		S
			A	240	V
370			Y		F
	340		L		L
			S		F
			R		P
			P		P
			S		K
			P		P
			F		K
378	347		D		D
	—			250	T
380	348		L		L
			F		M
	350		I		I
			R		S
			K		R
			S		T
			P		P
			T		E
			I		V
			T	260	T
390			C		C
			L		V
	360		V		V
			V		V
			D		D
			L		V
			A		S
			P		H
			S		E

(*continued*)

TABLE III (*Continued*)

	IgE		IgG
	K	270	D
400	G		P
	T		Q
	370 V		V
	N-[CHO]		K
	L		F
	T		N
	W		W
	S		Y
408	R		V
410	A	280	D
411	S		G
414	G		V
	380 K		Q
	P		V
	V		H
	N		N
	H		A
420	S		K
	T		T
	R	290	K
	K		P
	E		R
	390 E		E
	K		Q
	Q		Q
428	R		Y
430	N-[CHO]		N-[CHO]
433	G		S
	T		T
	L	300	Y
	T		R
	V		V
	400 T		V
	S		S
440	T		V
	L		L
	P		T
	V		V
	G		L
	T	310	H
	R		Q
	D		N
	410 W		W
	I		L
450	E		D
	G		G
	E		K
	T		E

(*continued*)

TABLE III (*Continued*)

	IgE		IgG	
		Y		Y
		Q	320	K
		C		C
		R		K
	420	V		V
		T		S
460		H		N
		P		K
		H		A
		L		L
		P		P
		R	330	A
		A		P
467		L		I
469	430	M		E
		R		K
		S		T
		T		I
		T		S
		K		K
		T		A
476		S	340	K
497		G		G
		P		Q
	440	R		P
500		A		R
		A		E
		P		P
		E		Q
		V		V
		Y		Y
		A	350	T
		F		L
		A		P
	450	T		P
510		P		S
		E		R
		W		E
		P		E
		G		M
		S		T
		R	360	K
		D		N
		K		Q
520	460	R		V
		T		S
		L		L
		A		T

(*continued*)

TABLE III (*Continued*)

	IgE		IgG
	C		C
	L		L
	I		V
	Q	370	K
	N		G
	F		F
530	470 M		Y
	P		P
	E		S
	D		D
	I		I
	S		A
	V		V
	Q	380	E
	W		W
	L		E
540	480 H		S
542	N		N
	G		E
	V	386	Q
	Q		—
	L		—
	P	387	P
	D		E
	A		N
550	R		N
	490 H		Y
	S		K
	T		T
	T		T
	Q		P
	P		P
	R		V
	K		L
	T		D
560	K	400	S
	500 G		D
562	S		G
565	G		S
	F		F
	F		F
	V		L
	F		Y
570	S		S
	R		K
	L	410	L
	510 E		T
	V		V

(*continued*)

TABLE III (*Continued*)

	IgE			IgG
		T		D
		R		K
		A		S
		E		R
		W		W
580		E		Q
		Q		Q
		K	420	G
	520	D		N
		E		V
		F		F
		I		S
		C		C
		R		S
		A		V
590		V		M
		H		H
		E	430	E
	530	A		A
		A	432	L
		S		—
		P	433	H
		S		N
		Q		H
		T		Y
600		V		T
		Q		Q
		R		K
	540	A	440	S
		V		L
		S		S
	543	V	443	L
		N		S
		P		P
		G		G
610		K		—

[a] The numbers in the first column correspond to the Kabat *et al.* (13) numbering scheme; those in the second column correspond to the Bennich numbering scheme (25); those in the third correspond to the EU numbering scheme (26). The domain boundaries are marked by dashed lines. The one-letter amino acid codes (27) are used here. Bound carbohydrate is indicated by [CHO].

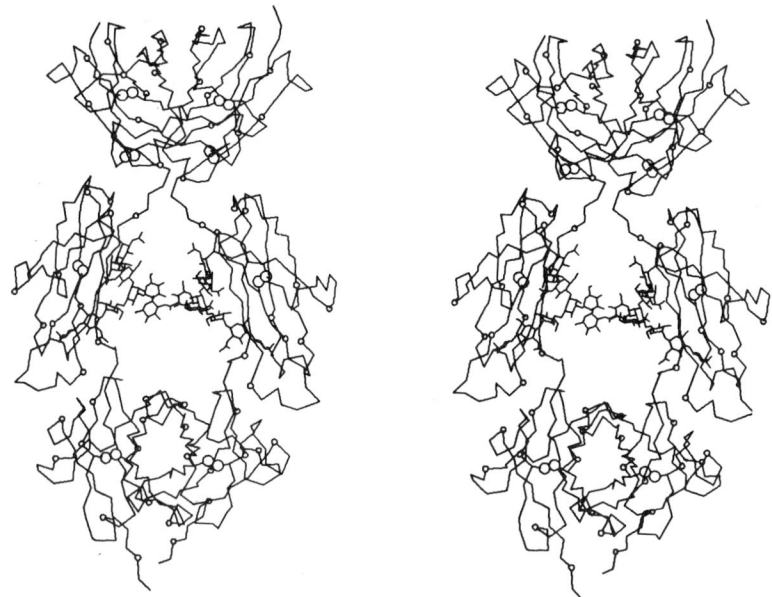

Fig. 3. Stereograph of the modeled alpha-carbon backbone of the Fc of human IgE. Small circles mark every tenth residue. Larger circles show the locations of the disulfide bridges. The CE2 domains are at the top. The carbohydrate positions are taken directly from Deisenhofer (3).

though it should be noted that the latter arrangement could not be ruled out but would simply require a more extensive rearrangement of the model in this region. In this regard it is worth noting that in IgM the symmetrical linkage is the only one possible as each mu chain possesses only one cysteine in this region.

In the absence of a crystal structure determination, a model such as this provides a useful three-dimensional framework for discussing the properties of the molecule. In particular it can be the basis for exploring the region of the molecule involved in binding to the Fc(epsilon) receptor (19–24).

REFERENCES

1. Davies, D. R., and Metzger, H. (1983). *Annu. Rev. Immunol.* **1**, 87–117.
2. Chothia, C., Novotny, J., Bruccoleri, R., and Karplus, M. (1985). *J. Mol. Biol.* **186**, 651–663.
3. Deisenhofer, J. (1981). *Biochemistry* **20**, 2361–2370.

4. Metzger, H., Alcaraz, G., Hohman, R., Kinet, J.-P., Pribluda, V., and Quarto, R. (1986). *Annu. Rev. Immunol.* **4**, 419–470.
5. Ishizaka, T., and Ishizaka, K. (1975). *Prog. Allergy* **19**, 60–121.
6. Hamburger, R. N. (1975). *Science* **189**, 389–190.
7. Stanworth, D. R., Kings, M., Roy, P. D., Moran, J. M., and Moran, D. M. (1979). *Biochem. J.* **180**, 665–668.
8. Bernstein, F. C., Koetzle, T. F., Williams, G. J. B., Meyer, E. F., Brice, M. D., Rogers, J. R., Kennard, O., Shimanouchi, T., and Tasumi, M. (1977). *J. Mol. Biol.* **132**, 535–542.
9. Saul, F., Amzel, L. M., and Poljak, R. J. (1978). *J. Biol. Chem.* **253**, 585–597.
10. Satow, Y., Cohen, G. H., Padlan, E. A., and Davies, D. R. (1986). *J. Mol. Biol.* **190**, 593–604.
11. Marquart, M., Deisenhofer, J., Huber, R., and Palm, W. (1980). *J. Mol. Biol.* **141**, 369–391.
12. Suh, S. W., Bhat, T. N., Navia, M. A., Cohen, G. H., Rao, D. N., Rudikoff, S., and Davies, D. R. (1986). *Proteins: Struct. Funct. Genet.* **1**, 74–80.
13. Kabat, E. A., Wu, T. T., Bilofsky, H., Reid-Miller, M., and Perry, H. (1983). "Sequences of Proteins of Immunological Interest." Department of Health and Human Services, Public Health Service, NIH, Washington, D. C.
14. Murata, M., Richardson, J. S., and Sussman, J. L. (1985). *Proc. Natl. Acad. Sci. U.S.A.* **82**, 3073–3077.
15. Chothia, C., and Janin, J. (1975). *Nature (London)* **256**, 705–708.
16. Alber, T., Grutter, M., Gray, T. M., Wozniak, J. A., Weaver, L. H., Baker, E. N., and Matthews, B. W. (1985). *UCLA Symp. Mol. Cell. Biol.: Protein Struct. Folding Des.* (in press).
17. Diamond, R. (1981). *In* "Biomolecular Structure, Function, Conformation and Evolution" (R. Srinivasan, ed.), Vol. 1, pp. 567–590. Pergamon, Oxford.
18. Hendrickson, W. A., and Konnert, J. H. (1980). *In* "Computing in Crystallography" (R. Diamond, S. Ramaseshan, and K. Venkatesan, eds.), pp. 13.01–13.23. Indian Acad. Sci. Int. Union Crystallogr., Bangalore.
19. Stanworth, D. R., Humphrey, J. H., Bennich, H., and Johansson, S. G. O. (1968). *Lancet* **2**, 17–18.
20. Ishizaka, K., Ishizaka, T., and Lee, E. H. (1970). *Immunochemistry* **7**, 687–702.
21. Perez-Monfort, R., and Metzger, H. (1982). *Mol. Immunol.* **19**, 1113–1125.
22. Baniyash, M., Eshnar, Z., and Rivnay, B. (1986). *J. Immunol.* **136**, 588–593.
23. Kenten, J., Helm, B., Ishizaka, T., Cattini, P., and Gould, H. (1984). *Proc. Natl. Acad. Sci. U.S.A.* **81**, 2955–2959.
24. Liu, F. T., Albrandt, K. A., Bry, C. G., and Ishizaka, T. (1984). *Proc. Natl. Acad. Sci. U.S.A.* **81**, 5369–5373.
25. Dorrington, K. J., and Bennich, H. H. (1978). *Immunol. Rev.* **41**, 3–25.
26. Edelman, G. M., Cunningham, B. A., Gall, W. E., Gottlieb, P. D., Rutishauser, U., and Waxdal, M. J. (1969). *Proc. Natl. Acad. Sci. U.S.A.* **63**, 78–85.
27. Dayhoff, M. O., ed. (1978). "Atlas of Protein Sequence and Structure," Vol. 5, Suppl. 3. Natl. Biomed. Res. Found., Washington, D. C.

13
Three-Dimensional Structure of a Lysozyme–Fab, Antigen–Antibody Complex

A. G. AMIT, G. BOULOT, G. BRICOGNE,* R. A. MARIUZZA,
S. E. V. PHILLIPS,[1] AND R. J. POLJAK

Département d'Immunologie
Institut Pasteur
74724 Paris, France
and
*Université Paris-Sud
and CNRS
91405 Orsay, France

INTRODUCTION

Our current understanding of the high-resolution, three-dimensional structure of antibodies derives from X-ray diffraction studies of human and murine myeloma immunoglobulins (reviewed in 1,2). The study of Fab and Fc fragments has been pursued in several laboratories. These fragments contain the molecular sites specific for different functions of the antibody molecule. Of particular interest is the site for antigen recognition which resides in Fab.

The three-dimensional structures of the Fab fragments of several myeloma immunoglobulins have been determined to high resolution by X-ray diffraction techniques (1,2). The structures of the complexes of two of these, Fab New (3) and Fab McPC 603 (4), with the haptenic groups vitamin K_1OH and phosphorylcholine, respectively, have also been determined. These studies have contributed to our understanding of the structure of antibody combining sites and their interactions with ligands. However, antigen–antibody interactions are more extensive and potentially more important for the elucidation of the physiological properties of antibodies and their induction. The introduction

---

[1] Present address: Astbury Department of Biophysics, University of Leeds, England.

of techniques to obtain lymphocyte hybrids that secrete antibodies of predetermined specificity (5) makes possible the production of large amounts of monoclonal antibodies and a systematic search for a crystalline antigen–antibody complex suitable for X-ray diffraction studies. Thus, Colman et al. (6) have reported crystals of the Fab fragment of a monoclonal anti-influenza virus neuraminidase antibody suitable for high-resolution X-ray diffraction analysis as well as attempts to crystallize the Fab–neuraminidase complex. More recently, Silverton et al. (7) have reported the crystallization of lysozyme–Fab complexes suitable for high-resolution X-ray diffraction studies.

In this chapter we describe our work on the selection and cloning of a hybrid cell line secreting an anti-hen eggwhite lysozyme (HEL) antibody, on the purification of its Fab fragment and the complex Fab–HEL, and on the crystallization and crystallographic study of this Fab–antigen complex. HEL was chosen as the protein antigen mostly because its three-dimensional (8) as well as antigenic structures have been analyzed extensively (9–15). Determination of the three-dimensional structure of this antigen–antibody complex can therefore be expected to provide precise information concerning the interactions of an antibody molecule with its antigen.

MATERIALS AND METHODS

Hybrid cell lines secreting anti-HEL antibodies were obtained (15a) by fusion of BALB/c immune spleen cells with the nonsecreting mouse hybrid myeloma cell line Sp2/0-Ag14 (16). BALB/c mice were immunized with one subcutaneous injection of 100 μg of purified HEL (Boehringer-Mannheim) in Freund's complete adjuvant followed by a second injection of antigen (100 μg) in Freund's incomplete adjuvant 15 days later; a third intravenous injection of protein (50 μg) in phosphate-buffered saline solution was administered on day 36 and the fusion was carried out 3 days later using polyethylene glycol 1000 as the fusion agent (17). Hybridoma cells were screened for antibody production by solid-phase radioimmunoassay (18) and cells producing anti-HEL antibodies were cloned by limiting dilution (19,20). Selected clones secreting anti-HEL antibodies were expanded in vivo to obtain ascitic fluid by injecting 3×10^6 hybridoma cells intraperitoneally into (BALB/c $\times$ C57BL/6)F_1 mice that had been primed with Pristane (Sigma) at least 14 days previously. Ascites fluids were stored frozen at $-20°$C in the presence of 0.01% (w/v) Na$_3$.

Monoclonal antibody D1·3 (IgG$_1$, kappa) was precipitated from ascitic fluid by addition of an equal volume of 100% saturated ammonium sulfate solution. The precipitate was redissolved in 0.1 M potassium phosphate (pH 7.2), dialyzed against 40 mM sodium phosphate (pH 8.0), and applied to a column of (diethylaminoethyl)cellulose (Serva) equilibrated with the same buffer. The antibody eluted in the run-through and was concentrated by precipitation at 80% saturated ammonium sulfate and dialyzed against 0.1 M potassium phosphate (pH 7.2). The Fab fragment was prepared by digestion with papain (from Worthington) (21) for 6 hours at 37°C at an enzyme to substrate ratio of 1:100 in the presence of 1.5 mM 2-mercaptoethanol and 1.25 mM EDTA. The reaction was terminated by the addition of 4 mM iodoacetamide followed by dialysis against 40 mM sodium phosphate (pH 8.0). The reaction mixture was applied to a (diethylaminoethyl)-cellulose column equilibrated with the same buffer. The Fab fragment eluted in the run-through and was concentrated by precipitation with ammonium sulfate at 80% saturation and dialyzed against 0.01 M potassium phosphate (pH 7.0).

The HEL–D1.3 Fab antigen–antibody complex was prepared by incubating a 1.25 M excess of HEL to D1.3 Fab overnight at 4°C (protein concentrations were calculated using 280-nm absorption coefficient $E_{1\,cm}^{1\%}$ of 25.5 for HEL and of 15.0 for Fab). Typically, 435 μl of a 9.2 mg HEL/ml solution in 0.01 M potassium phosphate (pH 7.0) was added to 1.0 ml of a 12.5 mg/ml solution of D1.3 Fab in the same buffer. The D1.3 Fab–HEL complex was separated from monomeric material by chromatofocusing. Briefly, the mixture was applied to a 15-ml column of PBE 94 (Pharmacia Fine Chemicals) equilibrated with 0.025 M diethanolamine acetate (pH 9.4), and proteins were eluted at their isoelectric points with Polybuffer 96 (Pharmacia Fine Chemicals)/acetate (pH 6.0). Three major peaks were obtained and identified by electrophoresis in 15% (w/v) polyacrylamide gels containing 0.1% (w/v) sodium dodecyl sulfate (22): peak I, eluting in the void volume, consisted for unbound HEL; peak II, eluting at pH 8.8, contained the Fab D1.3–HEL complex; and peak III, eluting at pH 8.2, corresponded to unbound Fab fragment. Peak II generally contained from two-thirds to three-fourths of the total Fab fragment applied to the column, as judged by visual inspection of column profiles and the relative intensities of protein bands in gel electrophoresis. Incubation of the material in peak III with a slight molar excess of HEL followed by rechromatography on PBE 94 as described above indicated no difference in its antigen-binding capacity from that of the original sample. This confirmed that the material in peak III was not denatured or nonspecific Fab fragment, and that our chromatographic procedure is a suitable method for the separation of complexed from uncomplexed Fab. The fractions corresponding to peak II were pooled and the protein separated from carrier ampholytes by precipitation with ammonium sulfate at 80% saturation followed by dialysis against 0.01 M potassium phosphate (pH 7.0).

CRYSTALLOGRAPHIC STUDIES

HEL–Fab complexes prepared from several anti-HEL monoclonal antibodies as described above (15a) were subjected to crystallization trials by vapor diffusion using the hanging drop technique (23). The Fab D1.3–HEL complex (10 to 15 mg/ml) crystallized at room temperature at polyethylene glycol 6000 concentrations of 20 to 22.5% (w/v), 0.1 M potassium phosphate (pH 6 to 7). The crystals from one such drop (containing about 50 μg protein) were collected and washed extensively with 20% polyethylene glycol, 0.1 M potassium phosphate (pH 6.0), and electrophoresed in a 15% polyacrylamide gel under nonreducing conditions alongside a sample of purified Fab D1.3–HEL complex; this confirmed that the two were essentially identical in terms of the relative ratios of HEL to Fab fragment in each, as judged by visual inspection of relative strengths of Coomassie blue-staining bands.

Crystals of the complex grew reproducibly as prisms or as prismatic plates terminated by bipyramidal faces and reached a size of up to 1.4 × 0.5 × 0.3 mm. The space group is monoclinic $P2_1$, with unit cell dimensions a = 55.6, b = 143.4, c = 49.1 Å, β = 120°30'. X-ray reflections corresponding to Bragg spacings of 2.5 Å and lower were observed on "stills" and on oscillation photographs taken with synchrotron radiation. The unit cell dimensions and symmetry, compared to those of other crystalline Fab fragments such as the anti-p-azophenyl-arsonate Fab R19.9 (24), indicate that the volume of the asymmetric unit is about 25% larger, as expected for an Fab–HEL complex. Thus, this crystalline material is very suitable for high-resolution X-ray diffraction studies.

The structure determination at 6 Å resolution has been reported (25). Three heavy atom isomorphous substitutions were obtained with K_2PtCl_4, $K_3F_5UO_2$, and p-hydroxymercuribenzene sulfonate (PHMBS). X-ray intensities were measured using a 4-circle automatic diffractometer. The conventional R-factors after refinement of heavy atom parameters, using a full-matrix least-squares method (26), were 49, 44, and 47%, respectively. The average figure of merit for the phases of the 1659 reflexions contained in the 6-Å sphere was 0.75. A balsa wood model based on the electron density map calculated with these reflexions showed the typical domain structure of Fab (27) and an additional globular region, corresponding to lysozyme, in close contact with Fab.

The map was subsequently displayed on an Evans and Sutherland PS 300 interactive graphics system using the program FRODO (28), together with the alpha-carbon backbones of the previously determined hen lysozyme (8) and Fab New (29) structures. The helical segments of lysozyme clearly corresponded to rods of electron density. The model was fitted to the map using only rigid body rotations and translations. The lysozyme chain crosses regions of lower density at positions 52–53 and 58–59 and at the turns 21–22 and 115–116, but the interpretation is unequivocal. No gross conformational change is evident at this resolution, but the two C-terminus residues appear to have moved and there may be a rearrangement at the 21–22 turn. Both these sites are involved in contacts with the antibody. The variable domains of Fab could be fitted well and the heavy (H) and light (L) chains distinguished by comparison with the Fab New model (27). In addition, the alternative fitting of the light and heavy chains was not as satisfactory. To fit the constant domain ($C_H1 + C_L$) it was necessary to rotate it about an axis through residues 103L and 117H, at the flexible switch regions, changing the angle between the constant and variable domains from 137° in Fab New to approximately 180°, closer to the 166° angle observed in Fab Kol (30).

DISCUSSION

The interactions between Fab and lysozyme extend over a large area (Fig. 1) and involve all six complementarity-determining regions (CDRs) (31) of the antibody. The most obvious contacts in lysozyme occur at positions 19–27 and 116–129, widely separated in its sequence. These residues do not strictly correspond to any of the four antigenic sites proposed from studies using polyclonal antibodies (9,10). Although side chains are not visible at this resolution, the model (Fig. 2) particularly suggests contacts to Asn 19, Gly 22, Ser 24, Gly 117, Asp 119, Gln 121, and Arg 125. Possible contacts to side chains such as those of Lys 13, Arg 14, and Arg 21 remain to be verified by a high-resolution analysis, which is currently under way. Biochemical experiments described below suggest that Gln 121 plays an important role in antigen–antibody complex formation. As shown

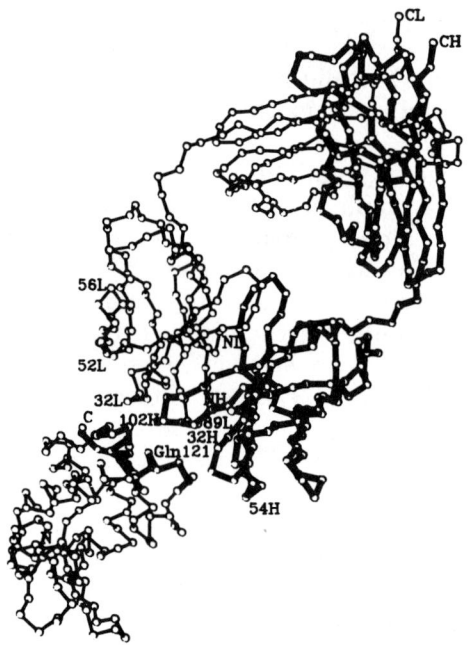

Fig. 1. Three-dimensional model of the Fab–HEL complex. Diagram of the alpha-carbon backbone obtained by fitting the structures of HEL (8) and of Fab New (29) to the 6-Å-resolution electron density map.

in Fig. 2, Gln 121 occupies a central position in the contacting region in close proximity to CDR1 and CDR3 of the light chain and to CDR3 of the heavy chain, its side chain penetrating the binding cleft of Fab D1.3.

To define the amino acid residues of lysozyme recognized by monoclonal antibody D1.3, the enzymatic activities of different avian lysozymes in the presence of Fab D1.3 were analyzed (Fig. 3). Chromatofocusing on HPLC established that those lysozymes that were inhibited in the assay formed complexes with Fab D1.3. Consequently, we take the enzymatic assay of Fig. 3 as an indicator of complex formation. Whether inhibition of activity is the result of steric hindrance or of an antibody-induced conformational change in the enzyme remains to be determined. Fab D1.3 inhibits the lytic activity of bobwhite quail lysozyme and of hen lysozyme but has little effect on that of California quail lysozyme. The quail lysozymes differ in their amino acid sequences (32, 33; see Fig. 4) only at positions 68 and 121. However, there is no apparent difference between the inactiva-

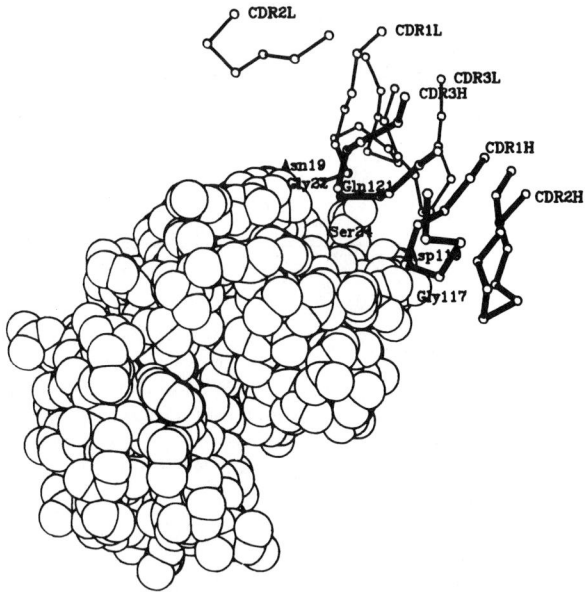

Fig. 2. Three-dimensional model of the Fab–HEL contact area. All atoms of lysozyme are shown as spheres of radius 1.7 Å; the orientation of the amino acid side chains is assumed to be the same as in crystalline HEL (8). Some of the residues most closely involved in contacts with the antibody are labeled. The complementarity-determining regions of the antibody are shown by their alpha-carbon backbones. The active site of lysozyme appears as a cleft to the lower right.

tion of hen lysozyme and bobwhite quail lysozyme, which differ at position 68 but share a Gln at position 121. This indicates that the replacement of Gln 121 by His in California quail lysozyme interferes with the formation of an antigen–antibody complex. Partridge (33) and turkey (34) lysozymes with a His at position 121 are also not inactivated by Fab D1.3 nor is Japanese quail lysozyme (35) with Asn at position 121, although this latter has additional sequence differences at positions 19 and 21, which make close contacts with Fab. These observations can be explained in terms of the three-dimensional structure presented here, in which Gln 121 occupies a central position in the antigen–antibody contact area. Monoclonal antibody D1.3 therefore recognizes amino acid replacements at position 121, as did polyclonal antisera in a microcomplement fixation assay (32,33). Thus, the effects of single amino acid substitutions on antigenic specificity and antigen recognition by the antibody are explained readily in terms of the three-dimensional model obtained in this study.

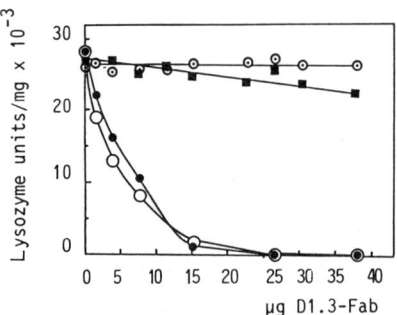

Fig. 3. Inactivation of different egg-white lysozymes by Fab D1.3. Samples of about 1.0 μg of different lysozymes were incubated with monoclonal Fab D1.3 in 0.1 M potassium phosphate, pH 7.0, at 37°C, for 90 min in a total volume of 100 μl. Enzymatic assays of the different lysozymes were performed using *Micrococcus lysodeikticus* cell walls (0.3 mg/ml in 0.1 M potassium phosphate, pH 7.0) as substrate. The assayed egg-white lysozymes are from hen (○), bobwhite quail (●), California quail (■), and partridge (◎).

The work reported above was continued to extend the resolution to 2.8 Å. X-ray intensities from 6 to 2.8 Å were measured using 10 native crystals, 10 PtCl$_4$-substituted crystals, 8 UO$_2$F$_5$-substituted crystals, and 9 PHMBS-substituted crystals. The native intensities were placed on the same relative scale (36) by overlaps with a data set obtained from a single crystal.

Heavy atom parameters (except the x, y, z coordinates) were refined for each heavy atom-substituted crystal independently, using a full matrix algorithm (26) which takes into account the implicit entrainment of the most probable phases by all the parameters. Difference-Patterson and difference-Fourier syntheses allowed us to locate two sites for PtCl$_4$ (one major), six sites for UO$_2$F$_5$, and eight sites for PHMBS. Table I gives some of the values obtained after refinement.

An electron density map to 2.8 Å was calculated with the refined phases. The location of the polypeptide chain backbone of lysozyme on this map agrees very closely with that obtained from the 6-Å-resolution map. Most of the path of the polypeptide chains can be followed without breaks in continuity. Many of the side chains of lysozyme, including the disulfide bonds, can be placed with orientations which are very close to those observed in native lysozyme. Examples are Trp 62, Trp 63, Met 12, and Phe 34. Other residues appear to have moved relative to the native conformation, such as 20, 21, and 22, in a loop that makes contacts with the antigen combining site on Fab and side chains such as that of Arg 14. An accurate account of these rela-

13. STRUCTURE OF AN ANTIGEN–ANTIBODY COMPLEX

```
            1    5    1    1    2    2    3    3    4    4    5    5    6    6
                      0    5    0    5    0    5    0    5    0    5    0    5
HEL    KVFGRCELAAAMKRHGLDNYRGYSLGNWVCAAKFESNFNTQATNRNTDGSTDYGILQINSRWWCN
PEL    ----------V------N-----------------------------------------------
CEL    -------------------------------------S---------------V-----------
BEL    -------------------------------------S---------------V-----------
JEL    --Y-------------K-Q----------------------------------------------
TEL    --Y-----------L--------------------------H-----------------------

                                              1    1    1    1    1    1    1
       6    7    7    8    8    9    9    0    0    1    1    2    2    2
       6    0    5    0    5    0    5    0    5    0    5    0    5    9
HEL    DGRTPGSRNLCNIPCSALLSSDITASVNCAKKIVSDGNGMNAWVAWRNRCKGTDVQAWIRGCRL
PEL    ------------------------------------------------------H--------
CEL    ------------------------T-------------------------------H--------
BEL    --K---------------------T----------------------------------------
JEL    -------------------------------------VH----------------N--------
TEL    -------K------------------------------A-G--------------H--------
```

Fig. 4. Amino acid sequences of different egg-white lysozymes. The full amino acid sequence of hen egg-white lysozyme is given in the one-letter code for amino acids. The sequences of the lysozyme from bobwhite (BEL), California (CEL), and Japanese (JEL) quails and from turkey (TEL) and partridge (PEL) are indicated only at the position in which they differ from HEL. The sequences are taken from references 32 and 33.

TABLE I
Phase Refinement

Resolution (Å)	No. of reflexions	$\langle m \rangle^a$	Phasing power		
			$PtCl_4$	UO_2F_5	PHMBS
5.50	1,947	0.74	1.31	5.80	4.49
3.71	4,491	0.67	1.67	7.03	3.04
3.19	3,505	0.58	2.71	2.60	2.13
2.80	4,179	0.45	2.69	2.05	3.21
Total	14,122	0.59	2.18	4.04	3.07

^{a} $\langle m \rangle$ is the average figure of merit. Phasing power is the rms heavy atom calculated structure factor divided by the rms residual.

tive movements can only be obtained after completion of model building and subsequent crystallographic refinement, which are currently under way.

ACKNOWLEDGMENTS

The work was supported by grants from the Centre National de la Recherche Scientifique, INSERM, and Institut Pasteur.

REFERENCES

1. Amzel, L. M., and Poljak, R. J. (1979). *Annu. Rev. Biochem.* **48**, 961–997.
2. Davies, D. R., and Metzger, H. (1983). *Annu. Rev. Immunol.* **1**, 87–117.
3. Amzel, L. M., Poljak, R. J., Saul, F., Varga, J. M., and Richards, F. F. (1974). *Proc. Natl. Acad. Sci. U.S.A.* **71**, 1427–1430.
4. Segal, D. M., Padlan, E. A., Cohen, G. H., Rudikoff, S., Potter, M., and Davies, D. R. (1974). *Proc. Natl. Acad. Sci. U.S.A.* **71**, 4298–4302.
5. Köhler, G., and Milstein, C. (1975). *Nature (London)* **26**, 495–497.
6. Colman, P. M., Gough, M. K. H., Lilley, G. G., Bagrove, R., Webster, R. G., and Laver, W. G. (1981). *J. Mol. Biol.* **152**, 609–614.
7. Silverton, E. W., Padlan, E. A., Davies, D. R., Smith-Gill, S., and Potter, M. (1984). *J. Mol. Biol.* **180**, 761–765.
8. Blake, C. C. F., Koenig, D. F., Mair, G. A., North, A. C. T., Phillips, D. C., and Sarma, V. R. (1965). *Nature (London)* **206**, 757–761.
9. Arnon, R. (1977). *In* "Immunochemistry of Enzymes and Their Antibodies" (M. R. J. Salton, ed.), pp. 1–28. Wiley, New York.
10. Atasssi, M. Z. (1978). *Immunochemistry* **15**, 909–936.
11. White, T. J., Ibrahimi, I. M., and Wilson, A. C. (1978). *Nature (London)* **274**, 92–94.
12. Kobayashi, T., Fujio, H., Kondo, K., Dohi, Y., Hirayama, A., Takagaki, Y., Kosaki, G., and Amano, T. (1982). *Mol. Immunol.* **19**, 619–630.
13. Smith-Gill, S. J., Lavoie, T. B., and Mainhart, C. R. (1984). *J. Immunol.* **133**, 384–393.
14. Meztger, D. W., Ch'ng, L. K., Miller, A., and Sercarz, E. E. (1984). *Eur. J. Immunol.* **14**, 87–93.
15. Benjamin, D. C., Berzofsky, J. A., East, I. J., Gurd, F. R. N., Hannum, C., Leach, S. J., Margoliash, E., Michael, J. G., Miller, A., Prager, E. M., Reichlin, M., Sercarz, E. E., Smith-Gill, S. J., Todd, P. E., and Wilson, A. C. (1984). *Annu. Rev. Immunol.* **2**, 67–101.
15a. Mariuzza, R. A., Jankovic, D. Lj, Boulot, G., Amit, A. G., Saludjian, P., Le Guern, A., Mazié, J. C., and Poljak, R. J. (1983). *J. Mol. Biol.* **170**, 1055–1058.
16. Shulman, M., Wilde, C. D., and Köhler, G. (1978). *Nature (London)* **276**, 269–270.
17. Galfré, G., Howe, S. C., Milstein, C., Butcher, G. W., and Howard, J. C. (1977). *Nature (London)* **266**, 550–552.
18. Metzger, D. W., Miller, A., and Sercarz, E. E. (1980). *Nature (London)* **287**, 540–542.

19. Kennett, R. H. (1979). In "Methods in Enzymology" (W. B. Jakoby and I. H. Pastan, eds.), Vol. 58, pp. 345–359. Academic Press, New York.
20. Galfré, G., and Milstein, C. (1981). In "Methods in Enzymology" (J. J. Langone and H. Van Vunkanis, eds.), Vol. 73, pp. 1–46. Academic Press, New York.
21. Porter, R. R. (1959). *Biochem. J.* **73**, 119–127.
22. Weber, K., and Osborn, M. O. (1975). In "The Proteins" (H. Neurath and R. L. Hill, eds.), 3rd ed., Vol. 1, pp. 179–223. Academic Press, New York.
23. Wlodawer, A., and Hodgson, K. O. (1975). *Proc. Natl. Acad. Sci. U.S.A.* **72**, 398–399.
24. Amit, A. G., Harper, M., Mariuzza, R. A., Saludjian, P., Poljak, R. J., Lamoyi, E., and Nisonoff, A. (1983). *J. Mol. Biol.* **165**, 415–417.
25. Amit, A. G., Mariuzza, R. A., Phillips, S. E. V., and Poljak, R. J. (1985). *Nature (London)* **133**, 156–158.
26. Bricogne, G. (1982). In "Computational Crystallography" (D. Sayre, ed.), pp. 223–230. Oxford Univ. Press, London and New York.
27. Poljak, R. J., Amzel, L. M., Avey, H. P., Becka, L. N., and Nisonoff, A. (1972). *Nature (London), New Biol.* **235**, 137–140.
28. Jones, T. A. (1978). *J. Appl. Crystallogr.* **11**, 268–272.
29. Poljak, R. J., Amzel, L. M., Avey, H. P., Chen, B. L., Phizackerley, R. P., and Saul, F. (1973). *Proc. Natl. Acad. Sci. U.S.A.* **70**, 3305–3310.
30. Matsushima, M., Marquart, M., Jones, T. A., Colman, P. M., Bartels, K., Huber, R., and Palm, W. J. (1978). *Mol. Biol.* **121**, 441–459.
31. Kabat, E. A., and Wu, T. T. (1971). *Ann. N.Y. Acad. Sci.* **190**, 382–393.
32. Ibrahimi, I. M., Prager, E. M., White, T. J., and Wilson, A. C. (1979). *Biochemistry* **18**, 2736–2743.
33. Ibrahimi, I. M., Eder, J., Prager, E. M., Wilson, A. C., and Arnon, R. (1980). *Mol. Immunol.* **17**, 37–46.
34. LaRue, J. N., and Speck, J. C., Jr. (1970). *J. Biol. Chem.* **245**, 1985–1991.
35. Kaneda, M., Kato, I., Tominaga, N., Titani, K., and Narita, K. (196). *J. Biochem. (Tokyo)* **66**, 747–749.
36. Fox, G., and Holmes, K. C. (1966). *Acta Crystallogr.* **20**, 886–891.

14

The Order–Disorder Paradox in Antigen–Antibody Union: Anti-Peptide Antibodies as a Probe for Structured Regions of Small Peptides

H. JANE DYSON, MARK RANCE, RICHARD A. HOUGHTEN,
PETER E. WRIGHT, AND RICHARD A. LERNER

Department of Molecular Biology
Research Institute of Scripps Clinic
La Jolla, California 92037

INTRODUCTION

Antibodies raised to short peptide fragments of proteins have been found to recognize with high frequency the cognate sequences in the intact proteins (1,2). This observation appears paradoxical in view of the accepted dogma that small peptides are not structured in water solution: How can an antibody molecule raised to a short peptide with no defined structure subsequently recognize the same sequence in a protein, in which the structure must be highly defined. This has been formulated as a "disorder–order phenomenon" (3) and probably means that proteins are more disordered and peptides more ordered than was previously supposed. We have hypothesized that this system—the peptide antigen and the anti-peptide antibody raised to it—could be used to detect regions of actual or incipient structure in peptides in water solution, in effect, a selection criterion for structure in peptides.

Two examples of immunogenic peptides have been found to be structured in solution (4,5). In both cases, unequivocal evidence for the structures was provided by high-resolution nuclear magnetic resonance (NMR) spectroscopy.

New developments in equipment and techniques have made ^{1}H-NMR the method of choice for structural studies of peptides and proteins in solution. High-field spectrometers of high sensitivity allow resolution of overlapped proton resonance signals. In addition, two-dimensional techniques recently developed are extremely useful in the rapid assignment of proton resonances to individual protons in peptides and proteins. The most useful two-dimensional techniques for the assignment of protons within individual amino acid residues have been two-dimensional correlated spectroscopy (COSY) (6–8) and double quantum spectroscopy (9,10), in which cross peaks are generated that correspond to protons which are connected through bonds (e.g., between C$^\alpha$H and NH of the same amino acid residue). Double quantum spectroscopy is particularly useful in the assignment of the protons of long side chains, such as in leucine, isoleucine, and lysine residues. Sequence-specific assignments are made using nuclear Overhauser effect spectroscopy (NOESY), from which information is obtained on through-space proximity of protons, in particular, that between the C$^\alpha$H of residue i and the NH of residue $i + 1$.

Evidence for the presence of structure in water solution is also provided by NMR studies, which are unique in that their information pertains specifically to the water solution used, unlike other techniques, such as crystallography, in which the information pertains to the solid state. Methods such as NOESY and temperature- and urea-dependence studies give information on solution structure.

NMR EVIDENCE FOR A TYPE II β-TURN

The first example found of a structured immunogenic peptide consisted of the immunodominant region of a larger peptide immunogen which formed a part of the trimer interface of influenza virus hemagglutinin (11). Its sequence is YPYDVPDYA (peptide I). Evidence for the presence of a β-turn was provided by two NMR experiments: the temperature dependence of the chemical shift of the amide resonances and the observation of nuclear Overhauser effects (NOEs) between certain protons in the peptide. The region of the NMR spectrum which includes the amide proton resonances of peptide I is shown in Fig. 1a for two temperatures. It is obvious from the figure

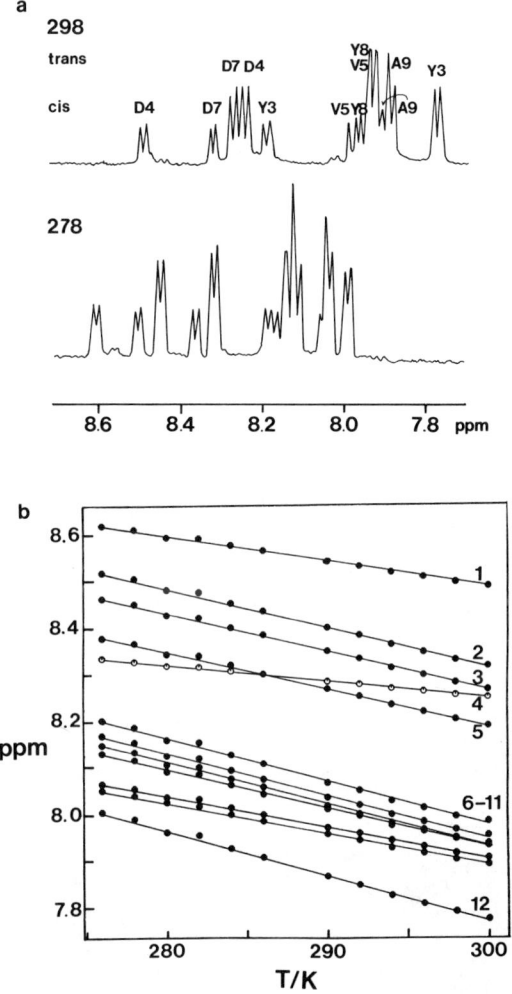

Fig. 1. (a) Amide proton region of ^{1}H-NMR spectrum of hemagglutinin peptide YPYDVPDYA (approximately 10 mM solution in 90% H$_2$O/10% D$_2$O, adjusted to pH 4.10 with 0.1 M NaOH and 0.1 M HCl). Resolution was enhanced by Lorentzian-to-Gaussian transformation. (b) Temperature dependence of the chemical shifts of the amide proton resonances of hemagglutinin peptide (I). (1) Asp 4 (cis); (2) Asp 7 (cis); (3) Asp 7 (trans); (4) Asp 4 (trans); (5) Tyr 3 (cis); (6) Val 5 (cis); (7) Tyr 8 (cis); (8) Val 5 (trans); (9) Tyr 8 (trans); (10) Ala 9 (cis); (11) Ala 9 (trans); (12) Tyr 3 (trans).

that there are two forms of the peptide and that the temperature dependencies of the amide proton resonances are not all the same. The two forms of the peptide arise through cis–trans isomerism around the Tyr 1–Pro 2 peptide bond. The presence of two forms of the peptide serves to complicate the spectra, but also provides an internal control: we find that the trans form of the peptide, which is the form in greater abundance, shows evidence of structure, whereas the cis form behaves like a normal peptide and has no structure. The temperature dependence of the amide proton resonances of peptide I (Fig. 1b) reveals that one proton, that of Asp 4 (trans), shows a reduced temperature dependence. This indicates that the proton is protected from water solvent to a considerable extent and is usually taken to mean the presence of a hydrogen bond involving this proton. The hydrogen bond is intramolecular, since the spectrum is not concentration dependent. The structure inferred for the peptide is shown in Fig. 2. This is confirmed by the observation of NOEs between the protons indicated in Fig. 2. The nuclear Overhauser effect arises from the "through-space" transfer of magnetization between two protons and can thus be used as a measure of the proximity of the protons; observation of NOEs in a situation such as this implies that the protons in-

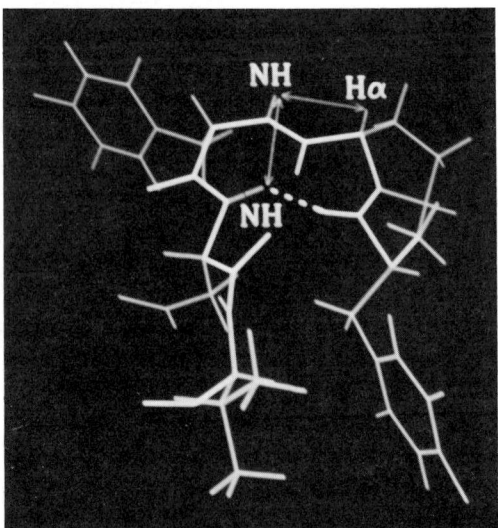

Fig. 2. Computer graphics-derived representation of the N-terminal portion of the hemagglutinin peptide I, showing the NOEs observed.

volved spend a significant proportion of their time in a conformation in which they are less than 3 Å apart.

NMR EVIDENCE FOR A HELICAL STRUCTURE

An immunogenic peptide with a sequence derived from the myohemerythrin of the marine worm *Themiste zostericola* has been found to contain a helical structure. The peptide, EVVPHKKMHKDFLEKIGGL (II), comprises the C-helix of the myohemerythrin molecule and has been found to be an immunodominant site for both antiprotein and anti-peptide antibodies. NMR studies of the C-helix peptide show that it contains a helical structure in water solution. Evidence for this is obtained from NOESY experiments (Fig. 3). It is found that the amide protons of certain of the residues in the peptide are in close proximity for a proportion of the time sufficient for NOEs to be observed between them. In addition, interresidue NOEs are observed between the $C^\beta H$ protons and amide protons of adjacent residues. This is summarized in Fig. 4.

In this context, it is significant that only some of the resonances are observed to behave this way; approximately half of the helix is apparently unstructured and shows no NH–NH NOEs. This provides a valuable internal control: the strong NOEs we observe for the structured half of the peptide must therefore arise from something other than random fluctuations bringing the protons into proximity.

The NMR evidence for the presence of considerable helical structure in this peptide is unequivocal, yet the peptide shows no evidence of such structure in the experiment used traditionally for its detection, circular dichroism (CD). This is an important observation, since it indicates that such traditional methods may not be sufficiently sensitive to observe structures in peptides which are easily detectable using more sensitive techniques. It also indicates that we are probably not observing in the NMR experiment a fixed helical structure of several turns. We interpret these results to show the presence of a "nascent" helix, that is, one in which short helical segments flicker in and out. This is highly significant, as it may well be the first direct observation of a potential initiation site for folding of a protein. Such structures have not been previously observed, for they are present too transiently in proteins under folding conditions and have not been detectable in small peptide fragments of proteins primarily because of a lack of sufficiently sensitive methods.

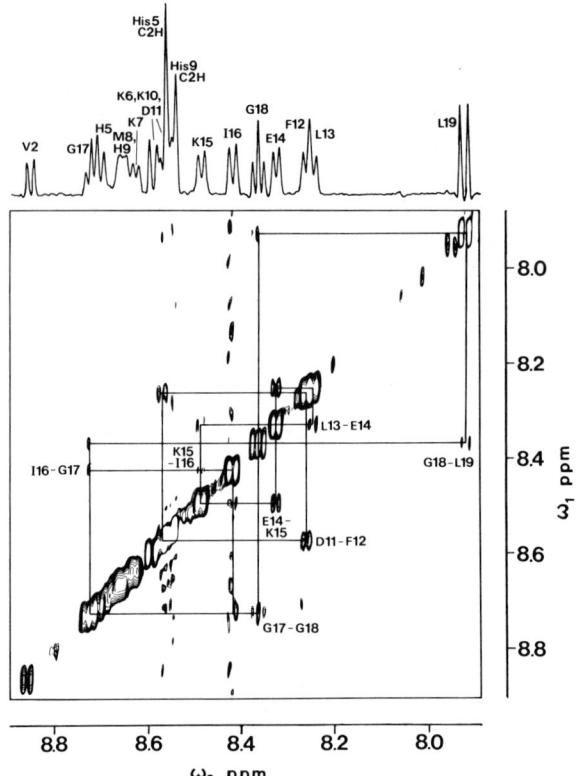

Fig. 3. NH–NH diagonal region of NOESY spectrum of C-helix peptide EV-VPHKKMHKDFLEKIGGL in 90% H_2O/10% D_2O solution at pH 5.1 and 278 K. The corresponding region of the one-dimensional spectrum is shown.

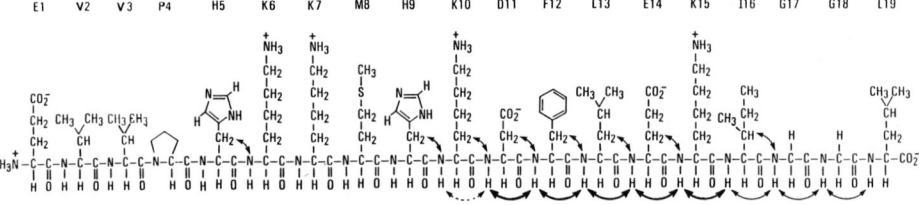

Fig. 4. A representation of the amino acid sequence of the C-helix peptide showing the NOEs observed in the NOESY spectrum.

DISCUSSION AND CONCLUSIONS

We have demonstrated the presence of secondary structure in water solutions of two short peptides which also happen to be immunodominant regions of proteins or larger antigens for anti-peptide antibodies. We propose that to raise antibodies which subsequently recognize the folded protein it is necessary for the peptide antigen to contain a high preponderance of a rudimentary structured form in its conformer population in solution. Similarly, for a protein to begin to fold, local structured regions must form which are presumably similar to those observed in these small peptides. It happens that the helical structure observed for the C-helix peptide corresponds to the structure observed in the native myohemerythrin (12), but there is no *a priori* reason why the earliest initiation structures should be present unchanged in the final folded form of the protein. For example, we observe the β-turn in the hemagglutinin peptide (I) at a position where no such turn is observed in the crystal structure of native hemagglutinin. Nevertheless, antibodies which recognize this immunodominant region of the peptide antigen will also recognize the native hemagglutinin, which implies a degree of flexibility in the protein, as has been previously suggested (13).

We conclude that the "disorder–order phenomenon" of anti-peptide antibodies bears closely on the problem of detection of early intermediates in protein folding. Work is continuing on other protein systems and in defining the conformation of the peptide when bound to the antibody.

ACKNOWLEDGMENTS

We thank Linda Tennant and Gail Donnan for technical assistance and Dan Bloch for help with computer simulations. We acknowledge financial support from NIH Grants A 119499-01 and CA27489.

REFERENCES

1. Lerner, R. A. (1982). *Nature (London)* **299**, 592–596.
2. Lerner, R. A. (1984). *Adv. Immunol.* **36**, 1–44.
3. Wright, P. E., Dyson, H. J., Rance, M., Ostresh, J., Houghten, R. A., Wilson, I. A., and Lerner, R. A. (1986). In "Vaccines 86" (R. A. Lerner, R. M. Chanock, and F. Brown, eds.), pp. 15–19. Cold Spring Harbor Lab., Cold Spring Harbor, New York.
4. Dyson, H. J., Cross, K. J., Houghten, R. A., Wilson, I. A., Wright, P. E., and Lerner, R. A. (1985). *Nature (London)* **318**, 480–483.

5. Dyson, H. J., Rance, M., Houghten, R. A., Wright, P. E., and Lerner, R. A. (1987). *J. Mol. Biol.* (submitted for publication).
6. Aue, W. P., Bartholdi, E., and Ernst, R. R. (1976). *J. Chem. Phys.* **64,** 2229–2246.
7. Bax, A., and Freeman, R. (1981). *J. Magn. Reson.* **44,** 542–561.
8. Rance, M., Sørensen, O. W., Bodenhausen, G., Wagner, G., Ernst, R. R., and Wüthrich, K. (1983). *Biochem. Biophys. Res. Commun.* **117,** 479–485.
9. Braunschweiler, L., Bodenhausen, G., and Ernst, R. R. (1983). *Mol. Phys.* **48,** 535–560.
10. Rance, M., and Wright, P. E. (1986). *J. Magn. Reson.* **66,** 372–378.
11. Wilson, I. A., Niman, H. L., Houghten, R. A., Cherenson, A. R., Connolly, M. L., and Lerner, R. A. (1984). *Cell (Cambridge, Mass.)* **37,** 767–778.
12. Hendrickson, W. A., and Ward, K. B. (1977). *J. Biol. Chem.* **252,** 3012–3018.
13. Tainer, J. A., Getzoff, E. D., Alexander, H., Houghten, R. A., Olson, A. J., Lerner, R. A., and Hendrickson, W. A. (1984). *Nature (London)* **312,** 127–133.

15
Characteristics of Protein Interfaces

WAYNE A. HENDRICKSON, JANET L. SMITH,
AND WILLIAM E. ROYER, JR.
Department of Biochemistry and Molecular Biophysics
College of Physicians and Surgeons
Columbia University
New York, New York 10032

INTRODUCTION

Many biological phenomena are mediated by direct interactions between macromolecular components, and protein–protein interactions often play a central role. In any case, attributes of the complexes between proteins are rather typical of macromolecular complexes generally. The structures of very few such complexes are presently known in atomic detail. However, the available results from protein crystallography do provide some insight into the nature of interfaces between associated proteins.

A comprehensive description of protein interfaces is beyond the scope of this brief overview. It is not clear how helpful such a detailed description might be, but, on the other hand, many of the general characteristics of these interfaces are so obvious that it is difficult to avoid being simplistic and trite. Nevertheless, we risk this overview in order to bring some general background to the various topics of this book that concern such interactions.

CATEGORIES OF PROTEIN INTERACTIONS

The best characterized of protein interfaces are those in oligomeric proteins. There are several such examples as many protein molecules naturally occur in associated form. The interactions between subunits have important functional consequences such as allostery. Since many protein molecules are dilute (micromolar levels) in physiological conditions, the interactions between protomers in oligomer interfaces must be quite strong to maintain the associated state. Even oligomers that occur in high concentrations, such as hemoglobin, tend to have rather tight associations.

The lattice contacts of crystalline proteins provide another important body of information about the structure of protein interfaces. These interfaces are in a sense artifacts of the solid state, but they are nonetheless exhibits of specific protein–protein interactions. In contrast to the interactions in oligomeric proteins, these are intrinsically rather weak. The individual lattice associations occur only as a consequence of the cooperative interplay of the three-dimensional array.

A third class of interactions is that in complexes between unlike protein molecules. This category includes such interesting examples as electron transfer complexes, antigen–antibody complexes, receptor–ligand complexes, multienzyme complexes, and enzyme–substrate or enzyme–inhibitor complexes. The strength of the interface between pairs of proteins in such unions is quite variable. Some, such as the trypsin–trypsin inhibitor complex, are essentially irreversible interactions (picomolar dissociation constants), whereas others, such as the actomyosin complex, must necessarily be quite labile. Antibodies can have a wide range of affinities against the same immunogen. One hopes that what has been learned from oligomer proteins and lattice contacts might profitably guide our thinking about the interface in this most important class of interactions.

COMPLEMENTARITY

Geometric complementarity of the surfaces that participate in a union between proteins is obviously necessary for complexation. It should also be clear that having topographical features that complement one another is not a sufficient condition for complexation. A striking illustration of this comes from the structure of clam hemoglobin (1). The globin subunits of the clam are very similar to those in human hemoglobin. Yet, as shown in Fig. 1, the oligomer interfaces comprise completely different surface elements in the two cases; those that are buried in human hemoglobin are completely exposed in clam hemoglobin.

What is essential for association is the detailed interdigitation of the

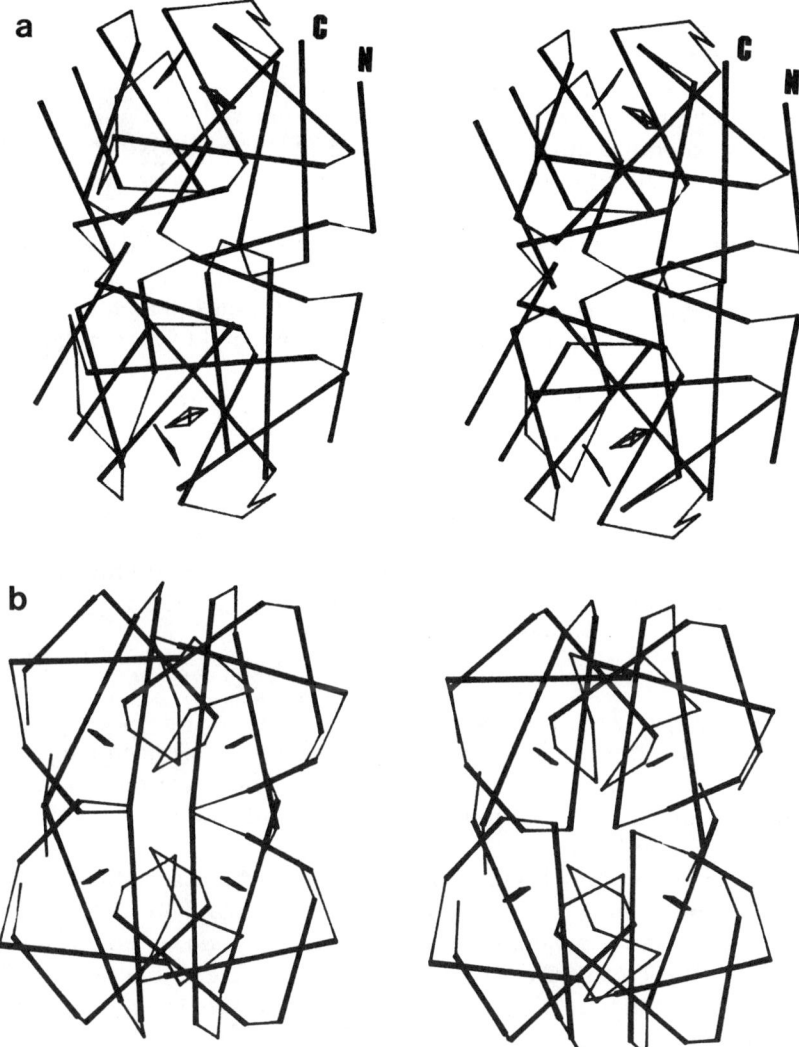

Fig. 1. Stereo drawings showing the distinctive subunit contacts in clam hemoglobin as compared with human hemoglobin. The molecular frameworks are shown as polypeptide tracings with the major helices (pre-A in clam; A,B,E,F,G, and H in both) drawn as heavy lines. Skeletal heme groups are also shown. (a) The clam tetramer is drawn with the molecular diad oriented horizontally in the plane of the paper. The pseudodiads relating subunits within dimers are also approximately in the plane of the paper. These cooperative dimers involve the E- and F-helices in the interfaces. The G–H surfaces that make up the mammalian $\alpha_1\beta_2$ interfaces are fully exposed. (b) The human CO tetramer is also drawn with the molecular diad horizontal in the plane of the paper. The pseudodiad relating α_1 and β_2 subunits is vertical in the plane. (Adapted from reference 1 with permission from *Nature*.)

surfaces in interfaces. At least in the cases of tight complexes, these fit as a hand into a glove. This can be seen in molecular graphics displays of cross sections through the interface of the superoxide dismutase dimer (E. Getzoff, personal communication) or through the trypsin–trypsin inhibitor complexes (R. Langridge, personal communication). Residues in such interfaces are found, on average, to be as close packed as are those in protein interiors (2). Complementarity of interfaces extends beyond steric factors to detailed chemical considerations such as hydrophobicity, charge pairing, and hydrogen bonding.

DETERMINANTS OF AFFINITY

The overall affinity of protein components for one another is measured by the association constant for the reaction of complex formation. This equilibrium is more commonly characterized for the dissociation reaction, and from this dissociation constant one can readily compute the free energy release on dissociation:

$$\Delta G = -RT \ln K_D$$

Tight associations, such as occur in dimeric proteins, typically have dissociation constants of 10^{-8} to 10^{-14} M. These correspond to free energies of 11 to 19 kcal/mol (R = 1.987 cal/mol/deg and T = 298° for 25°C).

The experimentally measurable free energy of association involves contributions from a number of components: (i) The direct interactions that can be observed in crystal structures do, of course, play a critical role. These van der Waals contacts, hydrogen bonds, and electrostatic interactions are attractive in favorably complementary interfaces, whereas they could be quite repulsive in mismatched surfaces. However, even though there are several dozen such pairwise interactions in a typical interface they individually are rather weak. For example, hydrogen bonds in an interface are usually compensated by hydrogen bonds to water in the exposed surfaces. It appears that other factors that are entropic in origin generally make greater energy contributions. (ii) There is an entropic gain from the liberation of organized water molecules on the associating hydrophobic surfaces. This hydrophobic effect adds greatly to the binding energy (3,4). (iii) A third qualitatively distinctive energy term arises from the entropic loss as a result of the reduction of translational and rotational freedom when a subunit is immobilized in a complex. Estimates from statistical thermodynamics suggest that this loss of the intrinsic entropy of a free subunit can represent a very large energy to be counterbalanced by positive binding contributions (2,3). (iv) Dynamic effects also have a significant role. Local mobility of surface groups or interdomain vibrational modes may be dampened by association (5). This would result in a further entropic loss. On the other hand, residual fluctuations in the interface will mitigate the loss of subunit entropy.

INTERFACES IN HUMAN HEMOGLOBIN

The hemoglobin molecule presents a well-worked, though still incompletely understood, model of protein interfaces. There are two rather different kinds of interfaces in the tetramer. One, the $\alpha_1\beta_1$ interface, has tighter binding and changes little if at all on oxygenation. The other, the $\alpha_1\beta_2$ interface, is more labile and it undergoes a major rearrangement on oxygenation (6). The dissociation constant for the fixed $\alpha_1\beta_1$ interface has recently been measured to be 1.5×10^{-12} M (7). This corresponds to a free energy change of -16.1 kcal/mol on association. The energy of association for these dimers into tetramers depends on ligand state and is -14.3 kcal/mol for deoxyhemoglobin and -8.0 kcal/mol for oxyhemoglobin (8). These interactions involve two $\alpha_1\beta_2$ interfaces as well as some $\alpha_1\alpha_2$ interactions. Taking into account the cooperative aspect of subunit entropy discussed below, we estimate the intrinsic dissociation constants for putative $\alpha_1\beta_2$ dimers to be no less than 4×10^{-4} M in deoxyhemoglobin and 8×10^{-2} M in oxyhemoglobin. Interestingly, lamprey hemoglobin is a weakly held dimer when deoxygenated but is monomeric when oxygenated (9). An $\alpha_1\beta_1$ type of contact is sterically precluded in this structure, but a $\alpha_1\beta_2$ type of dimer is possible (9,10).

The tight $\alpha_1\beta_1$ contact in hemoglobin is an extensive interface that buries approximately 850 $Å^2$ of surface on each subunit from solvent accessibility. The labile $\alpha_1\beta_2$ interfaces have approximately 650 $Å^2$ buried from each in the deoxy state and approximately 500 $Å^2$ in the oxy state (11). Chothia et al. (11) note that the fraction of the buried area contributed by polar (N, O, S) atoms is only slightly higher in the weaker contacts and suggest that the vast difference in binding strengths is due to hydrophobic free energy estimated as 25 cal/mol per $Å^2$. This analysis may be somewhat deceiving since a distinction between the fixed ($\alpha_1\beta_1$) and labile ($\alpha_1\beta_2$) contacts is in the character of side chains that contribute directly to the interactions. In the labile interfaces (particularly in the liganded state) more large polar or charged groups than large hydrophobic groups are involved, whereas the reverse is true for the fixed interface. Quite possibly, the water structures displaced from these two surfaces will differ in entropic content and hence in energy contribution. It is also noteworthy that the E- and F-helices in the extremely stable cooperative dimer interface in clam hemoglobin have outer surfaces that are markedly more hydrophobic than those in other hemoglobins (12).

SUBUNIT ENTROPY

Among the several factors contributing to the free energy of association, the entropy changes due to lost degrees of subunit freedom stand out as unique. Whereas the other factors—specific interactions, the hydrophobic effect, and internal mobility—are inherent properties of the surfaces in contact, this entropy factor depends in principle only on the immobilization of a free body and not on the niche that it comes to occupy. Thus, it is useful to separate the energy of association into an explicit bond energy (which itself has entropic as well as enthalpic components) and the subunit entropy (2,13).

Subunit entropy is also distinctive for its special role in the cooperativity of assembly. In the case of a cyclic oligomer there are N interfaces and thus N bond energy components but only $N - 1$ components of subunit entropy. That is, when the final subunit is put into place, two interface bonds are formed for the price of losing only one unit of subunit entropy. Thus, completion of the aggregate is strongly favored. This effect is also the secret of postnucleation crystal growth. Once a three-dimensional framework is laid, two or more bonds are made for each immobilization event. Interfaces of low intrinsic binding affinity that do not form in solution can do so in this cooperative manner. For example, dimeric clam hemoglobin has a crystal contact like the one that relates dimers within the tetrameric clam hemoglobin molecule (1). Such entropic considerations also pertain to other molecular events such as antibody binding to multivalent ligands or dimeric repressor binding to DNA.

Estimates of the actual magnitude of subunit entropy contributions to association differ widely. Chothia and Janin (2) suggest a loss of 20–30 kcal/mol for immobilizing typical subunits. However, Erickson and Pantaloni (13) find a reduced theoretical value of 13–15 kcal/mol if the subunits in the aggregate are allowed some vibrational motion (14). Consideration of the energetics of self-assembly of microtubules requires that the energy loss from subunit entropy must be less than 7 kcal/mol to fit observations (13). These disparities are large and might affect the conclusion of Chothia and Janin (2) that the hydrophobic effect is the major factor in stabilizing associated proteins. If subunit entropy is indeed rather low, then either the hydrophobic effect has also been overestimated, enthalpic interactions are greater, or perhaps entropic losses due to dampened mobility of the interface groups are substantial.

HYDRATION

Our considerations until now have focused on rather tight complexes. The subunit interfaces in protein crystals afford a look at associations that have much lower intrinsic bonding affinity. Typically these contacts involve less surface area than is in the major interfaces of oligomeric proteins, but the prevalence of water molecules in these interfaces is equally striking. An example from the structure of erabutoxin (15) is shown in Fig. 2. We have also seen similar hydrated interfaces in our well-refined structures of crambin (16), lamprey hemoglobin (17), and myohemerythrin (18).

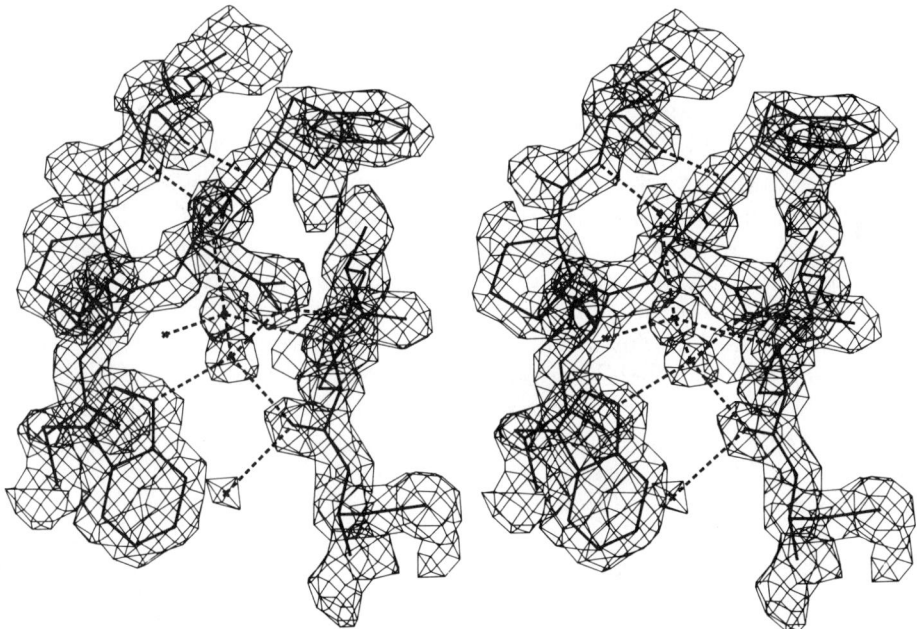

Fig. 2. Stereo drawing of a network of water-mediated hydrogen bonds (dashed lines) in a portion of a lattice interface in erabutoxin.

Localized water sites are an integral part of protein crystal structures. In the aforementioned well-refined structures we have located from 1.3 to 2.0 water sites per amino acid residue on average. Surfaces are covered by first and second shell water sites, and many of the highest occupancy and lowest mobility waters are at subunit interfaces. It may be that water molecules serve as facile adaptors to mold intrinsically noncomplementary surfaces to one another.

If waters are retained in an organized state in protein interfaces, then, in contrast to an anhydrous interface, a reduced entropic gain might be expected from the hydrophobic effect. However, the entropic cost of maintaining solvent structure in protein interfaces could be partially offset by disorder among multiple configurations for these waters (19). In addition, an enthalpic gain is expected from including water-mediated hydrogen bonds instead of leaving protein donors and acceptors uncompensated in an interface. Since the targets of antibodies and receptors are usually soluble, the target binding surfaces will often be rather hydrophilic. It seems likely from the crystal lattice results that water molecules may play an important role in such interfaces. Indeed, even the tight $\alpha_1\beta_1$ interface of hemoglobin includes a number of water molecules (14 in horse Met) (20).

MOBILITY

That protein molecules are flexible and dynamic is well established. Since the modes and amplitudes of vibration in a molecule depend on the various interatomic interactions that are involved, it is expected that these modes will be changed when proteins are bound in a complex. Similarly, a molecule that exists in a conformational equilibrium between substates might have this equilibrium modified in a complex. Mobility in the surfaces that are destined for an interface may affect the kinetics of binding and may broaden the spectrum of partners for a complex. However, in energetic terms, the reduction of flexibility in the bound interface corresponds to a decrease in entropy and a weaker bond.

Intrinsically rigid partners would be expected to form tighter complexes and it is argued that this accounts for the enhanced affinity of trypsin for trypsin inhibitor over free polypeptides. However, even here there is a reduction of mobility in the complex. As shown in Fig. 3, the inhibitory lysine shows very high group mobility (21). The mobility is greatly reduced for such lysine groups when they are bound in the specificity pocket of the enzyme (22). The mobility of chain segments on an enzyme surface can also be markedly reduced

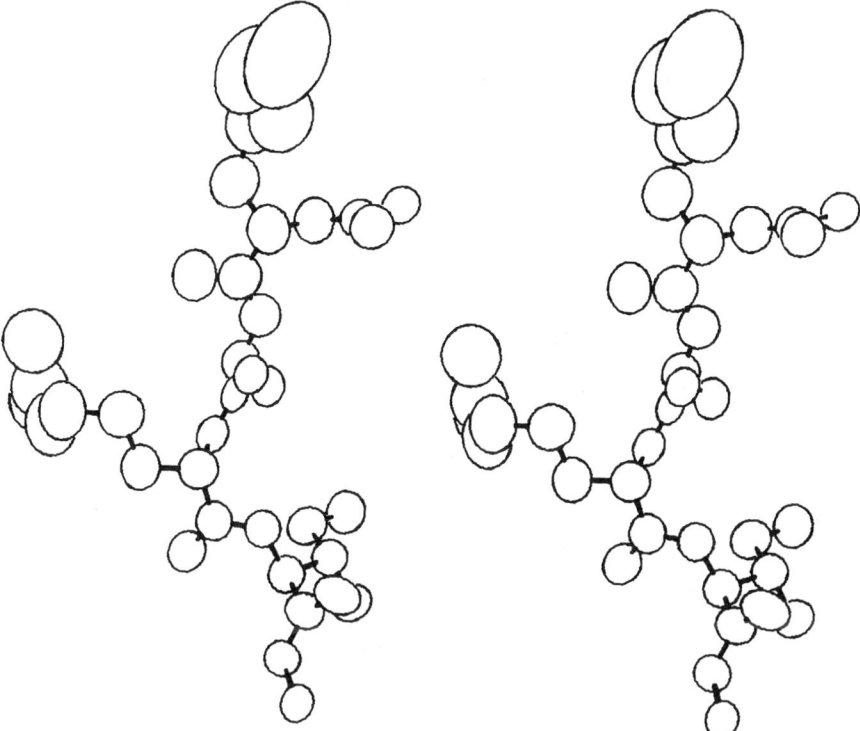

Fig. 3. Stereo drawing of a segment of bovine pancreatic trypsin inhibitor (Lys 15–Ala 16–Arg 17–Ile 18) that includes the inhibitory lysine side chain. Atoms are shown as 50% ellipsoids after subtraction of a lattice disorder component ($\Delta B = 5$ Å^2).

when inhibitors are bound as was shown for *Streptomyces griseus* protease A (23). Our own investigation of mobility of interface groups in hemerythrins also demonstrates the dampening impact of contacts on mobility (24). On the other hand, a surprisingly large fraction of the conformation disorder that we see in side chains occurs in lattice interfaces of the crystals (19).

Examples of conformational changes on complexation are less well documented. However, a striking example comes from studies of the binding of turkey ovomucoid inhibitor third domain to proteases. The binding interactions of the reactive-site loop of this inhibitor are virtually identical in complexes with α-chymotrypsin and with *S. griseus* protease B. However, because of differing interactions in more remote parts of the interface, the relative orientation of the reactive loop to the rest of the inhibitor differs markedly in the two complexes (R. J. Read, personal communication).

ACKNOWLEDGMENTS

This work was supported by grants from the National Science Foundation (PCM-8409658) and from the National Institutes of Health (GM-34102, GM-33954, and HL-34434.)

REFERENCES

1. Royer, W. E., Jr., Love, W. E., and Fenderson, F. F. (1985). *Nature (London)* **316**, 277–280.
2. Chothia, C., and Janin, J. (1975). *Nature (London)* **256**, 705–708.
3. Doty, P., and Myers, G. E. (1953). *Discuss. Faraday Soc.* **13**, 51–58.
4. Kauzmann, W. (1959). *Adv. Protein Chem.* **14**, 1–63.
5. Sturtevant, J. M. (1977). *Proc. Natl. Acad. Sci. U.S.A.* **74**, 2236–2240.
6. Perutz, M. F. (1970). *Nature (London)* **228**, 726–734.
7. Shaeffer, J. R., McDonald, M. J., Turci, S. M., Dinda, D. M., and Bunn, H. F. (1984). *J. Biol. Chem.* **259**, 14544–14547.
8. Ackers, G. K. (1980). *Biophys. J.* **32**, 331–346.
9. Hendrickson, W. A., and Love, W. E. (1971). *Nature (London), New Biol.* **232**, 197–203.
10. Honzatko, R. B., and Hendrickson, W. A. (1986). *Proc. Natl. Acad. Sci. U.S.A.* **83**, 8487–8491.
11. Chothia, C., Wodak, S., and Janin, J. (1976). *Proc. Natl. Acad. Sci. U.S.A.* **73**, 3793–3797.
12. Petruzelli, R., Goffredo, B. M., Barra, D., Bossa, F., Boffi, A., Verzili, D., Ascolig, F., and Chianconi, E. (1985). *FEBS Lett.* **184**, 328–332.
13. Erickson, H. P., and Pantaloni, D. (1981). *Biophys. J.* **34**, 293–309.
14. Steinberg, I. Z., and Scheraga, H. A. (1963). *J. Biol. Chem.* **238**, 172–181.
15. Smith, J. L., Corfield, P. W. R., Hendrickson, W. A., and Low, B. W. (1987). (Submitted for publication.)
16. Hendrickson, W. A., and Teeter, M. M. (1981). *Nature (London)* **290**, 107–113.
17. Honzatko, R. B., Hendrickson, W. A., and Love, W. E. (1985). *J. Mol. Biol.* **184**, 147–164.
18. Sheriff, S., Hendrickson, W. A., and Smith, J. L. (1987). (Submitted for publication.)
19. Smith, J. L., Hendrickson, W. A., Honzatko, R. B., and Sheriff, S. (1986). *Biochemistry* **25**, 5018–5027.
20. Fermi, G., and Perutz, M. F. (1981). "Haemoglobin and Myoglobin, Atlas of Molecular Structure in Biology," Vol. 2. Oxford Univ. Press, London and New York.
21. Yu, H.-A., Karplus, M., and Hendrickson, W. A. (1985). *Acta Crystallogr., Sect. B* **B41**, *191–201*.
22. Bolognesi, M., Gatti, G., Menegatti, E., Gauarneri, M., Marquart, M., Papamokos, E., and Huber, R. (1982). *J. Mol. Biol.* **162**, 839–868.
23. James, M. N. G., Sielecki, A. R., Brayer, G. D., Delbaere, L. T. J., and Bauer, C.-A. (1980). *J. Mol. Biol.* **144**, 43–88.
24. Sheriff, S., Hendrickson, W. A., Stenkamp, R. E., Sieker, L. C., and Jensen, L. H. (1985). *Proc. Natl. Acad. Sci. U.S.A.* **82**, 1104–1107.

16
Contributions of Individual Amino Acids to Protein Stability Determined by X-Ray Crystallographic Analysis of Selected and Directed Mutants of T4 Phage Lysozyme

TOM ALBER, TERRY M. GRAY, LARRY H. WEAVER,
JEFFREY BELL, JOAN A. WOZNIAK, KEITH WILSON,
SUN DAOPIN, AND BRIAN W. MATTHEWS
Institute of Molecular Biology and Department of Physics
University of Oregon
Eugene, Oregon 97403

The lysozyme from bacteriophage T4 is being used as a model system to determine the factors that influence the folding and stability of proteins. The three-dimensional structure of the protein is known and lysozymes with modified properties arising from single amino acid substitutions have been obtained by classical selection techniques as well as by site-directed mutagenesis. By rationalizing the stabilities of the mutant proteins in terms of their observed three-dimensional structures we attempt to quantitate the contributions that single amino acids make to protein stability.

INTRODUCTION

What does a single amino acid contribute to the stability of a globular protein? This is a problem that can be described in a general thermodynamic sense, i.e., one can write expressions for contributions arising from hydrogen bonding, hydrophobic interactions, electrostatics, chain entropy, and so on. Theoretical approaches to evaluating these expressions are hampered by a variety of problems, including inaccuracies in the potential functions, the complexity of solvent water, and the difficulty of modeling the denatured state of the polypeptide. Experimental studies of protein stability do not, in general, provide a method of factoring the observed results into the individual contribution of each amino acid in its unique environment in the native structure.

We are attempting to overcome these difficulties by comparing the structures and stabilities of mutants of phage T4 lysozyme that differ in sequence by a single amino acid. The mutants are of two distinct classes: selected and directed. Selected mutants are identified after "random" chemical mutagenesis using genetic screens to detect lysozymes that are temperature-sensitive (1) or have enhanced thermal stability (2). Directed mutants are obtained by oligonucleotide-directed mutagenesis, which allows the planned substitutions of selected residues with any of the naturally occurring amino acids.

STRUCTURE AND DYNAMICS OF WILD-TYPE LYSOZYME

The structure of the lysozyme from bacteriophage T4 has been determined crystallographically and refined at 1.7 Å resolution (3–5).

The crystallographic refinement gives information concerning the molecular dynamics of the structure in the crystal and, by inference, in solution as well. Figure 1 shows the apparent "thermal displacements" of the backbone of the molecule. Analysis of these individual motions indicates that the lower lobe in the figure undergoes a pronounced "hinge-bending" motion that opens and closes the active site cleft (5). X-ray analysis of the binding of mono-, di-, and trisaccharides to T4 lysozyme originally suggested the necessity of such motion for access of substrate to the "closed" active site cleft observed in the crystal structure (6).

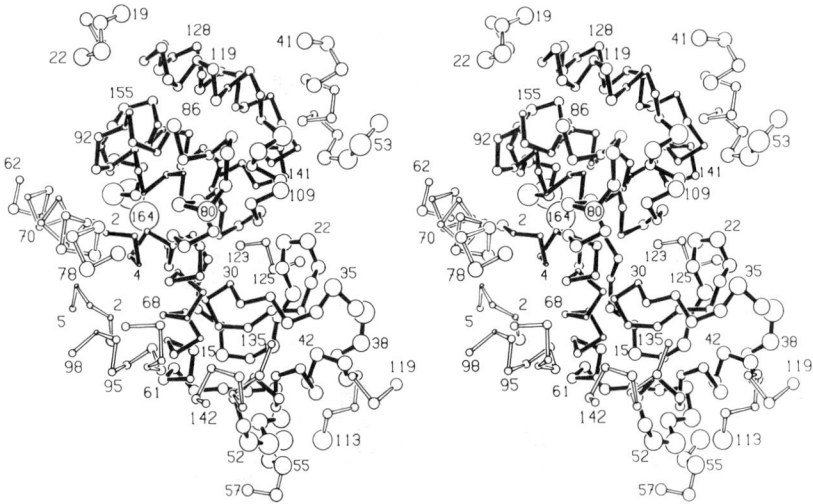

Fig. 1. Stereo drawing of the backbone of T4 phage lysozyme in its crystallographic environment. The circle at each α-carbon position has a diameter proportional to the refined crystallographic "thermal factor." One complete lysozyme molecule is drawn with solid bonds. Segments of neighboring molecules that surround the reference molecule and make contact with it are drawn with open bonds.

TEMPERATURE-SENSITIVE LYSOZYME MUTANTS

Using classical genetic methods developed by Streisinger and co-workers (1) we have obtained a number of temperature-sensitive (ts) lysozymes (7–10). Such ts mutants reveal amino acid substitutions that alter the stability of the lysozyme molecule yet still allow the protein to fold and be active at the permissive temperature. As discussed below, these sites are good candidates for further changes by site-directed methods. The temperature-sensitive lysozymes that we have studied in most detail are listed in Table I. The thermodynamics of unfolding of some of these lysozymes have been determined by Schellman and his co-workers (11).

Structural changes in two of the mutant lysozyme structures that are associated with temperature-sensitive mutations are shown in Fig. 2a and 2b. These figures are based on partially refined crystallographic coordinates. For the surface substitution Thr 157 → Ile there is a rearrangement in the crystallographically observable solvent structure associated with the amino acid substitution. In the case of the

TABLE I
Some Mutant Lysozymes Studied Using Crystallography

Lysozyme	Origin	Activity[a]	ΔT_m[b]	Resolution[c]	R[d]
Wild type	—	100	—	1.7	19.6
Arg 96 → His	ts	100	−14	1.9	19.5
Met 102 → Thr	ts	60	−13	2.1	18.1
Ala 146 → Thr	ts	55	−9	2.1	19.0
Gly 156 → Asp	ts	50	−5	1.7	17.8
Thr 157 → Ile	ts	90	−11	1.9	18.9
Thr 157 → Arg	Site specific	—	−5	1.7	—
Thr 157 → Asn	Site specific	—	−2	1.7	—
Thr 157 → His	Site specific	—	−8	1.7	—
Thr 157 → Ser	Site specific	—	−2.5	1.7	—

[a] Rate of hydrolysis of cell walls relative to wild type.
[b] Melting temperature (°C) relative to wild type, measured at pH 3. [From Alber et al. (10), Hawkes et al. (11), and personal communication from W. A. Baase, W. J. Becktel and S. Cook.]
[c] Resolution of crystallographic data (Å).
[d] Crystallographic residual at the present stage of refinement.

internal substitution Ala 146 → Thr, the larger side chain in the mutant structure causes neighboring groups to move outward to make room for the additional atoms. The indole of Trp 138 moves about 0.7 Å. This adjustment causes a repacking of the side chain of Met 102, which in turn largely fills a cavity that exists in the wild-type structure.

In general, our structural studies of ts mutant lysozymes are consistent with the view that each amino acid might contribute to stability via a variety of interactions, including hydrophobic contacts, hydrogen bonding, and ionic interactions. As discussed in the following section, site-directed mutagenesis can be used to help differentiate between these different contributions.

SITE-DIRECTED MUTAGENESIS

Because of the structures of the natural amino acids, the ts substitution Thr 157 → Ile is especially susceptible to analysis using site-directed mutagenesis. Thr 157 is located in a bend between two α-helices and is partly buried and partly exposed to solvent. The substitution Thr 157 → Ile was identied as a temperature-sensitive lysozyme with melting temperature (at pH 2) 11° lower than that of

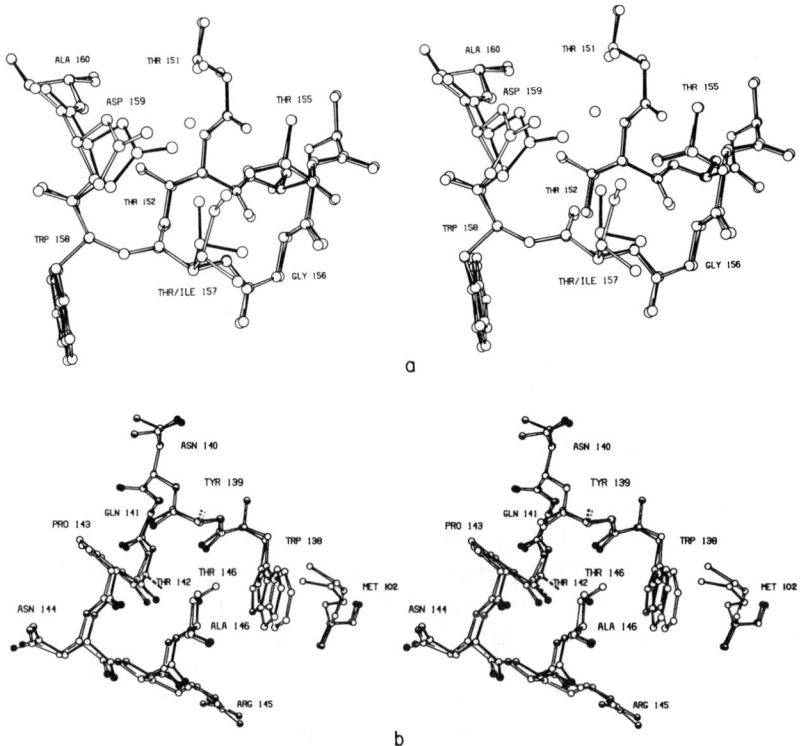

Fig. 2. Stereo drawings showing the structures of mutant lysozymes (open bonds) superimposed on the wild-type structure (solid bonds) in the vicinity of the amino acid substitution. (a) Temperature-sensitive mutant (Ile 157) and wild type (Thr 157). (b) Temperature-sensitive mutant (Thr 146) and wild type (Ala 146).

wild type. The consequences of this substitution are shown in detail in Fig. 2a and in schematic form in Fig. 3.

Several factors could underlie the decrease in stability. These include loss of van der Waals interactions of both the γ-carbon and the γ-hydroxyl, loss of hydrogen bonds (to Thr 155 and from the peptide NH of Asp 159), and changes in the structure of the surrounding solvent. By making a series of substitutions at position 157 (and at position 155 as well), we are attempting to discriminate between these contributing factors. For example, substitution with serine at position 157 should reveal the contribution of the hydrogen bonds to stability. Substitution of valine (or alanine) should show the importance of hydrophobic interactions, and so on. Of course one must also consider

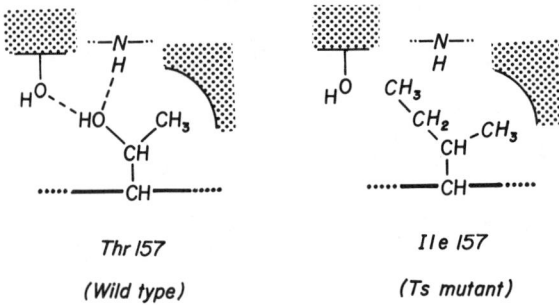

Fig. 3. Schematic drawing showing some of the interactions of threonine 157 in wild-type lysozyme and the consequences of replacing the threonine by isoleucine.

that these substitutions may change the free energy of the unfolded form of the protein.

By using a degenerate mixture of DNA primers it is possible to generate a number of different amino acid substitutions in a single experiment (10,12). Production of large quantities of the mutant lysozymes was facilitated by subcloning the genes into an *Escherichia coli* expression vector based on the *tac* and *lac UV5* promoters (13). The substitutions at position 157 obtained to date include Ser, Leu, Asn, His, and Arg (as well as Ile 157, the original ts mutant) (Table II, Fig. 4). Additional substitutions are in progress. Oscillation photography (14,15) is being used to measure diffraction data to 1.7 Å resolution for the mutant lysozymes. Sections of the difference density maps at 2.0 Å resolution for the two substitutions Thr 157 → Arg and Thr 157 → Asn are shown in Fig. 5.

TABLE II
Relative Stabilities of Lysozymes with Different Substitutions at Position 157

Amino acid at position 157	Melting temperature (°C ± 1°, measured at pH 2) relative to wild type
Threonine (wild type)	—
Asparagine	−2
Serine	−2.5
Leucine	−5
Arginine	−5
Histidine	−8
Isoleucine (ts mutant)	−11

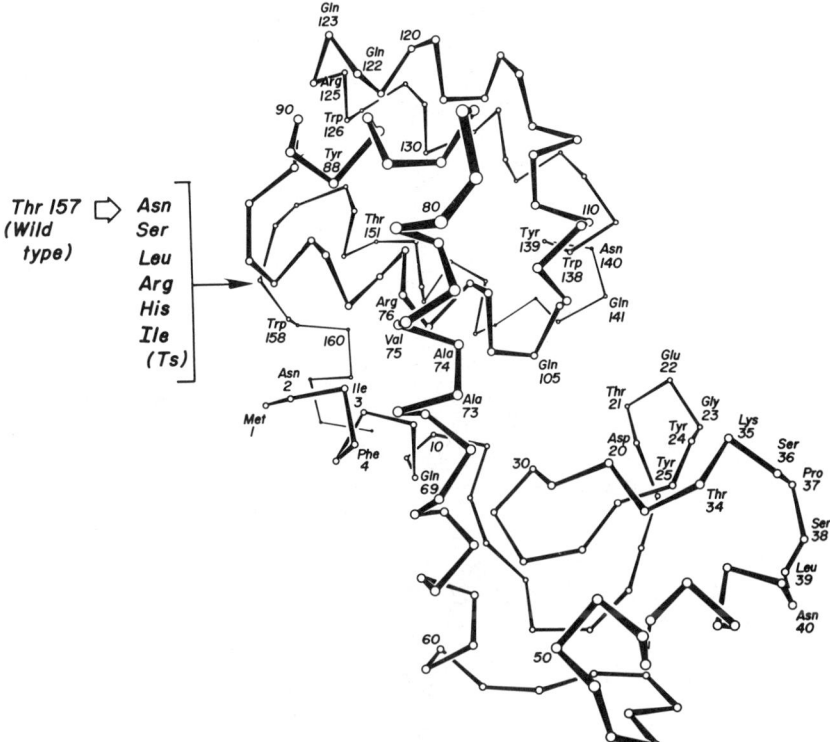

Fig. 4. The backbone of T4 phage lysozyme showing the location of residue 157. The site-directed substitutions shown are those that have been analyzed crystallographically at high resolution.

For the Thr 157 → Arg substitution the loss of the γ-hydroxyl of the threonine is clearly shown by the negative peak at this position. The γ-carbon of the arginine side chain occupies essentially the same position as the γ-carbon of the threonine and the distal end of the arginine side chain then curves around to form a salt linkage with Asp 159. The movement of this acid group toward the guanidinium of the arginine can be seen in Fig. 5a.

When Thr 157 is replaced by Asn, the oxygen of the Asn side chain occupies a position similar to that of the γ-hydroxyl of the threonine and appears to form essentially the same hydrogen bond network as present in the wild-type structure. This is achieved in spite of the difference in geometry of the threonine and asparagine side chains. In contrast to the Thr 157 → Arg substitution, the Thr 157 → Asn re-

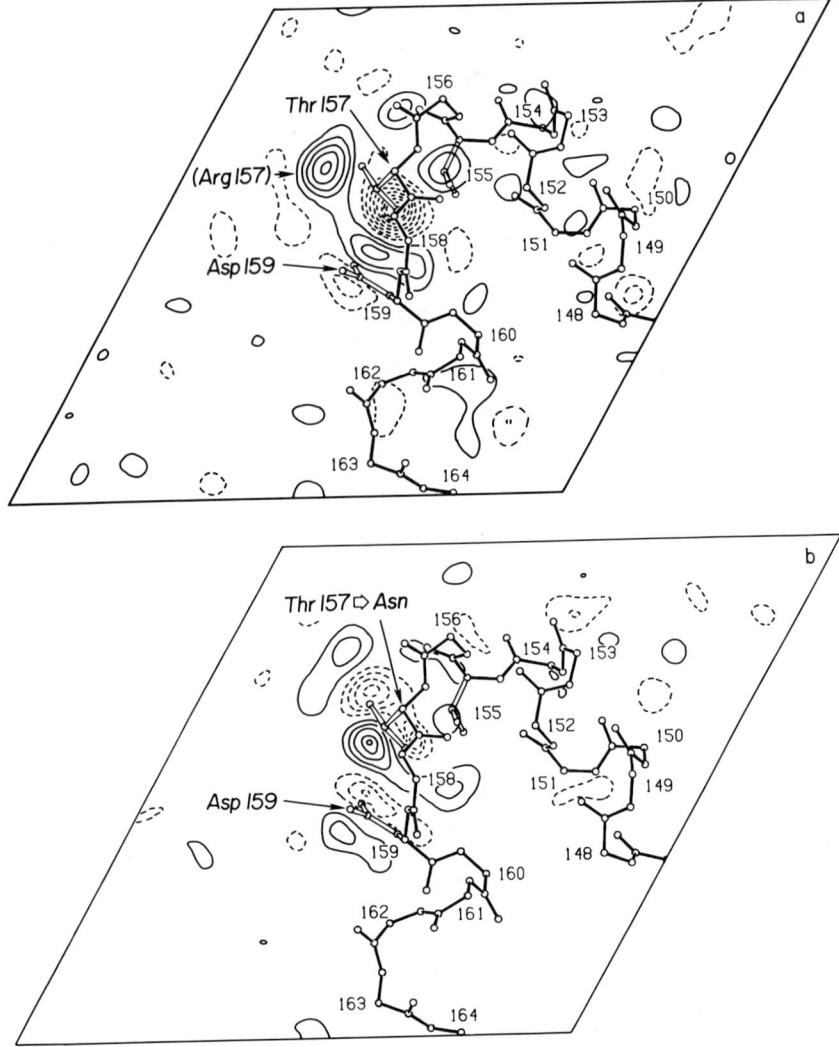

Fig. 5. The figure shows a single section taken from 2.0-Å-resolution difference electron density maps calculated for two site-directed amino acid substitutions. Superimposed on the difference electron density is the trace of the polypeptide backbone that lies near the section. Except for Thr 155, Thr 157, and Asp 159 (open bonds), all protein side chains have been omitted. (Residues 148–155 form part of one α-helix and residues 159–164 are in another helix.) The maps are calculated with amplitudes (F wild type − F mutant) and phases from the refined wild-type structure. Positive contours (solid) and negative contours (broken) are drawn at equal increments of 2σ, where σ is the standard deviation of the difference density throughout the unit cell. (a) Arg 157 minus Thr 157. (b) Asn 157 minus Thr 157.

placement causes Asp 159 to move away from the altered amino acid (Fig. 5b).

CONCLUSIONS

The analysis of the mutant lysozymes described here has a number of implications for protein stability in general.

1. *Thermal stability is global.* The temperature-sensitive mutations that we have characterized to date occur in separated locations. There is not one highly localized region with which temperature sensitivity is associated. The observations are consistent with the hypothesis that protein stability is global, with individual amino acids contributing additively to the overall stability of the protein.

2. *Local environment is important to stability.* In comparing the amino acid substitutions that occur in the different temperature-sensitive mutants, no simple pattern emerges (Table I). In different instances one sees charge changes, hydrophobicity changes, changes in solvent structure, changes in packing, and changes in hydrogen bonding. The consequence of a given mutation depends on the environment and interactions of the substituted amino acid within the folded protein and is not solely due to an effect on the stability of the unfolded state.

3. *Different types of interaction contribute to protein stability.* The variety of amino acid substitutions in different temperature-sensitive mutants shows that different effects, including hydrophobic interactions, electrostatic interactions, packing defects, solvent structure, and hydrogen bonding, all influence stability. Also some substitutions may destabilize the protein by differentially stabilizing the unfolded form.

The substitution Thr 157 → Ile seems to destabilize primarily through loss of hydrogen bonds involving O^γ. The alteration Ala 146 → Thr disrupts the packing within the molecule and introduces a new hydrogen bonding group that may not be easily accommodated by the wild-type structure. Replacement of Arg 96 by histidine causes a net loss of intramolecular hydrogen bonds and also an apparent loss of hydrophobic stabilization of the core of the molecule. We do not observe that temperature sensitivity is necessarily associated with the loss of ion pairs (16).

It has generally been assumed that hydrophobic interactions are of particular importance in thermal stabilization since such interactions are entropy driven (17,18). Chothia (19) has suggested that each

TABLE III
Side Chain Solvent Accessibility

Substitution		Side chain area (mutant-native) ($Å^2$)			
		Exposed to solvent[a]		Inaccessible to solvent[b]	
Nat	Ts	Hydrophilic	Hydrophobic	Hydrophilic	Hydrophobic
Arg 96 → His		−54.8	19.0	− 6.5	−7.5
Met 102 → Thr		4.6	−2.4	31.3	−72.5
Ala 146 → Thr		0	0	36.5	−3.5
Thr 157 → Ile		0	32.3	−36.5	28.0

[a] For a given amino acid the area exposed to solvent is for the side chain atoms and is calculated using the procedure of Lee and Richards (20).

[b] The area inaccessible to solvent is the difference between area exposed by the amino acid side chain when in a fully extended conformation and the area exposed in the folded protein. Hydrophilic surface is that corresponding to oxygen and nitrogen atoms and hydrophobic surface corresponds to carbon and sulfur (in methionine). Each entry is the difference between the surface area of the mutant side chain and the corresponding value for the wild-type side chain.

square angstrom of buried hydrophobic surface contributes 25 calories to the free energy of stabilization. Table III shows the changes in hydrophobic and hydrophilic surface area (20) in the temperature-sensitive mutants relative to the wild-type protein. In different cases the mutant lysozymes have either less hydrophobic surface buried or more hydrophobic surface exposed to solvent, or a combination of both. For one mutant there is little change in the hydrophobic surface, either buried or exposed. The surface area calculations do not support the idea that there is a direct correlation between buried hydrophobic surface area and the stability of the protein. For the substitutions listed in Table III there is also no clear correlation between changes in hydrophilic surface area and stability. It must be emphasized, however, that Table III constitutes a small sample and needs to be expanded before reliable generalizations can be made.

4. *Changes in stability need not be associated with large structural changes.* One of the striking results of this study is the close correspondence between the three-dimensional structures of the different temperature-sensitive mutants and the wild-type structure. For mutants Arg 96 → His and Thr 157 → Ile the amino acid substitutions are on the surface of the protein and the structural changes are quite localized. In the case of the internal substitution Ala 146 → Thr there are changes that propagate through neighboring residues, but such changes are, in general, not more than 0.5–0.8 Å (Fig. 2b). There is no

evidence that temperature sensitivity is necessarily associated with an obvious, dramatic change in the structure of the protein.

However, it should be noted that the crystallographic analyses (to date) are at room temperature, well below the melting temperature of wild-type lysozyme (54°C at pH 3). At room temperature we do not see any obvious changes in fluctuations or "looseness" of the mutant structures relative to the wild-type structure, as has been inferred from the compensating changes that are observed in the entropies and enthalpies of unfolding (11).

5. *Stability of secondary structure.* One way of stabilizing proteins could be to stabilize the folded conformations of individual α-helices and β-sheets, and perhaps turns as well (e.g., see ref. 21). Since T4 lysozyme has a high proportion of α-helices, this idea can be tested by evaluating the changes in helix stabilization associated with the amino acid substitutions in the temperature-sensitive mutants.

In Table IV we give two measures of helix propensity for the mutant lysozymes. The first measure, P_α, is the relative frequency of occurrence of a given residue in an α-helix as determined from a large number of known protein structures (22,23). The second measure, s, is the Zimm–Bragg helix-coil stability constant as determined by Scheraga and co-workers (24). For all temperature-sensitive mutations within helices the helical propensity either decreases or is ambiguous, whereas for the one example of a substitution not in a helix the helical propensity increases in the mutant.

TABLE IV
Changes in Helical Propensity for Mutant Lysozymes

Substitution	Location	P_α(mutant)/P_α(wild type)[a]		s(mutant)/s(wild type)[b]
		(A)	(B)	
Arg 96 → His	Helix	1.02	1.27	0.83
Met 102 → Thr	Helix	0.57	0.56	0.68
Ala 146 → Thr	Helix	0.58	0.64	0.77
Glu 128 → Lys	Helix	0.77	0.85	0.97
Thr 157 → Ile	Surface loop	1.30	1.18	1.39

[a] P_α is the relative frequency of occurrence in α-helices. (A) Data from Chou and Fasman (22). (B) Data from Levitt (23).

[b] s is the Zimm–Bragg helix-coil stability constant as determined by Scheraga and co-workers (24).

6. *Site-directed mutagenesis.* A series of different amino acid substitutions at a single locus can be used to probe the contributions of individual substituents to protein stability. We have initially investigated the role of Thr 157 of phage T4 lysozyme. The relative stabilities of six mutant lysozymes with substitutions at position 157 are given in Table II. Serine at this position stabilizes the protein more than isoleucine and raises the melting temperature to within 2.5° of the wild-type protein. Therefore the hydroxyl group of Thr 157 contributes more to the stability of the protein than does the methyl group.

The finding that the γ-OH of Ser 157 is sufficient to raise the T_m of T4 lysozyme by 8.5°C relative to Ile at this position suggests that hydrogen bonds can directly contribute to the stability of proteins. These results are not in agreement with the view that surface hydrogen bonds in the folded state are equivalent to those between the unfolded protein and water and consequently contribute to the specificity of folding but not to the stability of the native conformation (25). The contribution of a particular hydrogen bond to the free energy of the folded conformation may arise from an improvement in the average bond geometry or from an increase in the probability that the donor and acceptor in the folded protein are interacting. This predicts that each hydrogen bond in a protein makes a different contribution to overall stability. Donors and acceptors that are rarely or poorly paired in the folded state relative to their interactions with water in the unfolded state may even be destabilizing. All the substitutions made to date at position 157 have stabilities between that of the wild-type protein and the original temperature-sensitive mutant. The observation that no substitution is more stable than threonine suggests that in this case the wild-type protein has achieved optimum stability at this site. However, we believe it unlikely that proteins have evolved to have maximum stability. Indeed, it is possible to select lysozyme mutants that are more thermostable than the wild-type protein (2).

To determine the basis for the stability of each of the mutant lysozymes listed in Table II it will be necessary to determine their three-dimensional structures. This is in progress. In the case of Thr 157 → Arg (Fig. 5a), the arginine side chain loses van der Waals and/or hydrogen bonding interactions because it has no counterpart to the γ-oxygen of the wild-type threonine. However, it appears to compensate, in part, for this loss by the formation of a new salt bridge with Asp 159 (Fig. 5a). Although this salt bridge contains two hydrogen bonds between the side chain carboxyl and guanidinium moieties, it does not stabilize the protein to the same extent as the interactions of the wild-type threonine 157 side chain. Thus, the contribution of hydrogen bonds to protein stability apparently does not follow the pattern recently reported by Fersht et al. (26) for the free energies of hydrogen bonds involved in the binding of substrates to alanyl amino acyl tRNA synthetase from *Bacillus stearothermophilus*. These workers found that hydrogen bonds involving uncharged donors and acceptors contribute 0.5–1.5 kcal/mol of binding energy whereas those involving charged groups can contribute up to an additional 3 kcal/mol (26).

7. *Protein evolution.* One of the striking results of the site-directed mutagenesis is that most of the mutant lysozymes have stabilities closer to the wild-type (Thr 157) protein than to the temperature-sensitive mutant (Ile 157). In addition, amino acids that give rise to the most stable proteins need not be chemically or structurally similar. The protein must compensate in different ways for these chemically disparate substitutions. This suggests that proteins may have evolved to accommodate amino acid substitutions that may occur during evolution. Protein structures may be more resistant to destabilization by amino acid substitution than has been imagined, and mutations that lead to temperature sensitivity could be the exception rather than the rule.

ACKNOWLEDGMENTS

We are grateful to Drs. D. C. Muchmore and F. W. Dahlquist for providing the lysozyme expression system and thank these colleagues as well as Drs. W. J. Becktel, W. A. Baase, J. A. Schellman, and Mr. Lawrence McIntosh for discussions concerning the project in general.

This work was supported in part by grants from the National Institutes of Health (GM 21967, GM 20066), the National Science Foundation (PCM 8312151), and the Murdock Charitable Trust.

REFERENCES

1. Streisinger, G., Mukai, F., Dreyer, W. J., Miller, B., and Horiuchi, S. (1961). *Cold Spring Harbor Symp. Quant. Biol.* **26**, 25.
2. Alber, T., and Wozniak, J. A. (1985). *Proc. Natl. Acad. Sci. U.S.A.* **82**, 747.
3. Matthews, B. W., and Remington, S. J. (1974). *Proc. Natl. Acad. Sci. U.S.A.* **71**, 4178.
4. Remington, S. J., Anderson, W. F., Owen, J., Ten Eyck, L. F., Grainger, C. T., and Matthews, B. W. (1978). *J. Mol. Biol.* **118**, 81.
5. Weaver, L. H., and Matthews, B. W. (1987). *J. Mol. Biol.* **193**, 189–199.

6. Anderson, W. F., Grütter, M. G., Remington, S. J., Weaver, L. H., and Matthews, B. W., (1981). *J. Mol. Biol.* **147**, 523.
7. Grütter, M. G., Hawkes, R. B., and Matthews, B. W. (1979). *Nature (London)* **277**, 667.
8. Grütter, M. G., Weaver, L. H., Gray, T., and Matthews, B. W. (1983). *In* "Bacteriophage T4" (C. K. Matthews, E. Kutter, G. Mosig, and P. Berget, eds.), pp. 356–360. Am. Soc. Microbiol., Washington, D.C.
9. Grütter, M. G., Gray, T. M., Weaver, L. H., Alber, T., Wilson, K. and Matthews, B. W. (1987). *J. Mol. Biol.* (Submitted for publication.)
10. Alber, T., Grütter, M. G., Gray, T. M., Wozniak, J. A., Weaver, L. H., Chen, B.-L., Baker, E. N., and Matthews, B. W. (1986). *UCLA Symp. Mol. Cell. Biol. New Ser., Protein Struct. Folding Des.* (D. L. Oxender, ed.), Vol. 39, pp. 307–318. Alan R. Liss, New York.
11. Hawkes, R., Grütter, R., and Schellman, J. (1984). *J. Mol. Biol.* **175**, 195.
12. Alber, T., and Matthews, B. W. (1987). *In* "Methods in Enzymology," Vol. 154: Recombinant DNA, Part E (R. Wu and L. Grossman, eds.) (in press).
13. Muchmore, D. C., Russell, C. B., and Dahlquist, F. W. (1987). In preparation.
14. Rossmann, M. G. (1979). *J. Appl. Crystallogr.* **12**, 225.
15. Schmid, M. F., Weaver, L. H., Holmes, M. A., Grütter, M. G., Ohlendorf, D. H., Reynolds, R. A., Remington, S. J., and Matthews, B. W. (1981). *Acta Crystallogr., Sect. A* **A37**, 701.
16. Perutz, M. G., and Raidt, H. (1975). *Nature (London)* **255**, 256.
17. Kauzmann, W. (1959). *Adv. Protein Chem.* **14**, 1.
18. Brandts, J. F. (1967). *In* "Thermobiology" (A. H. Rose, ed.), p. 25. Academic Press, New York.
19. Chothia, C. H. (1974). *Nature (London)* **248**, 338.
20. Lee, B., and Richards, F. M. (1971). *J. Mol. Biol.* **55**, 379.
21. Argos, P., Rossmann, M. G., Grau, U. M., Zuber, H., Frank, G., and Tratschin, J. D. (1979). *Biochemistry* **18**, 5698.
22. Chou, P. Y., and Fasman, G. D. (1978). *Annu. Rev. Biochem.* **47**, 251.
23. Levitt, M. (1978). *Biochemistry* **17**, 4277.
24. Sueki, M., Lee, S., Powers, S. P., Denton, J. B., Konishi, Y., and Scheraga, H. A. (1984). *Macromolecules* **17**, 148.
25. Klotz, I. M., and Franzen, J. S. (1962). *J. Am. Chem. Soc.* **84**, 3461–3466.
26. Fersht, A. R., Shi, J.-P., Knill-Jones, J., Lowe, D. M., Wilkinson, A. J., Blow, D. M., Brick, P., Carter, P., Waye, M. M. Y., and Winter, G. (1985). *Nature (London)* **314**, 235–238.

PART IV

LIGHT-SENSITIVE PROTEINS

17
The Rhodopsins of *Halobacterium halobium*

WALTHER STOECKENIUS
*Cardiovascular Research Institute
University of California
San Francisco, California 94143*

INTRODUCTION[1]

Halobacteria live in concentrated brines and require NaCl concentrations near saturation for growth. They balance the high osmotic strength of the outside medium by maintaining a nearly equally high KCl concentration inside. Their main energy source is respiration, which generates an outside-positive electrochemical proton gradient across their cell membrane. This gradient drives locomotion, ATP synthesis, and an electrogenic Na^+/H^+ exchange. The resulting sodium gradient in turn drives amino acid uptake and probably other energy-requiring processes. Potassium is the most permeant ion and

---

[1] This is not a historical and comprehensive account of the discovery of the retinal proteins of halobacteria, it is rather an attempt to logically present the key findings that support our present view of their structure and function. The actual progress in these investigations followed a much more tortuous course than depicted here. Especially the early results in the investigation of halorhodopsin were confounded by the then unrecognized presence of slow rhodopsin. Since this review was written, the presence of a secondary sensory rhodopsin in the same membranes has been firmly established. It covers the blue–green region of the spectrum (18–22).
  Similar results were often obtained almost simultaneously by several laboratories, but led to different interpretations. Other groups involved in this work are those of Dieter Oesterhelt, Janos Lanyi, Tom Ebrey, Eilo Hildebrand, Toru Yoshizawa, Yonosuke Kobatake, and Yasuo Mukohata. Their conclusions do not always agree with the view presented here; a detailed discussion of the main discrepancies can be found in references 23–25.

is concentrated in the cell as a result of the outside-positive electric potential. The cells can use light as an alternative energy source because of the presence of bacteriorhodopsin (bR) in the cell membrane. This 26-kDa retinylidene protein has a broad absorption band in the green region of the spectrum and functions as a light-driven proton pump, so that it can substitute for respiration by generating an identical electrochemical force. Bacteriorhodopsin is well known and will not be further described here; for a review, see Stoeckenius and Bogomolni (1).

HALORHODOPSIN

For some time, light-driven proton translocations had been observed in halobacteria, which could not be attributed to bR unless some rather unlikely assumptions were made, e.g., misorientation of a substantial part of bR molecules and/or different effects of uncouplers on both sides of the membrane (2,3).[1] Illumination of an anaerobic cell suspension because of the activity of bR results in an acidification of the medium. As expected, the protonophore, carbonyl cyanide m-chlorophenyhydrozone (CCCP), increases the proton permeability of the membrane and in increasing concentrations reduces the acidification and finally abolishes it. However, at still higher concentrations, light-driven influx of protons is observed. MacDonald (4,5) demonstrated that this CCCP-facilitated, light-driven proton influx is a secondary effect and due to a light-generated, outside-positive membrane potential. He assumed that the primary translocated ion was Na^+, but Schobert and Lanyi (6) later showed that it is Cl^- transported into the cell. This implied that the cells in addition to the light-driven proton pump bR also contained a light-driven chloride pump, which was named halorhodopsin (hR). Action spectra for the CCCP-facilitated proton inflow were similar to those for bR-driven proton outflow, but showed a 10 to 20-nm red-shifted maximum. Inhibition of retinal synthesis by nicotine abolished both fluxes and addition of retinal restored them, indicating that hR like bR has a retinal chromophore. Its action spectrum and the effect of borohydride and hydroxylamine, which abolish its activity, suggested that the retinal also forms a protonated Schiff base with an amino group of the protein. The presence of hR in *Halobacterium halobium* had been overlooked for a long time because its concentration in the cell membranes of the halobacterium strains generally used is much lower, 10,000–20,000 molecules per cell compared to 150,000–250,000 for bR.

Since hR is a light-driven ion pump it must undergo a cyclic photoreaction and, as in bR, this can be observed by time-resolved absorption spectroscopy of whole cells or isolated cell membranes. The transient absorbance changes in the photocycle or the change in absorbance of hydroxylamine-bleached membrane on the addition of retinal can be used to obtain a more accurate approximation of the absorption spectrum than is possible using the proton fluxes. These experiments yield an absorption maximum at 578 nm (Fig. 1). Figure 2 compares the absorption of bR and hR 1 msec after exposure to a flash of green light. The differences are pronounced in the 500- to 600-nm region and the far blue-shifted intermediate M_{412} has no counterpart in the hR cycle. With some assumptions, a tenative photoreaction cycle has been constructed from such experiments (Fig. 3). While generally similar to the well-investigated photoreactions of bR, the lack of a short-wavelength intermediate suggests that, unlike bR, hR does not deprotonate its Schiff base during the photoreaction cycle. Also, in contrast to bR, both the absorption spectra and photocycle kinetics are sensitive to the concentration and species of anions present (7). Data given here are for ≥ 150 mM Cl$^-$ concentrations. One or more apparently deprotonated forms of the chromophore with an absorption maximum at 410 nm can arise at alkaline pH and/or exten-

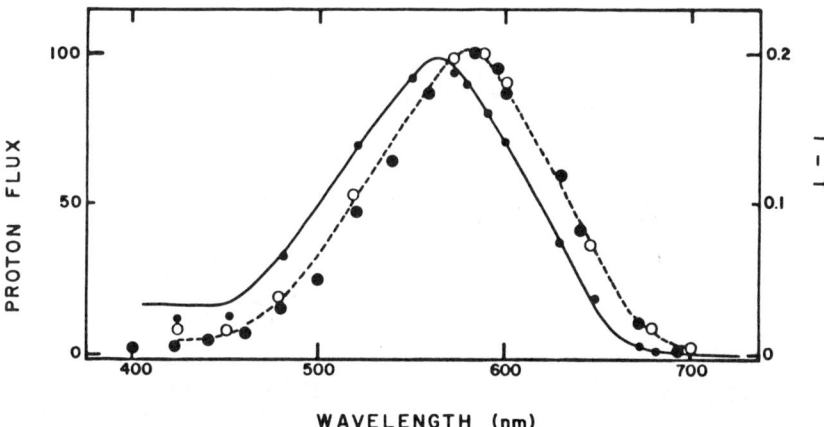

Fig. 1. A comparison of the action spectra for bR and hR. The solid line is the absorption spectrum of bR and the small dots indicate the absorbance changes as a function of actinic wavelength. The open circles are the corresponding data for hR and the solid circles indicate the extent of the light-driven, uncoupler-enhanced proton influx.

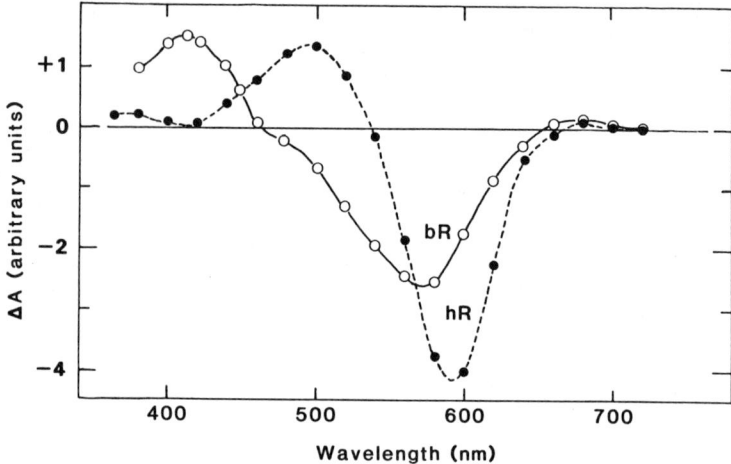

Fig. 2. Flash-induced difference spectra for hR (●) and bR (○) in membrane suspensions. The spectra 1 msec after the actinic flash were subtracted from the spectra before the flash.

sive long-wavelength illumination. Their physiological significance is still debated.

The transient absorbance changes caused by green light are a sensitive test for the presence of hR and we have used them to trace it through cell and membrane fractions and finally isolate it in essen-

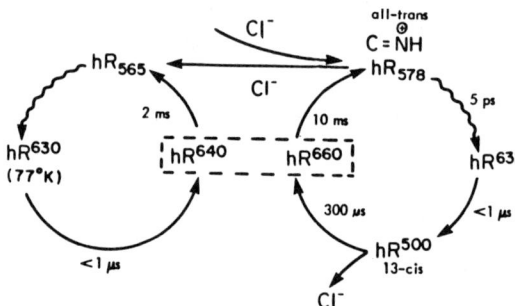

Fig. 3. Tentative photoreaction cycles for hR in the presence (right) and absence (left) of chloride. The times given are halftimes at room temperature. The subscripts indicate the absorption maxima and the superscripts the maxima of the difference spectra, i.e., the intermediate spectrum minus the initial state spectrum. Conformation of the retinal and protonation state of the Schiff base are indicated where known. It is possible that hR^{640} and hR^{660} are identical, which would yield a Cl^--pumping cycle. Note that hR^{632} has only been observed at 77°K but is likely to have a slightly blue-shifted absorption maximum.

tially pure form. The absorption spectrum of the isolated pigment, as expected, peaks at 578 nm and is in all other respects very similar to that of bR. Sodium dodecyl sulfate–polyacrylamide gel electrophoresis (SDS–PAGE) of the purified preparations shows a band migrating slightly faster than bR, indicating a molecular weight of ~25 kDa (8).

While the isolated pigment is apparently spectroscopically intact, it must still be tested for its competence to function as an ion pump. Incorporation of the purified pigment into lipid vesicles resulted in a preparation exhibiting light-driven, CCCP-facilitated proton efflux, which was abolished by addition of TPT, an electroneutral Cl^-/OH^- exchanging ionophore. Apparently the 25-kDa protein is functioning as the light-driven Cl^- pump, but it is incorporated into the lipid vesicles inside out, as is bR under the same conditions.

The most striking feature of hR is its similarity to bR. This is further emphasized by a comparison of their resonance Raman spectra. They indicate that both pigments in the ground state of their photoreaction cycle contain all-trans retinal linked by a protonated C=N anti-bond to the protein. Moreover, the distribution of fixed charges around the retinal must be very similar. A difference may be a slightly weaker hydrogen bond of the Schiff base to its counterion in hR (9). The apparent structural similarity between the two proteins makes their rather different functions difficult to understand and may force us to reexamine our models for proton transport in bR. A satisfactory model for Cl^- transport faces the difficulties that Cl^- has to be moved against a substantial electric potential and released into an approximately 4 M Cl^- solution.

The physiological role of hR is probably osmoregulation. A growing cell must take up anions to maintain its internal osmotic pressure, which requires a pump; the main cation K^+ can enter the cell driven by the electric potential. The question of how cells growing under aerobic conditions in the dark take up Cl^- remains to be answered.

SENSORY RHODOPSIN

Both bR and hR are light energy transducers; they use absorbed light energy to drive transport and other metabolic processes uphill. However, halobacteria also exhibit a light signal transducer function: they are phototactic. In darkness or constant light, they swim for a few seconds in a nearly straight line and then spontaneously reverse, changing direction by approximately 180°; this pattern continues indefinitely until the intensity and/or the wavelength of the light change. An increase in red or green light intensity suppresses reversals and a decrease induces reversals; the inverse holds true for blue and near-UV light. The cells, therefore, swim toward a source of long-wavelength light and away from short-wavelength light. A maximum in the action spectrum of the long-wavelength responses occurs between 550 and 600 nm and near 370 nm for the short-wavelength responses. Inhibition of retinal synthesis abolishes both phototactic responses and reconstitution with retinal or retinal analogues has shown that the receptor(s) must have retinal chromophores (10–12). Originally, bR or hR was thought to mediate the long-wavelength response and an unidentified retinal pigment, absorbing in the near UV, the short wavelength response. However, the isolation of mutants which lacked both bR and hR but retained both phototactic responses showed that this assumption was at least partially wrong (13).

It was obvious that the cells must contain at least one more retinal pigment. Flash spectroscopy of the bR^-hR^- mutants quickly revealed its presence, and reconstitution with retinal of membranes from retinal synthesis-inhibited cells or hydroxylamine-bleached membranes showed that its absorption maximum is at 587 nm (14). Its photoreaction cycle is characterized by an intermediate state with an absorption maximum at 373 nm (S_{373}), which rises in 250 μsec and decays with a halftime of ~800 msec. The intermediate, therefore, accumulates in significant amounts under physiological illumination conditions. S_{373} is converted back to the original state by blue and near-UV light in ~80 msec. Its photosteady-state concentration, therefore, depends strongly on the illumination intensity and wavelength; the pigment is photochromic. Preliminary results from labeling experiments and partial purification show that its molecular weight is nearly identical to that of hR. It is, however, present in much lower concentrations than the other two pigments: ~5000 molecules per cell compared to ~20,000 for hR and 250,000 for bR in the *H. halobium* strains with which most experiments have been carried out. The pigment has been named slow or sensory rhodopsin (sR).

The presence under physiological photosteady-state conditions of two forms of sR absorbing maximally at 587 and 373 nm suggests that sR may serve as the receptor for the phototactic responses to both short- and long-wavelength light. We assume that the generation or thermal decay of S_{373} causes an attractant response and its photoconversion a repellent response (Fig. 4). The model predicts that the sensitivity to short-wavelength light should depend on background illumination with long-wavelength light, and that background illumination with high intensity, short-wavelength light should convert an

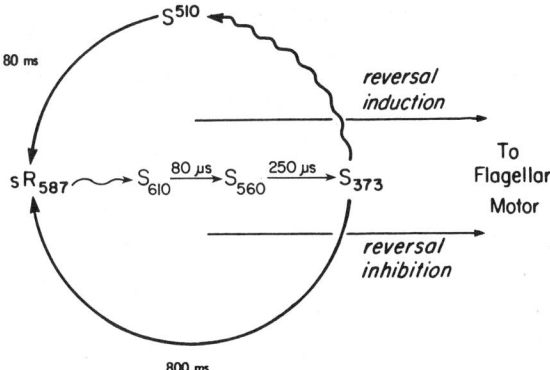

Fig. 4. The photocycle of sR and its effect on the direction of flagellar rotation. Subscripts and superscripts have the same meaning as in Fig. 3.

attractant long-wavelength stimulus into a repellent stimulus. The reason for the first prediction is that in the absence of light absorbed by the sR_{587} state, S_{373} does not exist. In the second case, practically all the pigment will be in the sR_{587} state, but any S_{373} formed when illumination with long-wavelength light is added will be immediately photoconverted to sR_{578} and therefore elicit a repellent response. Both predictions are borne out by experiment (15).

The phototactic response is very sensitive. On the basis of the model proposed, the molar absorption coefficient for sR obtained from reconstitution with retinal, and the sensitivity of the response, the photoconversion of one molecule of S_{373} is sufficient to trigger a reversal of swimming direction in a cell of *H. halobium*. The nature of the signal generated by the pigment is unknown. We have failed to detect any change in membrane potential when cells containing sR but not bR and hR were illuminated with visible or near-UV light, which strongly suggests that the receptor for the phototactic response is not a light-driven, electrogenic pump.

CONCLUSIONS

We thus find in halobacteria three retinal proteins which are apparently very similar in structure but have very different functions. All three are clearly designed to allow the use of light energy; two, bR and hR, by directly converting light energy into electrochemical gradients and the third, sR, by allowing the cell to move into an environment optimal for the light energy-converting function of the other two. It seems, therefore, unlikely that they have arisen independently.

It is not obvious, however, how a proton pump, consisting only of a 25-kDa protein, could be readily converted into a chloride pump or generate a nonelectrogenic signal for a tactic response. Further investigation on the relationship between these three proteins and the molecular mechanisms of their functions should prove interesting. Similar retinal proteins appear to be rather rare in nature, having been found only in the animal eye and a unicellular alga (16). The problem of a possible evolutionary relation to the visual pigments of animals is briefly discussed elsewhere (17).

ACKNOWLEDGMENTS

Work from our laboratory reported here has been supported by NIH Grant GM-27057 and NASA Grant NSG-7151. I thank Dr. Roberto Bogomolni for Figs. 1 and 2 and a critical reading of the manuscript.

REFERENCES

1. Stoeckenius, W., and Bogomolni, R. A. (1982). *Proc. Int. Congr. Biochem., 12th, 1982*, Abstracts, p. 296.
2. Kanner, B. I., and Racker, E. (1975). *Biochem. Biophys. Res. Commun.* **64**, 1054–1061.
3. Matsuno-Yagi, A., and Mukohata, Y. (1977). *Biochem. Biophys. Res. Commun.* **78**, 237–243.
4. Lindley, E. V., and MacDonald, R. E. (1979). *Biochem. Biophys. Res. Commun.* **88**, 491–499.
5. MacDonald, R. E., Greene, R. V., Clark, R. D., and Lindley, E. V. (1979). *J. Biol. Chem.* **254**, 11831–11838.
6. Schobert, B., and Lanyi, J. K. (1982). *J. Biol. Chem.* **257**, 10306–10313.
7. Stoeckenius, W., and Bogomolni, R. A. (1982). *Annu. Rev. Biochem.* **52**, 587–615.
8. Taylor, M. E., Bogomolni, R. A., and Weber, H. J. (1983). *Proc. Natl. Acad. Sci. U.S.A.* **80**, 6172–6176.
9. Smith, S. O., Marvin, M. J., Bogomolni, R. A., and Mathies, R. A. (1984). *J. Biol. Chem.* **259**, 12326–12329.
10. Spudich, J. L., and Stoeckenius, W. (1979). *Photobiochem. Photobiophys.* **1**, 43–53.
11. Dencher, N. A., and Hildebrand, E. (1979). *Z. Naturforsch., C: Biosci.* **34C**, 841–847.
12. Sperling, W., and Schimz, A. (1980). *Biophys. Struct. Mech.* **6**, 165–169.
13. Spudich, E. N., and Spudich, J. L. (1982). *Proc. Natl. Acad. Sci. U.S.A.* **79**, 4308–4312.
14. Bogomolni, R. A., and Spudich, J. L. (1982). *Proc. Natl. Acad. Sci. U.S.A.* **79**, 6250–6254.

15. Spudich, J. L., and Bogomolni, R. A. (1984). *Nature (London)* **312**, 509–513.
16. Foster, K. W., Saranak, J., Patel, N., Zarilli, G., Okabe, M., Kline, T., and Nakanishi, K. (1984). *Nature (London)* **311**, 756–759.
17. Stoeckenius, W. (1985). *Trends Biochem. Sci.* **10**, 483–489.
18. Takahashi, T., Tomioka, H., Kamo, N., and Kobatake, Y. (1985). *FEMS Microbiol. Lett.* **28**, 161–164.
19. Wolff, E. K., Bogomolni, R. A., Scherrer, P., Hess, B., and Stoeckenius, W. (1986). *Proc. Natl. Acad. Sci. U.S.A.* **83**, 7272–7276.
20. Sundberg, S. A., Alam, M., and Spudich, J. L. (1986). *Biophys. J.* **50**, 895–900.
21. Spudich, E. N., Sundberg, S. A., Manor, D., and Spudich, J. L. (1986). *Proteins: Struct. Funct. Genet.* **1**, 239–246.
22. Scherrer, P., McGinnis, K., and Bogomolni, R. A. (1986). *Proc. Natl. Acad. Sci. U.S.A.* **84**, 402–406.
23. Dencher, N. A. (1983). *Photochem. Photobiol.* **38**, 735–767.
24. Hildebrand, E., and Schimz, A. (1986). *Trends Biochem. Sci.* **11**, 402.
25. Stoeckenius, W. (1986). *Trends. Biochem. Sci.* **11**, 402–403.

18
Structure and Function in Photosynthetic Reaction Centers

STEVEN G. BOXER
Department of Chemistry
Stanford University
Stanford, California 94305

INTRODUCTION

The initial photochemical steps in photosynthesis take place in a membrane-bound pigment protein complex known as the reaction center (RC). RC preparations are available from many photosynthetic organisms, and those isolated from the photosynthetic bacteria are the simplest and best defined chemically and structurally. Recently, Michel succeeded in crystallizing the RC from *Rhodopseudomonas viridis* (1), and a full X-ray structural analysis is in progress (2,2a). Most physical and mechanistic studies have been performed on RCs from a related species, *Rhodopseudomonas sphaeroides*, R-26 (3), for which there are, as yet, no structural data. The chemical compositions of the RCs from the two species are related, but there are significant differences. Both contain four bacteriochlorophylls, two Mg-free bacteriopheophytins, two quinones, one nonheme iron, and three polypeptides. The chromophores in *R. sphaeroides* are a-type bacteriochlorophylls and both quinones are ubiquinone, whereas the chromophores in *R. viridis* are b-type and one of the quinones is menaquinone (4).

It is generally believed that the mechanism of the initial photoinduced charge separation in both species is closely related. Two of the bacteriochlorophylls are closely associated as a dimer (5) and together serve as the primary electron donor, generally denoted P. The initial electron acceptor is less clear, however, it is likely that one monomer bacteriochlorophyll and one bacteriopheophytin are involved. The electron acceptor is called I (intermediate acceptor) and the formation of the initial charge-separated ion pair, $P^{+\cdot}I^{-\cdot}$, occurs within, at most, a few picoseconds following photo-excitation of P (6,7). The electron on $I^{-\cdot}$ moves on within several hundred picoseconds to reduce the primary quinone, Q_A [ubiquinone in *R. sphaeroides*, menaquinone in *R. viridis* (4)]. From here, the electron moves on to a second quinone, Q_B (ubiquinone in both species), which accumulates two electrons and two protons before entering the quinone pool. At the same time, $P^{+\cdot}$ is reduced by bound cytochrome, completing the photocatalytic cycle.

The X-ray structural data published to date for *R. viridis* (2,2a) both clarify and complicate this simplified mechanism. The dimeric nature of P is clearly seen, and the quinone is found near the opposite side of the protein complex from P, presumably on the other side of the membrane. A monomeric bacteriochlorophyll and monomeric bacteriopheophytin are found arrayed in sequence between the dimer and the quinone, with the bacteriochlorophyll closer to the dimer. Unexpectedly, this arrangement is duplicated, so that two nearly equivalent electron transfer pathways, related by a pseudo-C_2 axis, are evident. Why the charge separation occurs along one pathway rather than the other is mysterious as of this writing; however, the picture may become much clearer when the full protein structure is completed and refined.

The initial charge separation steps are summarized schematically in Fig. 1, where no attempt is made to chemically identify I or to consider the two possible electron transfer pathways. It is simple to remove successive electron acceptors in the electron transport chain (3), blocking charge separation at that point and permitting measurements of the recombination times, which are also shown in Fig. 1. The striking result is that each forward charge separation rate exceeds the recombination rate by several orders of magnitude, leading to an overall high efficiency of the sequence of forward reactions. Although kinetic spectroscopy provides essential data on the time scale of various transformations in the scheme (6,7), it does not provide insight into the energetics of intermediate states and often gives ambiguous results on the identities of intermediates. Furthermore, as will be discussed below, the kinetics in blocked RCs are not simple because of the presence of electron spins on the intermediates.

Based on theoretical considerations originating with Marcus nearly

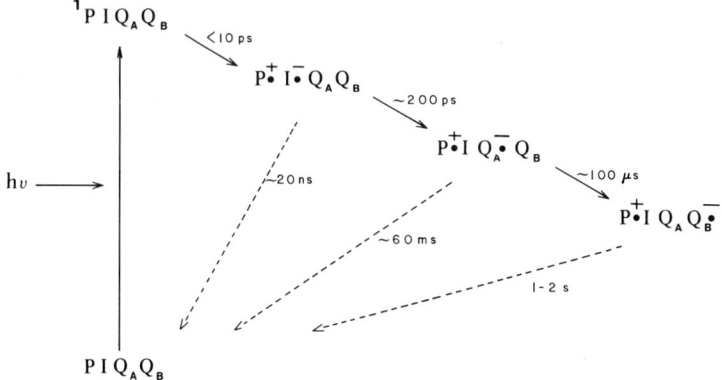

Fig. 1. Kinetic scheme for the primary intermediates in bacterial photosynthesis. P is the primary electron donor (bacteriochlorophyll dimer), I is the intermediate acceptor (monomeric bacteriochlorophyll and/or bacteriopheophytin), Q_A is the strongly bound quinone, and Q_B is the weakly bound quinone.

30 years ago (8) and numerous experimental tests in model systems (9,10), we expect that the kinetics of electron transfer will be closely related to the free energy change for each step. Other key factors which determine the rate of electron transfer include the distance between donor and acceptor, their mutual orientation, changes in the structure of the donor, acceptor, and nearby solvent (primarily the protein) when charge is separated, and the electronic coupling between the donor and acceptor through the intervening protein medium. When comparing the rate of charge separation and recombination for the same donor and acceptor, say, P and I, most of these factors are nearly constant, with the exception of the free energy changes in the forward and reverse directions. We have developed methods for studying these kinetics and energy gaps using the effects of magnetic fields on the reaction scheme (11) and this approach is outlined below.

MAGNETIC FIELD EFFECTS IN PHOTOSYNTHETIC REACTION CENTERS

PHYSICAL ORIGINS OF THE EFFECTS

To study the initial charge separation and recombination steps, electron transfer from $I^{\pm}$ to Q_A can be blocked either by prereducing Q_A to Q_A^- or by removal of Q_A altogether. We prefer the latter approach because it eliminates the complication of spin–spin interaction between Q^- and I^- in the intermediate $P^+I^-Q^-$, an interaction which we have shown to cause complications in the analysis because of strong interactions between the spin on Q_A^- and the high-spin Fe(II) which is close to Q_A (12). With Q_A removed, the reaction scheme simplifies; however, as shown in Fig. 2, RCs blocked at this step form the triplet state of P, ^{3}P. Since the initial charge separation step to form P^+I^- from the singlet excited state of P has 100% quantum efficiency, the formation of ^{3}P cannot be occurring by ordinary intersystem crossing and must arise during the charge recombination step. Both P and I are closed-shell molecules, so the ions P^+ and I^- are radical ions with unpaired spins. Since the intermediate P^+I^- originates from the singlet excited state of P by a very fast electron transfer reaction, the two spins do not initially have a random phase relationship, rather they are correlated in a net singlet spin state, denoted $^1(P^+I^-)$ or the singlet radical ion pair. This is illustrated in the classical precession diagram at high magnetic field (H > 1 kG) in Fig. 3. If the spins on P^+ and I^- had identical precession frequencies, then their phase relationship would continue to be singlet until recombination to the singlet ground state (rate constant k_S), a spin-allowed reaction. However, we expect that the precession frequencies of the two spins will be different because P^+ and I^- are chemically distinct species (i.e., they have different g-factors). As a result, the phase relationship between the spins will evolve as one spin precesses faster than the other, and the spin multiplicity of the radical pair will evolve to the triplet phase relationship illustrated on the right of Fig. 3. This is the T_0 magnetic sublevel, and only this sublevel of the three possible

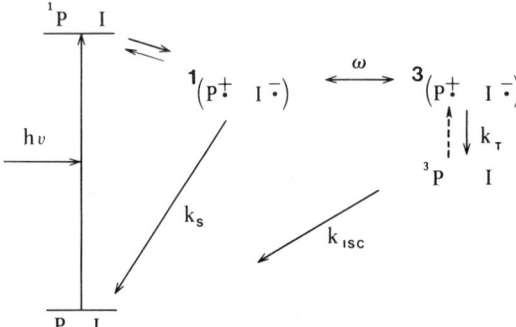

Fig. 2. Kinetic scheme for the initial photochemistry in quinone-depleted *Rhodopseudomonas sphaeroides*, R-26 RCs. The triplet state of the primary electron donor, ^{3}P, is formed by spin-allowed recombination from the triplet radical ion pair, $^3(P^+I^-)$.

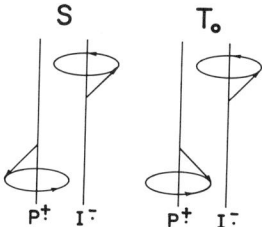

Fig. 3. Vector model for precession of spins in a radical pair at high magnetic field strength. The spin on I^- precesses faster than that on P^+. The $S \to T_0$ conversion requires no spin flips, but only a change in the relative phase of the two electron spins.

triplet sublevels will be populated by this simple precession mechanism known as the Δg effect (13).

Once the radical pair is in the triplet state, spin-allowed recombination to form the excited triplet state, ^3P, becomes possible (rate constant k_T), so it is seen that singlet–triplet mixing in the radical pair is the key to explaining the formation of ^3P. The triplet state formed in RCs, as detected by epr spectroscopy, was shown by Thurnauer et al. (14) to be predominantly in the T_0 state as predicted by this mechanism. From the vector diagram, it is also seen that the evolution of the phase relationship between the spins will continue indefinately, oscillating between S and T_0 at a characteristic frequency, ω, until interrupted by a recombination reaction. Since the spin precession frequency increases linearly with the strength of an applied magnetic field, the frequency of S-T_0 interconversion due to the difference in procession frequency will increase with magnetic field. If triplet recombination is fast relative to this interconversion rate, then an increase in the interconversion rate will increase the yield of ^3P. This is the origin of the magnetic field effect on the yield of ^3P at high field, and a measurement of this field effect can be related to the magnetic properties of P^+I^- and the decay kinetics. A proper quantum mechanical description of this problem is straightforward and is discussed elsewhere in detail (11).

The actual situation is a great deal more complicated and interesting. The spins on P^+ and I^- are not isolated as described above, rather they interact with each other and with the nuclear spins (mostly protons) that are present in the molecules comprising the radicals. Species believed to be P^+ and I^- can be trapped separately and studied by epr and endor spectroscopy (15), so a great deal is known about the electron spin–nuclear spin interactions. The nuclear spins create an internal magnetic field, which is felt by the electron spins. Many nuclear spins are involved on each radical, and we assume that the nuclear spin states are populated at random. In simplified terms it is least probable that all nuclear spins are up or down in a particular radical (a situation which would produce the largest internal magnetic field), and it is most probable that half of the spins are up and half are down, leading to roughly zero internal field. From this, it is evident that not all radical pairs have the same internal magnetic field, that is, the sample has an intrinsic heterogeneity with respect to magnetic environments, unlike the oversimplified vector diagram in Fig. 3. This heterogeneity can be incorporated into the theory by summing over the distribution of nuclear states. Physically, the net result is that at a given field some radical pairs evolve faster than others from the S to T_0 state, that is, there is a distribution of ω's.

At zero and low magnetic field (H < 1 kG), the simple precession diagram fails altogether. The g-factor difference plays no role in causing spin multiplicity evolution, and only the electron–nuclear hyperfine interaction is effective. In contrast to the high-field case in which only S and T_0 levels are considered, all three magnetic sublevels of the radical pair triplet state must be considered. It is likely from the distances between $P^{\pm}$ and $I^{\mp}$ in the *R. viridis* crystal structure (2,2a) that spin–spin interactions should be small (this is confirmed by the data), so the three magnetic levels in $^3(P^+I^-)$ are degenerate or nearly so at zero field. As a field is applied, the levels split off, as illustrated in Fig. 4. The electron–nuclear hyperfine interaction is capable of mixing S with all three magnetic levels; however, the effectiveness of mixing with the T_+ and T_- levels will decrease as they split off. In crude terms, all three triplet levels mix with S at zero field, whereas only one triplet level mixes with S at higher fields. Thus, ω decreases on application of a small field, and this will lead to a reduction in the

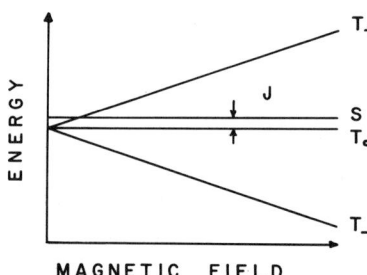

Fig. 4. Energy levels for a radical pair as a function of applied magnetic field strength in the absence of electron–dipole electron–dipole interactions, but with a small isotropic exchange interaction, J, leading to a small singlet–triplet splitting.

yield of ^3P with increasing field. Again, a proper theoretical treatment of the problem is straightforward (11,16,17); and the zero-field quantum yield of ^3P, the amount of reduction in yield with applied field, and the shape of the fall-off can be related to magnetic and kinetic properties of P^+I^-.

DATA AND ANALYSIS

An example of the type of data which we have obtained for the field dependence of the ^3P yield is shown in Fig. 5. As predicted qualitatively above, the yield drops on application of a small field due to the loss of S-$T_\pm$ degeneracy and increases at higher field as the g-factor difference between the radicals becomes important. It can be seen that these effects are very large, involving an increase of more than 100% in the ^3P yield when the field is increased from 1 to 50 kG. Measurements of the triplet yield can be complimented with mea-

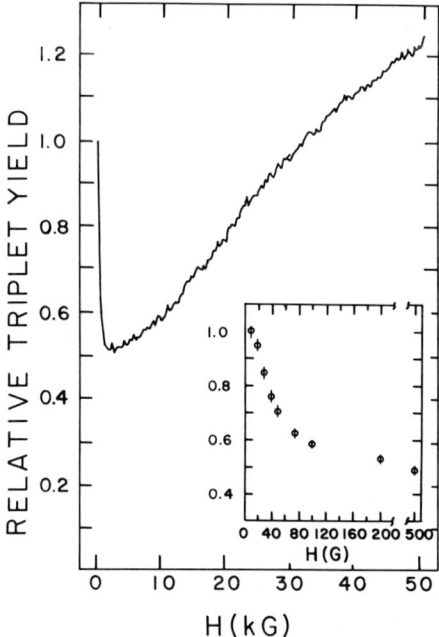

Fig. 5. Relative triplet yield in quinone-depleted RCs at 293°K between 0 and 50 kG. Inset: expansion of the low field region from 0 to 500 G. For the latter, error bars are the standard deviation determined from five experiments on two samples; field strengths are accurate to within 2 G.

surements of the radical pair lifetime [^{1}(P†I†) and ^{3}(P†I†) cannot be distinguished optically so their combined decay is measured]. Referring to the scheme in Fig. 2, it is simple to predict the effect of an applied field on the lifetime depending on the relative magnitudes of k_S and k_T. Since ω decreases going from zero field to a few hundred gauss, the radical pair will live longer if $k_T > k_S$. Further application of a field increases ω; if $k_T > k_S$ then the lifetime should get shorter again. Exactly the opposite effect of field on the lifetime is predicted if $k_T < k_S$. The radical pair lifetime was measured in collaboration with Holten and Kirmaer (18); the lifetime is about 13 nsec at zero field, about 17 nsec at 1 kG, and drops to 9 nsec at 50 kG, as predicted if $k_T > k_S$. Quantitative analysis of this data combined with the triplet yield data using theoretical expressions discussed in Boxer et al. (11) gives quite precise values for the recombination rate constants: $k_S \sim 5 \times 10^7$ sec^{-1} and $k_T \sim 3-5 \times 10^8$ sec^{-1}.

REACTION CENTER ENERGETICS

As discussed in the Introduction, it is expected that the kinetics for an electron transfer reaction depends on the free energy change for the reaction, among many factors. The data discussed above and previous picosecond kinetic studies provide three rate constants for the electron transfer reaction between P and I: ^{1}PI $\rightarrow$ ^{1}(P†I†), $k_{ET} \sim 3 \times 10^{11}$ sec^{-1} (6,7); ^{1}(P†I†) $\rightarrow$ PI, $k_S \sim 5 \times 10^7$ sec^{-1}; ^{3}(P†I†) $\rightarrow$ ^{3}PI, $k_T \sim 3-5 \times 10^8$ sec^{-1}. The magnetic field effects also demonstrate that the singlet and triplet radical pairs have nearly the same energy (17,18). Earlier work from Shuvalov and Parson (19) provides an estimate for the ^{3}P energy above the ground state. Thus, we need either the ^{1}PI − P†I† or the P†I† − ^{3}PI energy gap for a complete set of related energies and kinetics (see Fig. 2).

Quite by accident, we discovered that the ^{3}P decay rate depends on magnetic field (20). Based on what has been discussed above, this immediately suggests that ^{3}P decays to some extent through the radical pair from which it was born in the first place. This is indicated in the scheme in Fig. 2 with a vertical dotted arrow between ^{3}PI and ^{3}(P†I†), and amounts to an exact reversal of the pathways discussed thus far. We can predict the magnetic field effect on the ^{3}P decay rate with the same physical picture and principles used above for the ^{3}P yield and the radical pair lifetime. Assume that ^{3}P decays by two competing mechanisms as shown in Fig. 2: ordinary intersystem crossing to the ground state, k_{isc} ($k_{isc} \sim 10^4$s^{-1}), and reformation of

$^3(P^+I^-)$. Since S-T mixing (ω) and k_S are much faster than k_{isc}, the route $^3PI \rightarrow {}^3(P^+I^-) \xrightarrow{\omega} {}^1(P^+I^-) \rightarrow PI$ competes with k_{isc} even though most reformed $^3(P^+I^-)$ immediately decays back to ^3P (k_T is fast relative to ω except at very high field). On going from zero to a small magnetic field, $\omega(H)$ decreases, consequently decay through the radical pair route is less effective, and ^3P decays more slowly. As the field is increased further, $\omega(H)$ increases, ^3P decay through the radical pair route is more effective, and the ^3P decay rate should increase. These are precisely the magnetic field effects which we observe. Since the ^3P decay pathway through the radical pair is activated, we expect that the ^3P lifetime should be strongly temperature dependent and that the magnetic field effect will disappear at low temperature (k_{isc} is not expected or found to depend on temperature or field). Both predictions are confirmed by experimental results; the temperature-dependent lifetime is shown in an Arrhenius plot in Fig. 6.

The theoretical treatment of the magnetic field effect on the triplet decay rate is very much like that for the triplet yield, except the initial state is now pure triplet instead of pure singlet. This introduces several assumptions which were not needed when treating the singlet-born radical pair. There are three triplet sublevels in ^3P, and we assume that all are involved in reforming $^3(P^+I^-)$ from ^3PI. The assumption that nuclear spin states are randomly populated is not likely to be

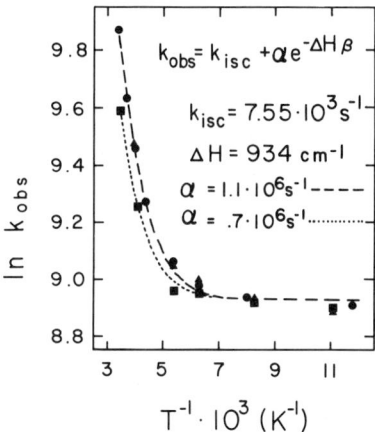

Fig. 6. Arrhenius plots of ^3P decay rate as a function of temperature at several magnetic field strengths in quinone-depleted RCs. Data points taken at zero field are donated ●; at 5 kG, ■; at 45 kG, ▲. The data were fit to the equation in the figure, which is the theoretical expression derived from a full quantum mechanical treatment of the problem (20).

true, because the nuclear spin configuration in ^{3}PI is that which was most effective in forming ^{3}PI in the first place. Thus, $\omega(H)$ in the reverse direction may not be the same as in the forward direction, but we assume it to be (21). With these assumptions [and others discussed in detail in Chidsey et al. (20)], we can fit the temperature-dependent lifetime data to the equation shown in Fig. 6 to extract the energy gap for the ^{3}PI → ^{3}(P†I^{-}) step: about 0.12 eV. Combined with the ^{3}P energy, this gives an energy gap of about 0.28 eV for the initial photochemical step in RCs: ^{1}PI → ^{1}(P†I^{-}).

With this information, we can compare the energy gaps and kinetics for all reactions involving P and I summarized in Fig. 7: ^{1}PI → ^{1}(P†I^{-}), $k_{ET} \sim 3 \times 10^{11}$ sec^{-1}, energy gap ~0.28 eV; ^{1}(P†I^{-}) → PI, $k_S \sim 5 \times 10^7$ sec^{-1}, energy gap ~1.1 eV; ^{3}(P†I^{-}) → ^{3}PI, $k_T \sim$ 3–5 $\times 10^8$ sec^{-1}, energy gap ~0.12 eV. The largest rate is observed for an intermediate energy gap, whereas the rate falls off for larger and smaller energy gaps. A fall-off in rate for large exothermicity is an important prediction of the Marcus theory (8). This may be the reason why the recombination rates are much slower than the forward reaction rate. In addition, the rate for the primary charge separation step is known to be nearly activationless. We note that for activationless electron transfer processes, such as ^{1}PI → ^{1}(P†I^{-}), the energy gap is equal to the total reorganization energy. Our estimated value of about 0.28 eV is quite small and is comparable to estimates for the inner-sphere reorganization energy associated with electron transfer between rigid macrocycles such as those which participate in the initial steps of photosynthesis. Therefore, it appears that the outer-sphere reorganization energy is very small, suggesting minimal change of the protein environment accompanying the initial electron transfer step, PI → P†I^{-}. This, in turn, suggests that residues in the environment of the donor and ac-

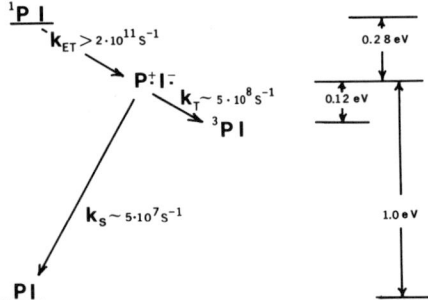

Fig. 7. Summary of kinetics and energetics for electron transfer between P and I.

ceptor may be organized in order to minimize reorganization following charge separation. It may well be that the difference in reorganization energy for electron transfer for the two potential electron transfer pathways revealed by the R. viridis X-ray structure (2,2a) controls the relative rates along each pathway.

ACKNOWLEDGMENTS

The author thanks his students whose work was outlined in this paper: Dr. M. G. Roelofs, Dr. C. E. D. Chidsey, L. Takiff, and R. Goldstein. This research is supported by NSF Grant DMB-8607799.

REFERENCES

1. Michel, H. (1982). J. Mol. Biol. **158**, 567–572.
2. Deisenhofer, J., Epp, O., Miki, K., Huber, R., and Michel, H. (1984). J. Mol. Biol. **180**, 385–398.
2a. Deisenhofer, J., Epp, O., Miki, K., Huber, R., and Michel, H. (1985). Nature (London) **318**, 618–624.
3. Feher, G., and Okamura, M. Y. (1978). In "The Photosynthetic Bacteria" (R. K. Clayton and W. R. Sistrom, eds.), pp. 349–86. Plenum, New York.
4. Shopes, R. J., and Wraight, C. A. (1985). Biochim. Biophys. Acta **806**, 348–356.
5. Norris, J. R., Druyan, M. E., and Katz, J. J. (1973). J. Am. Chem. Soc. **95**, 1680–1682.
6. Martin, J. L., Breton, J., Hoff, A. J., Migus, A., and Antonetti, A. (1986). Proc. Natl. Acad. Sci. U.S.A. **83**, 957–961.
7. Breton, J., Martin, J. L., Migus, A., Antonetti, A., and Orszag, A. (1986). Proc. Natl. Acad. Sci. U.S.A. **83**, 5121–5125.
8. Marcus, R. A. (1956). J. Chem. Phys. **24**, 966.
9. Miller, J. R., Calcaterra, L. T., and Closs, G. L. (1984). J. Am. Chem. Soc. **106**, 3047–49.
10. Wasielewski, M. R., Niemczyk, M. P., Svec, W. A., and Pewitt, E. B. (1985). J. Am. Chem. Soc. **107**, 1080–1082.
11. Boxer, S. G., Chidsey, C. E. D., and Roelofs, M. G. (1983). Annu. Rev. Phys. Chem. **34**, 389–417.
12. Roelofs, M. G., Chidsey, C. E. D., and Boxer, S. G. (1982). Chem. Phys. Lett. **87**, 582–588.
13. Lepley, A. R., and Closs, G. L., eds. (1973). "Chemically Induced Magnetic Polarization." Wiley, New York.
14. Thurnauer, M. C., Katz, J. J., and Norris, J. R. (1975). Proc. Natl. Acad. Sci. U.S.A. **72**, 3270–74.
15. Hoff, A. J. (1979). Phys. Rep. **54**, 77–200.
16. Werner, H.-J., Schulten, K., and Weller, A. (1978). Biochim. Biophys. Acta **502**, 255–268.
17. Haberkorn, R., and Michel-Beyerle, M. E. (1979). Biophys. J. **26**, 489–498.

18. Chidsey, C. E. D., Kirmaier, C., Holten, D., and Boxer, S. G. (1984). *Biochim. Biophys. Acta* **766**, 424–437.
19. Shuvalov, V. A., and Parson, W. W. (1981). *Proc. Natl. Acad. Sci. U.S.A.* **78**, 957–961.
20. Chidsey, C. E. D., Takiff, L., Goldstein, R., and Boxer, S. G. (1985). *Proc. Natl. Acad. Sci. U.S.A.* **82**, 6850–6854.
21. Goldstein, R., and Boxer, S. G. (1987). *Biophys. J.* (in press).

19

The Structural Basis of Photosynthetic Light Reactions in Bacteria

ROBERT HUBER
Max-Planck-Institut für Biochemie
8033 Martinsried, West Germany

In photosynthesis light energy is absorbed and subsequently converted into chemical energy stored in energy-rich compounds. Photosynthesis occurs in certain bacteria and green plants in specialized photosynthetic membranes which either are infoldings of the cell membrane (in prokaryotes) or are localized in chloroplast organelles (in eukaryotes). The primary events of the photosynthetic process consist of light absorption and charge separation. These occur in two functional units, the light-harvesting complexes and the reaction centers, respectively. The light-harvesting complexes funnel the energy of the absorbed light to the reaction centers with very high efficiency. In the reaction centers the charge separation across the photosynthetic membrane occurs (Fig. 1).

A vast amount of functional and structural data of components of the photosynthetic apparatus of bacteria has been accumulated (1) but a detailed understanding requires knowledge of the three-dimensional structures of the macromolecular complexes involved in these reactions at the atomic level. This goal has been achieved recently for both light-harvesting complexes and reaction centers by crystallographic analyses.

The main component of the light-harvesting organelles of the cyanobacterium *Mastigocladus laminosus* was analyzed by Schirmer *et al.* (2). The reaction center of the purple bacterium *Rhodopseudomonas viridis* was crystallized by Michel (3) and its crystal structure was determined by Deisenhofer *et al.* (4–6).

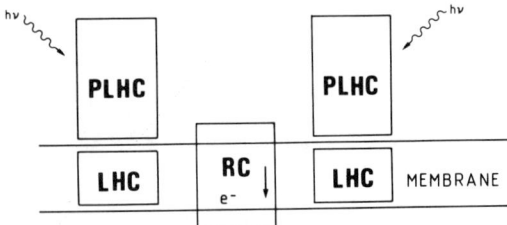

Fig. 1. Components of the photosynthetic apparatus of bacteria. RC, reaction center of the photosystem; PLHC, peripheral light-harvesting complex; LHC, membranous light-harvesting complex; membrane, photosynthetic membrane.

The light-harvesting organelles of cyanobacteria (the phycobilisomes) form antennalike structures that are assembled from stacked hollow double disks. The disks are formed from trimers of ($\alpha\beta$) protein pigment complexes. The pigments are open-chain tetrapyrrole systems to which these protein complexes owe their deep blue color. The pigments cover the wall of the antenna and stick into the central channel. They absorb the light and conduct light energy probably by a mechanism of inductive resonance within a few picoseconds along the antenna to the photosynthetic membrane. Energy transfer within the phycobilisomes occurs almost without loss by fluorescence or other mechanisms of energy dissipation.

The folding of the individual protein subunits is largely α-helical. Very remarkably, the arrangement of the helical segments in the globular protein bodies resembles the fold first observed in the globin family of proteins. This may point to a very distant evolutionary relationship of biliproteins and globins (Fig. 2).

The reaction center of *Rhodopseudomonas viridis* is a huge protein pigment complex assembled from four different proteins, the H, L, and M subunits and the cytochrome. *In vivo* it is embedded in the photosynthetic membrane and in the crystals it is associated with detergent. The central part of the complex is made of the L and M subunits. The L and M subunits have similar shapes and foldings and resemble dishes to hold the essential chromophores, four bacteriochltorophylls, two bacteriopheophytins, an inorganic iron, and a quinone. These chromophores are arranged in a very special way to form two similar branches of aromatic systems in close contact. The branches meet at two closely associated bacteriochlorophylls, the special pair. This is the primary electron donor which, on excitation by light, releases an electron. The electron is conducted via a bacteriochlorophyll and a bacteriopheophytin to the quinone (Qa), the primary electron acceptor. This process of charge separation occurs in

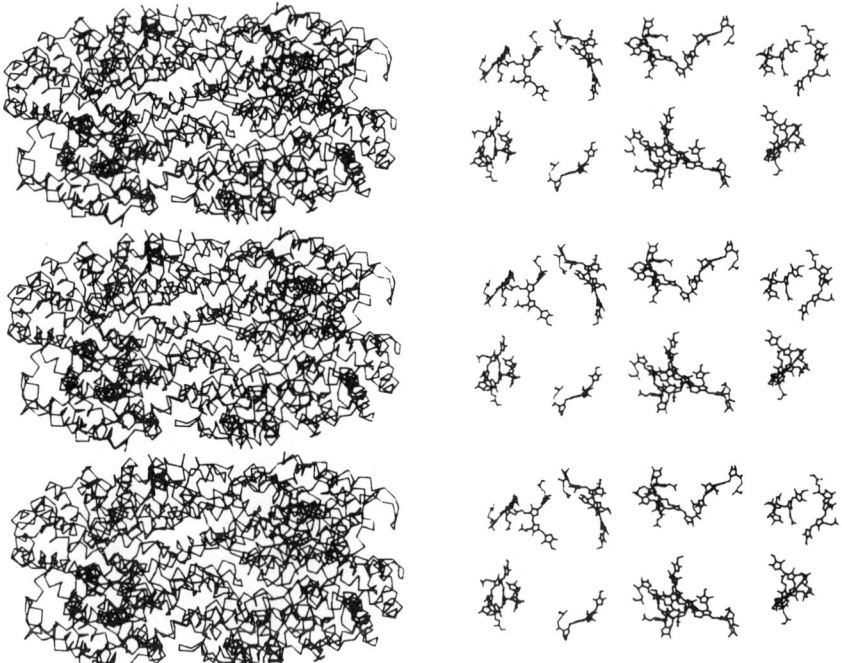

Fig. 2. Three hexamers of *Agmenellum quadruplicatum*. C, phyocyanin polypeptide chain chromphores.

the picosecond range. The unique arrangement of the chromophoric groups is necessary to prevent back reaction. The special pair is located close to the periplasmic side of the photosynthetic membrane and the quinone is on the cytoplasmic side. The establishment of an electrical charge across the membrane is followed by further reactions to transfer the electron from Qa to a secondary quinone (Qb) and to generate a pH gradient. The oxidized special pair is reduced by the cytochrome component of the reaction center complex and the circuit thus closed. The cytochrome has four heme groups arranged in a row along which fast electron transfer seems to be possible. The cytochrome forms the periplasmic and the H subunit the cytoplasmic cover of the LM pigment complex, which is integrated in the photosynthetic membrane (Fig. 3).

The arrangement of the chromophores in the phycobilisomes and the reaction center indicate a unique pathway for the light energy traveling along the antenna and the photoinduced electrons crossing the membrane, respectively.

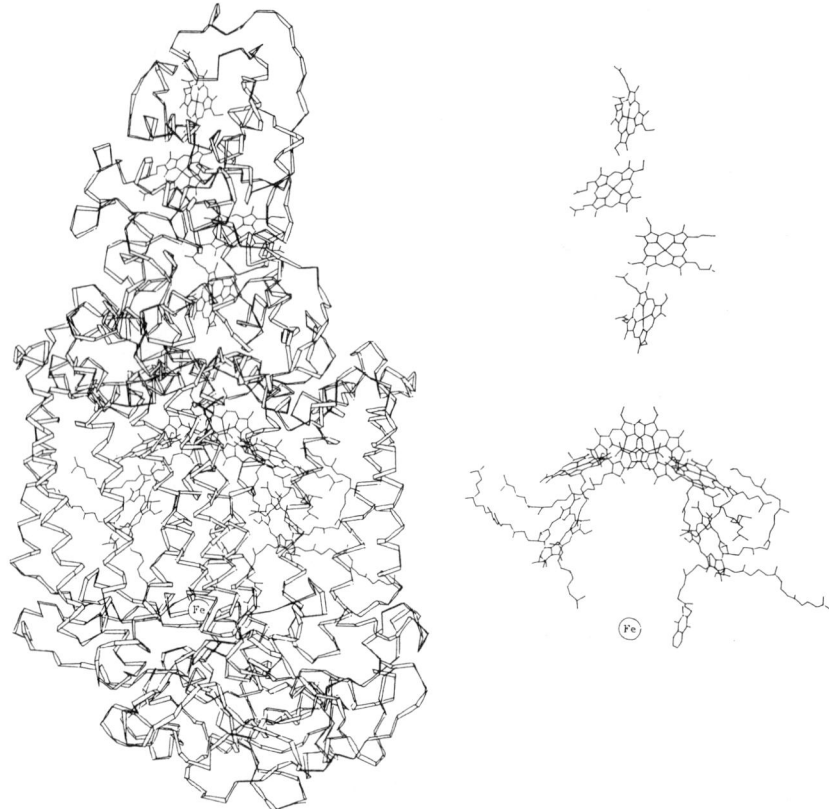

Fig. 3. Reaction center of *Rhodopseudomonas viridis*. Polypeptide chain and chromophores of cytochrome and the L, M, and H subunits.

REFERENCES

1. Clayton, R. K., and Sistrom, W. R., eds. (1978). "The Photosynthetic Bacteria." Plenum, New York.
2. Schirmer, T., Bode, W., Huber, R., Sidler, W., and Zuber, H. (1985). *J. Mol. Biol.* **184**, 257–277.
3. Michel, H. (1982). *J. Mol. Biol.* **158**, 567–572.
4. Deisenhofer, J., Epp, O., Miki, K., Huber, R., and Michel, H. (1984). *J. Mol. Biol.* **180**, 385–398.
5. Deisenhofer, J., Michel, H., and Huber, R. (1985). *Trends Biochem. Sci.* **10**, 243–248.
6. Deisenhofer, J., Epp, O., Miki, K., Huber, R., and Michel, H. (1985). *Nature (London)* **318**, 618–624.

PART V

MEMBRANE PROTEINS AND SIGNALING

20
X-Ray Analysis of Deaminooxytocin: Conformational Flexibility and Receptor Binding

J. E. PITTS, S. P. WOOD, I. J. TICKLE, A. M. TREHARNE,
Y. MASCARENHAS,[1] J. Y. LI,[2] J. HUSAIN, S. COOPER,
T. L. BLUNDELL

Laboratory of Molecular Biology
Department of Crystallography
Birkbeck College
London WC1E 7HX, England

V. J. HRUBY

Department of Chemistry
The University of Arizona
Tucson, Arizona

H. R. WYSSBROD, A. BUKU, and A. J. FISCHMAN

Department of Physiology and Biophysics
Mount Sinai Medical Center
City University of New York,
New York, New York 10029

---

[1] Present address: Instituto de Fisica e Quimia de Sao Carlos, Universidade De Sao Paulo, Campus De Sao Carlos, CEP 13.560, Sao Carlos (SP), Brazil.

[2] Present address: Institute of Biophysics, Academia Sinica, Beijing, People's Republic of China.

INTRODUCTION

Two crystal structures of deaminooxytocin have been determined at 1.1 Å resolution from isomorphous replacement and anomalous scattering X-ray measurements. The two closely related conformers with disulfide bridges of different chirality which occur in each crystal may be important in receptor recognition and activation.

The neurohypophyseal hormones, oxytocin (I) and vasopressin, are closely related nonapeptides, and have very different physiological effects.

$$H_2N-Cys-Tyr-Ile-Gln-Asn-Cys-Pro-Leu-Gly-NH_2$$
(I)

Oxytocin elicits smooth muscle contraction causing milk ejection and uterine contractions in mammals, whereas vasopressin is involved in water balance control. The various vertebrate hormones differ at positions 2, 3, 4, and 8; the similarity in the rest of the molecule is reflected in a degree of cross-reactivity between the variety of target tissue receptors and associated biological effects.

The synthesis of oxytocin (1) provided the basis for many subsequent structure function studies. Much of the earlier work concerned the relation of systematic changes in primary structure to biological activity, but more recent studies have highlighted the additive nature of modifications which favor certain pharmacological effects. This has resulted in the design of highly selective, long-acting superagonists and antagonists of therapeutic potential (2,3). One synthetic analogue of particular interest is deaminooxytocin (1β-mercaptopropionate oxytocin), which was the first to be found more active in most tests than the natural hormone (4).

Spectroscopic studies, such as nuclear magnetic resonance (NMR), laser Raman, and circular dichroism, have shown that oxytocin can exist in several differing conformations, although certain well-defined intramolecular hydrogen bonds characterize most members of the population of conformers in solution (5,6). In view of this inherent flexibility, it is necessary to examine the conformation and dynamics of the hormone and many analogues not only in aqueous conditions but also in other environments, which may be better models for the hormone in its complex with the receptor. In fact conditions in the crystal where solvent availability is limited and multiple weak intermolecular interactions exist may be of more significance in this respect than the structural studies in solutions of dimethyl sulfoxide (DMSO) which are popular with spectroscopists.

Although crystals of oxytocin were first reported in 1952 (7), the crystal structure has proven elusive. In 1965 preliminary crystal data for deaminooxytocin were reported by Low and Chen (8) and an active 6-selenodeaminooxytocin analogue was later purified and crystallized (9). No further progress has been reported, although other crystals of oxytocin have been described and a structure analysis has been completed for two COOH-terminus peptides (10,11).

Here we present the three-dimensional structures of two crystal forms of deaminooxytocin and a further related crystal form of 6-selenodeaminooxytocin defined by X-ray analyses at resolutions between 1.09 and 2.1 Å. We compare the crystal structures to conformations proposed from NMR and other spectroscopic studies in DMSO and water. We show that even in the crystals the deaminooxytocin is flexible and that there are at least two well-defined conformers. We argue that the partially hydrophobic environment in the crystal may have features in common with the receptor and that the flexibility seen in the deamino analogues may be required for full agonist rather than antagonist activity.

STRUCTURE DETERMINATION

In view of the lack of success encountered earlier (8,9), we decided to synthesize fresh materials, grow new crystals, and recollect X-ray data. Crystal data for deaminooxytocin, air-dried deaminooxytocin, and 6-selenodeaminooxytocin are summarized in Table I. The similarity in cell parameters of the air-dried C2 crystals contrasts with a large β angle change previously reported for dried deaminooxytocin crystals (8).

The absence of visible pseudosymmetry in the dried crystals and the close isomorphism with the seleno analogue prompted the use of data sets from these two crystals for the initial phase determination. The sulfur positions in the C2 crystal form were located using selenium isomorphous difference, selenium anomalous difference, sulfur anomalous difference Patterson maps, and an anomalous difference Fourier map (12) (see Fig. 1) (13–15). The selenium single isomorphous replacement with anomalous differences (SIRAS) technique alone did not provide an interpretable map, but when supplemented with higher-quality anomalous differences due to sulfur in the native structure, a "best" phased Fourier map at 2.1 Å resolution provided a rough model of the structure. In view of some serious ambiguities in the initial model, it was necessary to refine with unrestrained recipro-

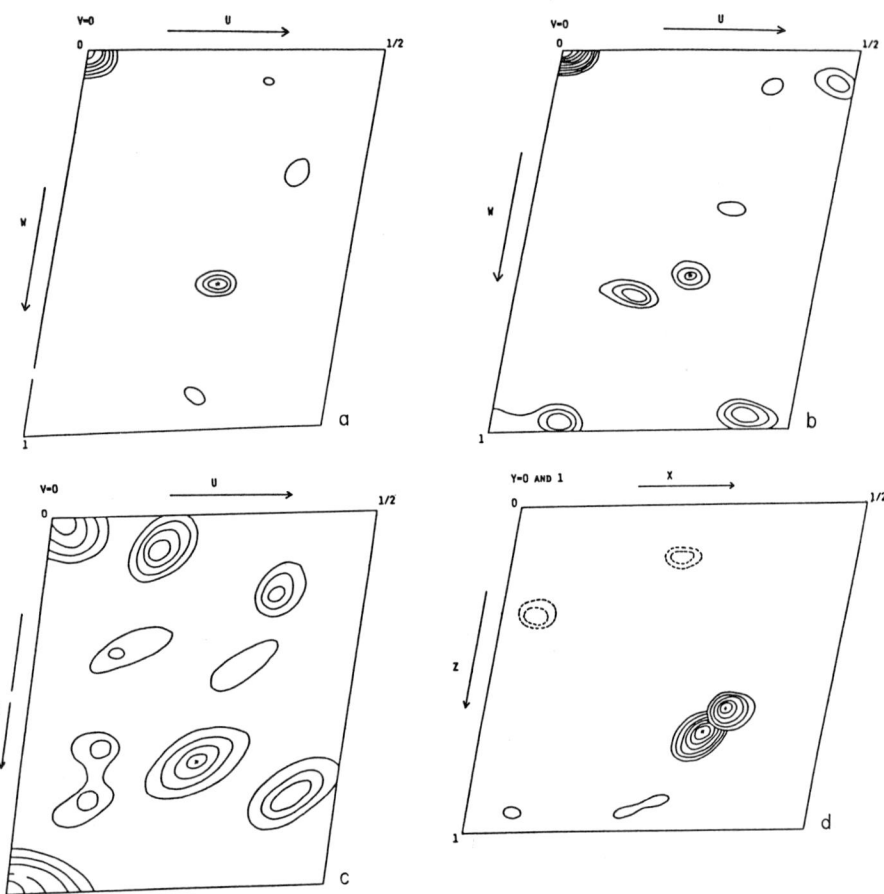

Fig. 1. For the C2 crystal forms, an isomorphous difference Patterson (a) map with coefficients $(Fs-s - Fs-se)^2$ located the selenium atom and this was confirmed in a noisy anomalous difference Patterson (b) with coefficients $[Fs-se(+) - Fs-se(-)]$. An anomalous difference Patterson map with coefficients $[Fs-s(+) - Fs-s(-)]$ (c), showing an extended peak at the same location as the selenium Harker vector, indicates the direction of the bridge and the quality of the sulfur anomalous data. The S1 position was finally located from an anomalous difference Fourier (d, composite of three sections) map in which the imaginary part of the structure factor alone is used to produce a map of anomalous scatterers present. The Fourier coefficients were the anomalous differences due to the sulfur atoms and phases were calculated by single isomorphous replacement by the method of Blow and Crick from data in (a) and (b) above. Final phases were calculated using the isomorphous contribution due to the selenium replacement and its associated anomalous scattering, together with the anomalous scattering due to sulfur. In this calculation, the data from deaminooxytocin can be considered to have served both, as "native" and as "derivative" with zero isomorphous contribution. This was achieved by providing real form factors for sulfur of zero at all angles and selenium

TABLE I
Crystal Data For Deaminooxytocin, Air-Dried Deaminooxytocin, and 6-Selenodeaminooxytocin[a]

	a (Å)	b (Å)	c (Å)	$\beta°$
"Wet" crystals P2$_1$	27.27	9.06	23.04	102.24
4681 reflections to 1.09 Å				
R_{sym} = 3%				
"Dry" crystals C2	27.08	9.06	22.98	102.06
1840 reflections to 1.2 Å				
R_{sym} = 5.7%				
(1507 ano's)				
Seleno crystals C2	27.26	9.19	23.01	102.72
425 reflections to 2.1 Å				
R_{sym} = 5.2%				
(307 ano's)				

[a] All crystals were grown by slow cooling of aqueous or 20% aqueous ethanol solutions of peptide at 40 mg/ml from 35–40°C to 18°C. Solutions and containers were purged with nitrogen gas for the seleno analogue to minimize oxidation. The P2$_1$ crystals were mounted in sealed Lindemann tubes with mother liquor, whereas the C2 crystals were produced by air-drying a previously wet-mounted crystal. Data were measured with ω/θ scans on a Hilger and Watts four-circle diffractometer with CuK$_\alpha$ radiation. All reflections were measured in small shells for each crystal.

cal space methods. The power of computer graphics facilities in simultaneous presentation of multiple maps and difference maps in color was most important in properly defining the turn in residues 2–5, the disulfide bridge, and in placing waters. These rebuilding steps provided major improvements in the refinement. Re-

---

minus sulfur factors. Imaginary form factors were as supplied in the International Tables. The mean figure of merit was 0.68 at 2.1 Å resolution.

Improvement of the initial model was carried out using unrestrained reciprocal space refinement with all the data to 1.2 Å. Rebuilding was carried out using maps calculated with $2 W_{sim} \cdot F_{obs} - F_{calc}$ and $W_{sim} \cdot F_{obs} - F_{calc}$ and α_{calc} as coefficients, where W_{sim} is the weight recommended by Sim (14). Interpretation of maps was performed with an Evans and Sutherland Picture System II and an extensively modified version of the FRODO program (15). Movement of the main chain nitrogen atom of Gln 4 to the carbonyl position of Ile 3 and vice versa prompted rebuilding of the initial model type I turn to type II. Instability in the disulfide region and high thermal parameters were resolved by defining two conformers with quite different S1 positions. For anisotropic refinement and inclusion of hydrogens the observation/parameter ratio was maintained by using restrained refinement (13). For the continued refinement in P2$_1$ with 1Å data, the two molecules of the asymmetric unit were generated with the distinct disulfide chiralities seen in the C2 cell. Water molecules were not initially included.

strained refinement (13) was employed in later stages when refining nonhydrogens anisotropically and hydrogens isotropically gave $R = 10.0\%$ at 1.2 Å resolution for the C2 crystal form. The resulting parameters were then used to initiate the refinement of the crystal form with space group $P2_1$ and data to 1 Å resolution after an appropriate origin shift to the coordinates. The final residual was 9% after several cycles of modeling from difference maps and refinement. Restrained refinement was also carried out for the selenooxytocin at 1.9 Å resolution in order to maximize the observations/parameters ratio at this resolution. Details of the structure analyses of these three crystals will be published elsewhere. The general features of the oxytocin molecules are similar in all three crystals examined, although the disorder observed in the C2 cell (in the disulfide bridge, Ile 3, etc.) is partially although not completely resolved in the $P2_1$ cell.

CRYSTAL STRUCTURE DESCRIPTION

The crystal structure of deaminooxytocin shows the 20-membered ring closed by Cys 6 and Cys 1 in a disulfide bridge. There are two β-turns in the structure, a type II turn between Tyr 2 and Asn 5 stabilized by two hydrogen bonds between the amide and carbonyl of both residues, and a type III turn between Cys 6 and Gly 9 stabilized by a hydrogen bond between the Cys 6 CO and Gly 9 peptide NH. Both turns are shown in Fig. 2. The type II turn was initially assigned type

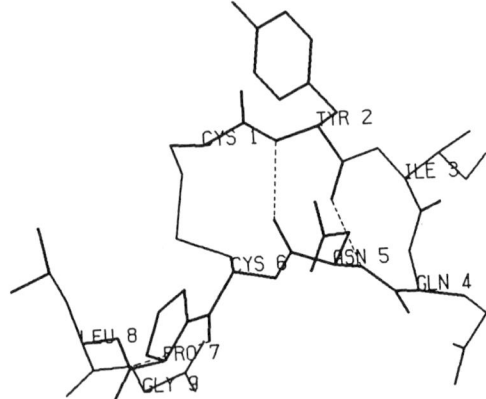

Fig. 2. The three intramolecular hydrogen bonds found in deaminooxytocin are illustrated and (C2 cell) have ON distances and N–H–O angles of (1) Tyr 2 CO – Asn 5 NH, 2.93 Å, 156.31°, (2) Tyr 2 NH – Asn 5 CO, 3.09 Å, 162.62°, and (3) Cys 6 CO – Gly 9 NH, 3.37 Å, 134.19°.

I in view of the bulky corner residues (Ile 3, Gln 4) but the refinement would not proceed with this geometry. Other small cyclic peptides with bulky corner residues have been described with type II turns (16,17). The main chain torsion angles of the 20-membered ring are such that the ring is curved (see Figs. 2 and 3), with Tyr 2 and Asn 5 side chains on the top side. The curved shape is completed by the acyclic tail of the molecule, which is oriented away from the ring with a trans Cys 6–Pro 7 peptide. The top surface of the curve is also made up of Ile 3, the α and β carbon of Gln 4, the ring of Pro 7, and Leu 8. The underside of the curve is devoid of side chains, the terminal amide of Gly 9 and side chain amide of Gln 4 being stationed at the extremities of the underside cleft. In the crystal there are at least two possible conformations for the disulfide bridge region which differ in S1 position. In the C2 cell one conformation results in a disulfide torsion angle of $+76°$ with right-handed chirality and the other with a torsion angle of $-101°$ and left-handed chirality (see Fig. 4). The two S1 positions are about 2 Å apart and equally populated. The alternate positions for S1 can be reached with only small differences of other nonhydrogen atoms because of the parallel nature of the $c\beta6-S6$ and $C\alpha_1-C\beta_1$ bonds, a situation which is displayed in the familiar "boat" and "chair" conformations of cyclohexane. It is not possible to define any differences in the disordered C2 cell at this resolution. In the $P2_1$ cell both molecules exhibit the S1 disorder but different chiral forms, similar to those seen in the C2 cell, are predominantly populated (70/30) in each molecule. The torsion angles $[C\beta(1)-S\gamma(1)-S\gamma(6)-C\beta(6)]$ are $\underline{+77}$, -87 and -94, $\underline{+83}$, the dominant isomer being underlined. Furthermore, the dominant conformers are associated with distinct differences in the main chain torsion angles and in the $C\alpha-C\beta$ rotamers of the Ile 3 side chain. Small differences are also apparent in the side chains of Tyr 2 and Asn 5. Together these differences in structure between the two molecules of the $P2_1$ cell correlate well with those atoms showing high thermal parameters in the C2 cell. Figure 5 shows anisotropic thermal ellipsoids for the C2 structure which can be compared with an overlaid comparison of the two molecules in the $P2_1$ asymmetric unit seen in Fig. 4.

All the hydrogen bonding potential of the molecule is satisfied by either intramolecular interactions or bonding with water and other oxytocin molecules. There are seven water positions in the C2 cell, two of which lie close to crystallographic two-fold axes and are disordered. The water molecules are found in two groups (Fig. 6). Four waters are found in a complex cage network around another water close to the twofold axis and form a hydrophilic zone in the crystal near the underside "cleft" with potential hydrogen bonds to Gln 4

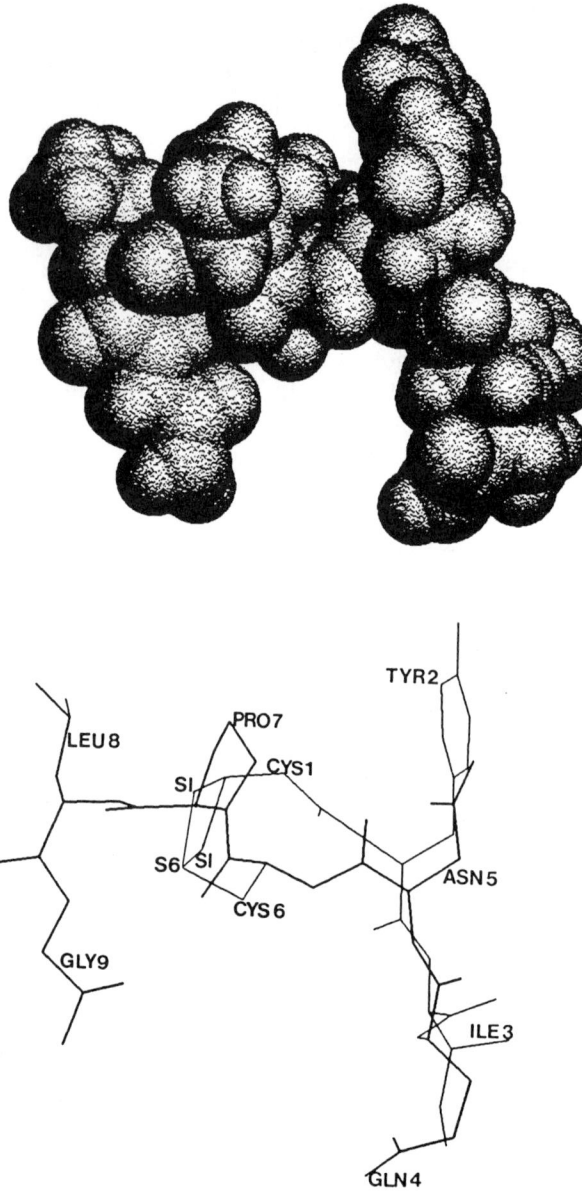

Fig. 3. The space filling and equivalent stick model for deaminooxytocin showing the "pleat" of the 20-membered ring and the accentuation of this form by the "tail" peptide and the Gln 4 side chain dispositions. The main chain torsion angles $(\phi,\pi)^{RES}$ for one molecule of the $P2_1$ cell are (-, 101)[1]. (-126, 164)[2]. (-65, 125)[3]. (56, 29)[4]. (-158, 66)[5]. (-128, 98)[6]. (-73, -12)[7]. (-77, -33)[8]. (-176, 168)[9].

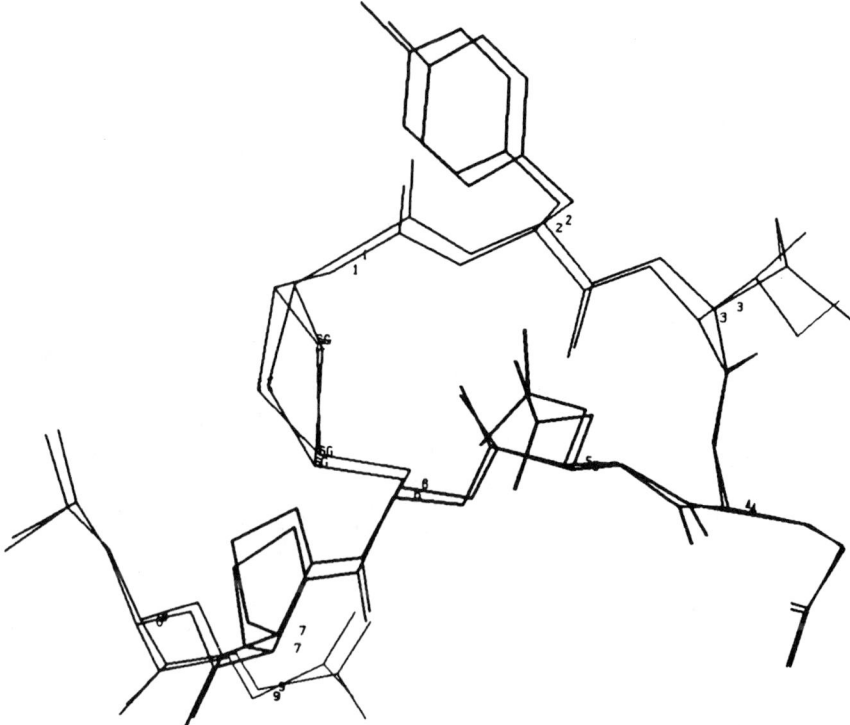

Fig. 4. The two molecules of the P2₁ asymmetric unit are shown superimposed to demonstrate the two conformations in the S1 region, the associated Ile 3 rotamers, and other conformational differences.

CO, Gln 4 NH, Gln 4 OE$_1$, Gln 4 NE$_2$, Asn 5 OD$_1$, Gly 9 CO, and the C-terminus amide NH$_2$ of several oxytocin molecules. The other two waters, one of which again is disordered close to the two-fold axis, are in the region of Tyr 2 (with hydrogen bonds to Tyr 2 OH, Asn 5 CO, Cys 1 CO) and interrupt an otherwise hydrophobic zone in the crystal formed by Tyr 2, Leu 8, Pro 7, and S1 from symmetry-related molecules. In the P2$_1$ cell the water distribution is generally the same but there are differences in location, numbers, and disorder. The waters close to the pseudosymmetry axes are no longer disordered. There is one water molecule more in the P2$_1$ cell and this water is in close contact with Asn 5 CO. Its removal may explain the change in space group on drying.

There are intermolecular hydrogen bonds between Ile 3 NH and Pro 7 CO, Asn 5 ND$_2$ and Leu 8 CO, Cys 6 NH and Gln 4 OE$_1$, Cys 1 CO and Asn 5 ND$_2$, and Ile 3 CO and the C-terminus amide NH$_2$. The

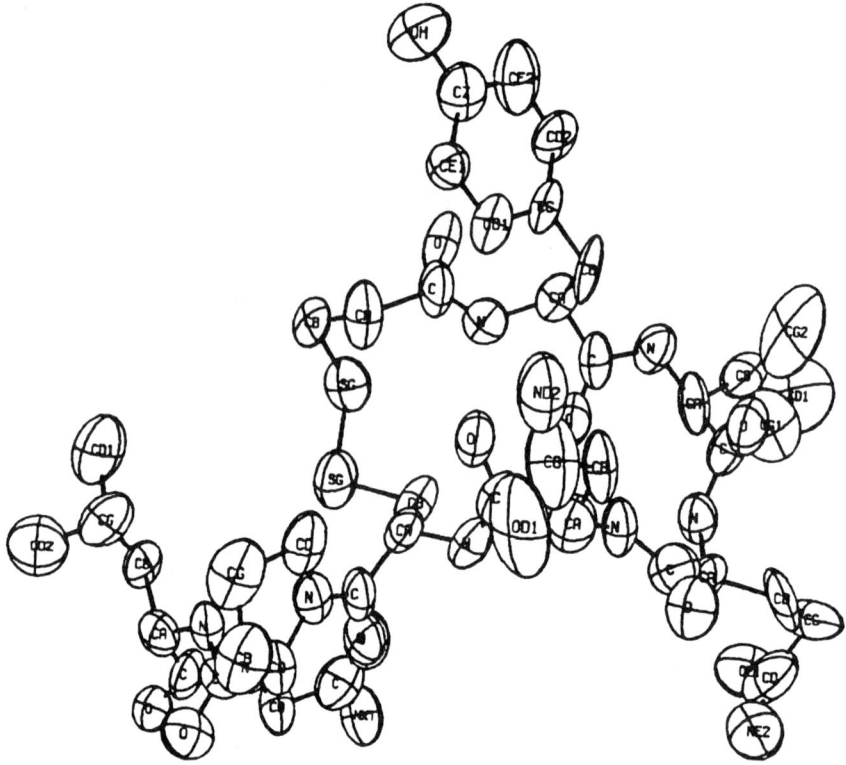

Fig. 5. Anisotropic thermal ellipsoids for the C2 structure of deaminooxytocin. Only one disulfide chirality is shown.

packing of the molecules around the two-fold axis brings the residues 1 of two molecules close together. If the terminal amino group were in position then electrostatic repulsion would not favor this packing and this may explain why oxytocin itself is not crystallized in this form.

The structure of the 6-seleno analogue at 2.1 Å resolution is very similar except in the region of the disulfide bridge. The torsion angles [Cβ(1)–Sγ(1)–Sγ(6)–Cβ(6)] are distorted to +79° and −87° and the relative occupancies are shifted to 62% for the LH and 38% for the RH conformer. Again the S1 positions are about 2 Å apart and the disorder cannot be detailed away from the bridge.

SOLUTION STRUCTURE

The conformations of oxytocin and its deamino analogue have been the subject of extensive spectroscopic studies, and views on the struc-

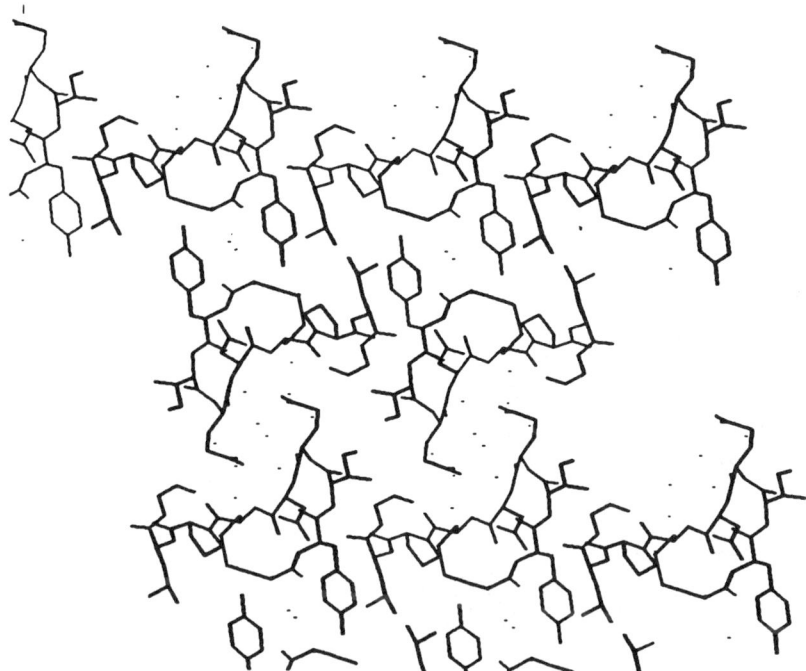

Fig. 6. When viewed down the two-fold axis of the C2 cell the packing of deaminooxytocin molecules shows distinct hydrophobic and hydrophilic zones, the former being interrupted by two water molecules in the vicinity of Tyr 2.

ture have evolved with advances in theory and technique. The original model of Walter et al. (18–21) was based on an interpretation of the NMR of oxytocin and deaminooxytocin in dimethyl sulfoxide (DMSO) as well as on evidence from hydrogen exchange, circular dichroism, and other studies. While subsequent interpretations of the NMR data became more cautious and proton assignments in water (22), Raman spectroscopy (23–25), and energy calculation data (26–29) were reported, the turn between residues 2 and 5 with a hydrogen bond between Tyr 2 CO and Asn 5 NH remained a major conformational feature for oxytocin in DMSO and this is now confirmed in the amphipathic environment of the crystal. There is general agreement that in water there is considerable conformational flexibility, particularly in the tail region, with less good evidence for intramolecular hydrogen bonds.

Although the initial definition of the turn in DMSO was not explicit, the bulky corner residues of the 2/5 turn were subsequently considered more consistent with a type I turn. A type I turn is generally preferred unless Gly or D-Ala are located at residue $i+2$. However,

Glickson (5) suggested that unique conformational constraints due to the 20-membered ring might stabilize a type II turn and this is confirmed by the characteristic $J_{N\alpha}$ of Ile 3 and the similarity of the spectra of oxytocin and Gly 4 oxytocin (30). It is therefore probable that a type II turn exists in DMSO solution and in crystals. An additional hydrogen bond between Asn 5 CO and Tyr 2 NH has been indicated by NMR for deaminooxytocin in DMSO solution and this is also observed in the crystal structure.

The chirality of the disulfide bridge in solution has also been the subject of much discussion. A right-handed screw with a dihedral angle close to $+90°$ was proposed on the basis of the sign and rotatory strength of the near-UV circular dichroism of oxytocin in water (20,31), but the validity of this interpretation has been seriously questioned (23). Although laser Raman spectroscopy indicates that most molecules in solution have dihedral angles close to $±90°$, the preponderance of a particular screw cannot be determined. It is in fact possible that both conformers exist in equilibrium and that the population of each conformer, in solution as in the crystals, depends on the analogue and the environment. This would explain changes in the circular dichroism on titration of the α-amino group and on addition of organic solvent to aqueous solutions of oxytocin and differences between the spectra of oxytocin, deaminooxytocin, selenodeaminooxytocins, and other analogues (31,32). It is likely that chemical differences would affect flexibility of the tocin ring and that environment changes would favor different conformers offering different accessibilities of hydrophobic and hydrophilic groups.

There is evidence for the maintenance in DMSO solution of the type III β-turn between residues 6 and 9 with a hydrogen bond between the Cys 6 CO and the Gly 9 peptide NH, and the existence of a trans Cys 6–Pro 7 peptide, both of which are observed in the crystals. However, the tocin ring and acyclic tail were proposed in DMSO to fold over so that a hydrogen bond is formed between the Asn 5 amide carbonyl and the Leu 8 NH. This appears to be replaced by intermolecular hydrogen bonds in the crystal structure which lead to a more open structure. In fact such an open conformation is consistent with the NMR in water (6), where there is no evidence for such intramolecular hydrogen bonds. In any case, it is likely that there is conformational flexibility between the tocin ring and the β-turn of the acyclic tail and this is confirmed by conformational energy calculations which suggest a number of equi-energy conformers (26–29).

In conclusion, the solution data taken together with the crystal structures indicate that the two β-turns are stable features of oxytocin and its analogues. Both crystal structure and solution data are consistent with flexibility of the disulfide and it is probable that both conformers exist, but in different proportions in different environments. Finally, there may be considerable flexibility in the spatial reltionship between the tocin ring and acyclic tail, but it is probable that where intermolecular hydrogen bond donors and acceptors are available—in water and in the crystal, for example—there are no intramolecular hydrogen bonds between the two parts. Similar conformational preferences, although perhaps more flexibility, are evident in arginine and lysine vasopressins (6). To understand the flexibility of oxytocin further we recently carried out a normal mode analysis for deaminooxytocin and calculated 1000 psec of molecular dynamics simulation both *in vacuo* and in the crystal lattice, which will be described elsewhere. Similar work has recently been presented by Hagler *et al.* (33).

RECEPTOR BINDING AND BIOLOGICAL RESPONSE

In the absence of a purified and characterized oxytocin receptor complex, it is necessary to use hormone analogues to define those molecular features which are responsible for biological activity. For each analogue we must consider the structural consequences of chemical modifications, especially in such a small peptide where flexibility is expected and modifications might cause drastic redistributions in conformer populations. In this respect it is fortunate that spectroscopic investigations have been reported for many analogues (5,6). It is also important to differentiate the molecular features responsible for receptor affinity from those critical for eliciting a biological response (2). Unfortunately such data are not always available for oxytocin analogues as many were prepared before development of receptor affinity assays and agonism or antagonism could be detected by competitive bioassay. Although hormone tracer of high specific activity has not been easily available as iodination of oxytocin is inappropriate, tritiated hormone of moderate specific activity has been used to characterize receptors with K_d in the nanomolar region.

In spite of these difficulties, models for oxytocin and vasopressin that are important for binding and activation have been proposed. The first and most comprehensive model, proposed by Walter and coworkers, was based on the solution conformation in DMSO as defined by NMR together with an assessment of the analogue assay data available for the uterine target (35,36). Ile 3, Gln 4, Pro 7, and Leu 8 were held to constitute the binding message whereas the proximity of Tyr 2

OH and Asn 5 carboxamide was important for activation. In this model the 20-membered tocin ring was seen as essentially planar with an almost featureless underside and a topside dominated by Tyr 2 and Asn 5 in a cleft between ring and acyclic tail. Some features of this model may be correct, but the X-ray analysis shows that the ring is pleated like part of a β-sheet with Tyr 2 and Asn 5 being close together on one side and the Gln 4 protrudes on the other. These features may be maintained in solution and at the receptor. The crystal structure also shows that the tail is disposed away from the tocin ring although this could move in the hormone receptor complex.

In the absence of specific receptor affinity data, antagonists have played a major role in understanding the binding process. Using such an approach Hruby (6) proposed a general "dynamic" model of peptide hormone activity which complements the ideas of the "cooperative" model of Walter (35). The model highlights the importance of flexibility in agonist activity and stresses the use of conformationally restricted analogues such as 1-L-penicillamine oxytocin to probe binding, activation events, and energy barriers between such states (6,37,38). 1-Penicillamine derivatives of oxytocin, which have the two hydrogens at the β-carbon of Cys 1 replaced by methyl groups, are potent antagonists at the uterine receptor. Circular dichroism and Raman and NMR spectroscopy (6,28,37,38) show that they have much reduced flexibility in the ring moiety, a preferred disulfide torsion angle of between 110° and 115°, and a conformation which is said to exclude the Tyr 2 ring rotamer which would place the aromatic group over the tocin ring, a major feature for agonism in the "cooperative" model of Walter (35).

The existence of two conformers in the crystal structures of both deaminooxytocin and its 6-seleno analogue are evidence of conformational flexibility. We now propose a model in which these conformers are required for receptor recognition and transduction of the biological response. We assume that the receptor is also flexible and can adopt two conformations. In the absence of hormone, it adopts a conformation which binds the oxytocin conformers with right-handed disulfide chirality. This inactive state is then stabilized by rigid, right-handed disulfide-containing analogues such as 1-penicillamine derivatives which act as antagonists (37,38). On the other hand, strong agonists stabilize the receptor in a conformation complementary to the left-handed disulfide oxytocin conformer. An efficient transduction of the biological response will then depend on both a low-energy barrier between conformers as in deaminooxytocin (39) and a strong interaction with the receptor in the active form.

Our model predicts that agonist activity will be enhanced by either kinetic effects involving flexibility and interconvertibility of the two conformers or thermodynamic effects which stabilize the left-handed conformer or increase its interactions with the receptor. Modifications at side chains, for example, Phe 2→Tyr and Gln 4→Thr, which increase potency (6,40,41), may often be due to optimization of direct interactions with the receptor—in this case formation of hydrogen bonds. However, substitutions at the same positions may also affect flexibility and have kinetic effects. Thus, replacement of Gln 4 by glycine may result in a decrease of activity due to loss of a direct interaction with the side chain but a compensatory increase of potency due to increased flexibility; there is evidence that the existence of Gly 4 leads to rearrangements in the tocin ring similar to those in deaminooxytocin.

Our model does not imply a necessary interaction of the disulfide with the receptor as we have shown that small conformational changes occur in the main chain of the tocin ring. (Larger changes also occur in the side chains of Tyr 2 and Ile 3, but it needs to be established by further structure analyses that these are related to the conformation of the disulfide rather than crystal packing.) Thus our model is consistent with the activity of derivatives in which the sulfur atoms of the cysteine are replaced by methylene (42) or selenium atoms (43).

Circulating in the bloodstream, oxytocin will spend much time in an aqueous environment. However, at the receptor, water will be excluded from much of the surface of the hormone. Experimental data from analogues indicate that hydrophobic and hydrogen-bonded interactions, possibly through the free main chain CO and NH groups as well as side chains, are likely to dominate the receptor interaction. We suggest that the molecule binds as a neutral species so that the deprotonated amino terminus can form a hydrogen bond to the receptor. The ability to form such a hydrogen bond without deprotonation would explain the enhanced activity of analogues in which the amino terminus is replaced by a hydroxyl group (6). If there were a charge on the receptor neither these analogues nor deaminooxytocin would be expected to bind the receptor strongly. The receptor binding may also require the formation of the two hydrogen bonds between residues 2 and 5 as found in deaminooxytocin and analogues with glycine at residue 4. All these observations indicate that the conformation of the neutral deaminooxytocin in an environment which excludes water from much of the hormone—as in the crystal—might reveal important aspects of the conformation at the receptor. In the crystal multiple hydrogen-bonded and hydrophobic interactions between molecules form·as flexible molecules from aqueous solution associate in the growing lattice. These might simulate the receptor environment.

Both oxytocin and vasopressin are synthesized as much larger single-chain precursors which fold before being processed to give a two-chain neurophysin/neurohypophyseal hormone complex (44,45). These complexes form dimers which probably serve to stabilize the hormone in storage. On release into the circulation the complexes dissociate and the hormones eventually find their way to membrane receptors. Studies of the interaction of hormones and analogues with neurophysin have been reviewed (45–47). The principal sites on the hormone for interaction are residues 1, 2, and 3 as shown by binding studies of hormone analogues and short peptides, some of which bind with up to 60% of the binding energy of the whole hormone. Deaminooxytocin apparently does not bind at all to neurophysins and the loss of binding with change of pH confirms the importance of the amino terminus of the hormone in forming a salt bridge with a charged carboxylate on the neurophysin. The single tyrosine (48) residue on the protein is also implicated in binding. Although there is evidence from CD for perturbation of the hormone disulfide bridge geometry or environment on binding as well as a conformational change in the protein, no general dependence on the overall hormone conformation has been demonstrated, with some totally inactive analogues binding as well as the hormone itself. The crystal structure of deaminooxytocin shows the disposition of residues 1–3 from which anticomplimentary surfaces might be generated, but details of the interaction must await structure analysis of neurophysin–hormone complexes. Some progress has been made with the structure of neurophysin–dipeptide complexes but suitable crystals containing the whole hormone have not yet been obtained (49). It is possible that the existence of the hormone precursor makes the cyclization of the tocin ring with its relatively unfavorable conformation more efficient, the appropriate conformer of the oxytocin portion being directed by the precursor "fold."

REFERENCES

1. Du Vigneaud, V., Ressler, C., Swan, J. M., Roberts, C. W., Katsoyannis, P. G., and Gordon, S. (1953). *J. Am. Chem. Soc.* **75**, 4879.
2. Rudinger, J. (1972). *Recent Prog. Horm. Res.* **28**, 131.
3. Manning, M., Lowbridge, J., Haldar, J., and Sawyer, W. H. (1977). *Fed. Proc., Fed. Am. Soc. Exp. Biol.* **36**, 1848.
4. Ferrier, B. M., Jarvis, D., and Du Vigneaud, V. (1965). *J. Biol. Chem.* **240**, 4264.

5. Glickson, J. D. (1975). In "Peptides: Chemistry, Structure, Biology" (R. Walter and J. Meienhofer, eds.), p. 787. Ann Arbor Sci. Publ., Ann Arbor, Michigan.
6. Hruby, V. J. (1981). In "Topics in Molecular Pharmacology" (A. S. V. Burgen and G. C. K. Roberts, eds.), p. 100. North-Holland Publ., Amsterdam.
7. Pierce, J. G., Gordon, S., and Du Vigneaud, V. (1952). *J. Biol. Chem.* **199**, 929.
8. Low, B. W., and Chen, C. C. H. (1966). *Science* **151**, 1552.
9. Chiu, C. C., Swartz, I. L., and Walter, R. (1969). *Science* **163**, 925.
10. Rudko, A. D., Lovel, F. M., and Low, B. W. (1971). *Nature (London), New Biol.* **232**, 18.
11. Reed, L. L., and Johnson, P. L. (1973). *J. Am. Chem. Soc.* **95**, 7523.
12. Kraut, J. (1968). *J. Mol. Biol.* **37**, 225.
13. Morfew, A., and Moss, D. S. (1982). *Comput. Chem.* **6**, 1.
14. Sim, G. A. (1960). *Acta Crystallogr.* **13**, 511.
15. Jones, T. A. (1978). *J. Appl. Crystallogr.* **A34**, 863.
16. Chaing, C. C., Karle, I. L., and Wieland, T. (1982). *Int. J. Pept. Protein Res.* **20**, 414.
17. Karle, I. L., and Chaing, C. C. (1984). *Int. Congr. Crystallogr.*, *13th*, Absr. 03.2-15.
18. Urry, D. W., and Walter, R. (1971). *Proc. Natl. Acad. Sci. U.S.A.* **68**, 956.
19. Urry, D. W., Ohnishi, M., and Walter, R. (1970). *Proc. Natl. Acad. Sci. U.S.A.* **66**, 111.
20. Urry, D. W., Quadrifoglio, F., and Walter, R. (1968). *Proc. Natl. Acad. Sci. U.S.A.* **60**, 967.
21. Johnson, L. F., Shwartz, I. L., and Walter, R. (1969). *Proc. Natl. Acad. Sci. U.S.A.* **64**, 1269.
22. Brewster, A. I. R., and Hruby, V. J. (1973). *Proc. Natl. Acad. Sci. U.S.A.* **70**, 3806.
23. Maxfield, F. R., and Sheraga, H. A. (1977). *Biochemistry* **16**, 4443.
24. Hruby, V. J., Deb, K. K., Fox, J., Bjarnason, J., and Tu, A. T. (1978). *J. Biol. Chem.* **253**, 6060.
25. Tu, A. T. (1979). *J. Biol. Chem.* **254**, 3272.
26. Brewster, A. I. R., Hruby, V. J., Glasel, J. A., and Tonelli, A. E. (1973). *Biochemistry* **12**, 5294.
27. Kotelchuck, D., Sheraga, H. A., and Walter, R. (1972). *Proc. Natl. Acad. Sci. U.S.A.* **69**, 3629.
28. Honig, B., Kabat, E. A., Katz, C., Levinthal, C., and Wu, T. T. (1973). *J. Mol. Biol.* **80**, 277.
29. Nikiforovich, G. V., Leonova, V. I., Galaktianov, S. G., and Chipens, G. I. (1979). *Int. J. Pept. Protein Res.* **13**, 363.
30. Wyssbrod, H. R., Ballardin, A., Gibbons, W. A., Roy, J., Schwartz, I. L., and Walter, R. (1975). In "Peptides: Chemistry, Structure, Biology" (R. Walter and J. Meienhofer, eds.) p. 815. Ann Arbor Sci. Publ., Ann Arbor, Michigan.
31. Beychok, S., and Breslow, E. (1968). *J. Biol. Chem.* **243**, 151.
32. Walter, R. (1968). In "Peptides" (E. Bricas, ed.), p. 50. North-Holland Publ., Amsterdam.
33. Hagler, A. T., Osguthorpe, D. J., Dauber-Osguthorpe, P., and Hempel, J. C. (1985). *Science* **227**, 1309.
34. Soloff M. S., Schroeder, B. T., Chakraborty, J., and Pearlmutter, A. F. (1977). *Fed. Proc., Fed. Am. Soc. Exp. Biol.* **36**, 1861.
35. Walter R. (1977). *Fed. Proc., Fed. Am. Soc. Exp. Biol.* **36**, 1872.
36. Walter R., Schwartz, I. L., Darnell, J. H., and Urry, D. W. (1971). *Proc. Natl. Acad. Sci. U.S.A.* **68**, 1355.
37. Hruby V. J., Rockway, T. W., Viswanatha, V., and Chan, W. Y. (1983). *Int. J. Pept. Protein Res.* **21**, 24.

38. Mosberg, H. I., Hruby, V. J., and Meraldi, J. P. (1981). *Biochemistry* **20**, 2822.
39. Smyth, D. G. (1970). *Biochim. Biophys. Acta* **200**, 395.
40. Hruby, V. J. (1986). *In* "Biochemical Actions of Hormones" (G. Litwack, ed.). Vol. XIII, pp. 191–241. Academic Press, New York.
41. Hruby, V. J. (1986). *In* "Oxytocin: Clinical and Laboratory Studies." Elsevier/North-Holland, New York (in press).
42. Fric, I., Kodicek, M., Jost, K., and Blaha, K. (1974). *Collect. Czech. Chem. Commun.* **39**, 1271.
43. Walter, R., and Chan, W. (1967). *J. Am. Chem. Soc.* **89**, 3892.
44. Land, H., Grez, M., Ruppert, S., Schmale, H., Rehbein, M., Richter, D., and Schutz, G. (1982). *Nature (London)* **302**, 342.
45. Land, H., Schutz, G., Schmale, H., and Richter, D. (1982). *Nature (London)* **295**, 299.
46. Breslow, E. (1979). *Annu. Rev. Biochem.* **48**, 251.
47. Blumenstein, M., Hruby, V. J., and Viswanatha, V. (1983). *In* "Biomolecular Stereodynamics" (R. H. Serma, ed.), Vol. 2, p. 353. Acadenine Press, Guilderland, New York.
48. Chaiken, I. M., Abercrombie, D. M., Kanmera, T., and Sequeira, R. P. (1983). *In* "Peptide and Protein Reviews" Vol. I., (M. T. W. Hearn, ed.), Vol. 1, p. 139. Decker, New York.
49. Pitts, J. E., Wood, S. P., Hearn, L., Tickle, I. J., Wu, C. W., Blundell, T. L., and Robinson, I. C. A. F. (1980). *FEBS Lett.* **121**, 41.

21
The Acetylcholine Receptor: What the Three-Dimensional Structure Tells Us about Ion Conductance

ROBERT MICHAEL STROUD AND JANET FINER-MOORE
Department of Biochemistry and Biophysics
University of California
San Francisco, California 94143-0448

INTRODUCTION

Nicotinic acetylcholine receptors (AChR) are the best understood of the neurochemical receptors. The most studied AChR's are those found at the junctions between nerve and muscle cells, or in the electrocytes of electric fish or eels. Located in the plasma membrane of the target cell, they translate the binding of the neurotransmitter acetylcholine into conductivity through an ion-conducting channel (1–5). Essentially all the modulation or regulation of function is seen in terms of the closing rate after an event. The channels are a binary logical element of the nervous system, being always either fully open or fully closed. We determined the three-dimensional structure of this molecule from acetylcholine-rich membrane fractions prepared from the electric fish *Torpedo californica* (6). The structure, at 25 Å resolution, was obtained by a combination of electron microscope studies on tubular crystalline arrays of the native acetylcholine receptor (7) and by X-ray diffraction of AChR-rich membranes (8).

Recently tubular arrays of extensively proteolyzed AChR from *T. marmorata* have been analyzed by frozen hydrated electron microscopy to give a 25-Å-resolution three-dimensional structure (9,10). The structure appears to be more fivefold symmetric than previous views of the receptor, particularly that portion of the structure on the extracellular side of the membrane (9,10). Because these *T. marmorata* samples were proteolyzed, the striking fivefold symmetry of the three-dimensional reconstruction may reflect symmetry of the subunit protein cores.

The *T. marmorata* receptors pack in a different lattice than the lattices we have observed for *T. californica* (7,9). The arrays that we observe for the nondegraded *T. californica* AChR fall into two classes (7). In one class the cell dimensions are $a = 92 \pm 2.2$ Å, $b = 85 \pm 1.7$ Å, and $\gamma = 125 \pm 2°$ (measured clockwise from the a axis). In the second lattice type, the unit cells are approximately 20% smaller than in the first, the a axis is parallel to the tube axis, and its length is 84.5 ± 0.5 Å. The *T. marmorata* receptors form a P2 plane lattice with $a = 90.0$ Å, $b = 160.8$ Å, and $\gamma = 120.2°$ (9). The area occupied per molecule is approximately the same as in our type I arrays. For close packing of pentamers, or pentagonal shapes, the P2 lattice is approximately 10% more efficient, in terms of the ratios of the smaller amount of unit cell occupied by lipid to that occupied by protein, than a P1 lattice would be.

The P2 lattice is formed by packing double rows of receptors in a manner similar to what is seen in quick-frozen, deep-etched postsynaptic membranes of electroplaques (11). The oblique *T. californica* AChR lattice is quite clearly not constructed from such double rows of receptors. Proteolysis of the *T. marmorata* receptors may be responsible for the more regular-looking P2 lattice, though we have never observed this lattice with proteolyzed *T. californica* AChR. In spite of the fact that the frozen hydrated sample preparation presumably ensures quite good specimen preservation, the phospholipid bilayer is not located in the analysis of the P2 arrays (10), suggesting that lipolysis may also be a factor in generating the more efficiently packed pentamers. In our attempts to grow tube structures from *T. californica* AChR *without* inhibition of proteases, lipases, or bacterial growth, we have not yet observed the rather neat double rows of receptors.

Our own analysis of three-dimensional crystals of AChR after solubilization in detergent shows that the molecular envelope is similar to that seen in the native receptor tubes, thus we aim for three-dimensional X-ray and electron crystallography from these crystals. The unit cell dimensions of detergent crystals are much larger than those for the tube membranes [$a = 135$ Å, $b = 145$ Å, and c is currently undetermined since we have so far analyzed only the (001) plane by electron crystallography—this is fortuitously the plane most easily seen in fragments—and $\gamma = 117°$].

THE IONIC CHANNEL LIES BETWEEN FIVE QUASI-EQUIVALENT SUBUNITS

AChR consists of five transmembranous subunits which transpose the membrane as shown by the binding of antibodies or by susceptibility to proteolysis from either side of the membrane (4,5). The center of the molecule, which is densely stained by negative stains such as uranyl acetate, contains these electron-dense uranyl ions throughout 80–90% of the entire length of the AChR, in a cylinder which is 7 Å in

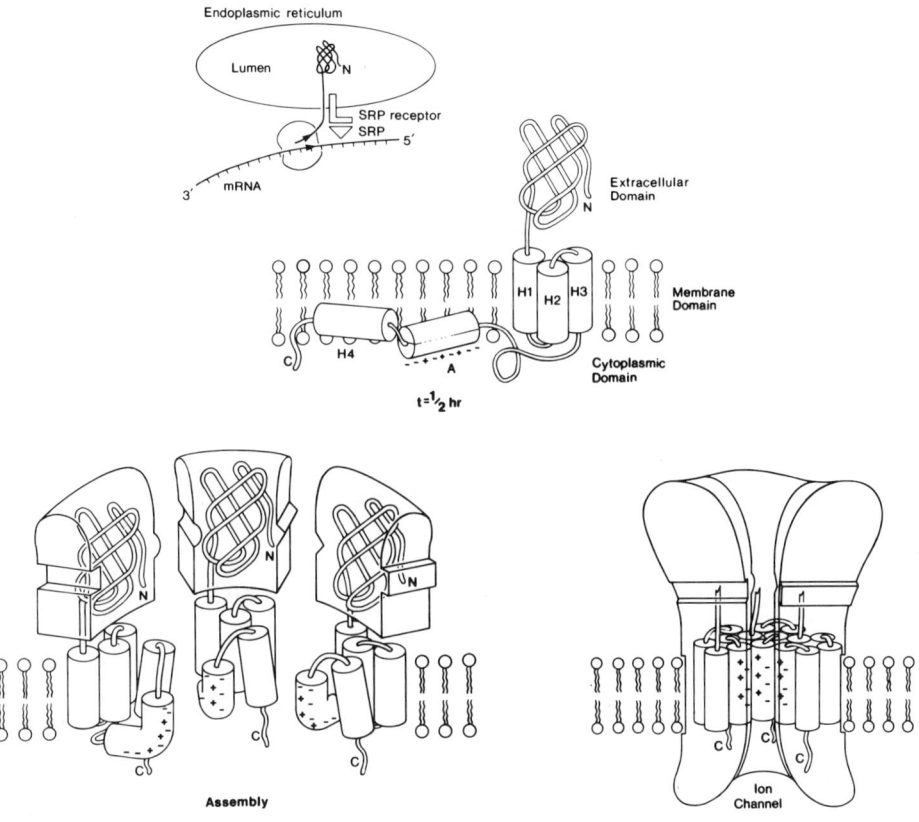

Fig. 1. Drawing of the three-dimensional structure of AChR derived from electron microscopy and X-ray diffraction is indicated alongside our proposal for assembly of the AChR from translated subunits. The disordered cytoplasmic domain must be larger than seen in this EM structure. The subunit locations are tentative though it is clear that the α subunits are not adjacent to one another. The ion channel probably contians ions in the resting state based on Tb^{3+} binding experiments (12). Cys 208/209 lie at the top crest, where toxin and cholinergic ligands bind (13).

diameter in the narrowest transmembranous part of the molecule (6) (Fig. 1) (12,13). By X-ray diffraction it was shown that neurotoxins which compete with acetylcholine for binding sites primarily located on the α-chains are located on the top crest of the AChR (14). Thus ligands binding on the top crest must signal their effect through at least 60 Å to the entrance to the ion channel.

X-ray diffraction shows that the AChR contains between 12 and 30 α-helices predominantly oriented perpendicular to the membrane plane (8). Alpha-helices provide a means of protecting hydrogen-bonding carbonyl and amide groups during cotranslational insertion into the membrane. There is precedent for α-helices which interface between protein and lipid. They are, for example, seen in the photoreaction center complex (15) and are inferred for HMG-CoA reductase, bacteriorhodopsin, and many other proteins (16) (Fig. 2) (17).

TOPOGRAPHY OF SEQUENCES CONFIRMS EXISTENCE OF AN AMPHIPATHIC MEMBRANE CROSSING

Amino acid sequences, deduced from the sequences of cDNA clones, show a homology between all four different gene (α, β, γ, δ) products from N to C terminus (18). Each sequence contains four very hydrophobic sequences of ~27, 26, 26, and 22 amino acids which bear all expected characteristics of those that transpose the membrane. The topology of each subunit is most probably identical in light of this strong sequence homology. Using antibodies generated against peptides contained within the sequence of the α- and β-chains, we were able to establish that residues around 392 and around 500 lie on the cytoplasmic side of the membrane, whereas the sequence 1–228 lies on the synaptic surface (see Fig. 3) (19). The latter sequence contains the amino terminus, generated by cleavage of the signal sequence in the lumen of the endoplasmic reticulum, and it contains regions involved in the binding of ligands (13) as well as the sites of N-glycosylation (18) (Fig. 1). The antibody experiments imply an even number of crossings between residues 392 and 500. In each subunit this region contains one highly hydrophobic sequence M4 and the amphipathic unusual sequence which presents an alternation of charge and hydrophobic residues, with a periodicity expected for sequences in α-helical structure.

Periodicities in sequences can be analyzed by Fourier transformation of the hydrophobicities of amino acids (20). Such Fourier transforms quantitatively demonstrate the degree of amphipathic quality

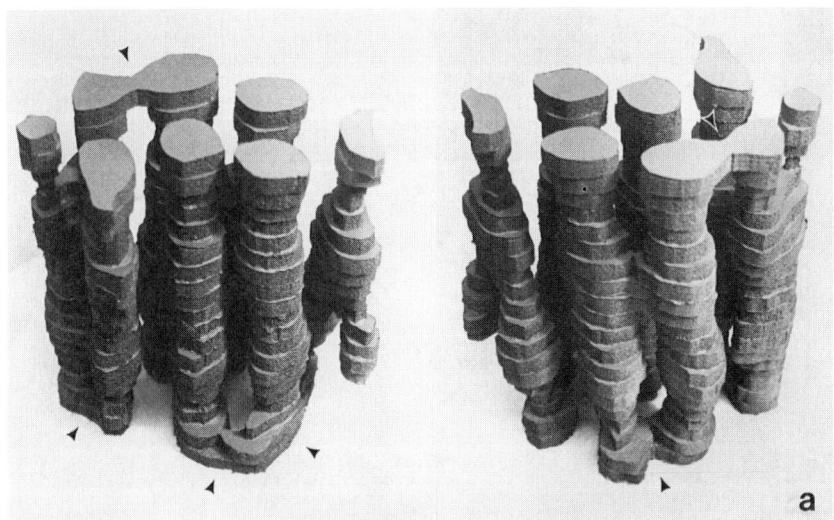
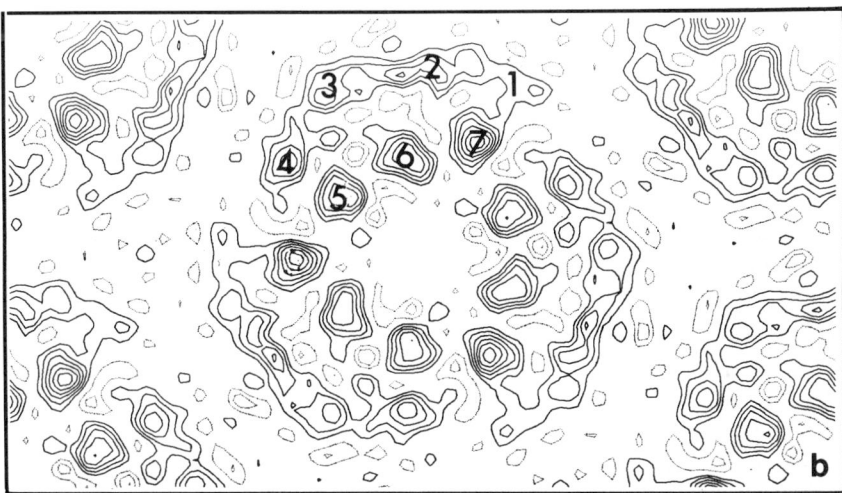

Fig. 2. (a) The three-dimensional structure of bacteriorhodopsin refined by Agard and Stroud (16) shows predominantly helical rods perpendicular to the membrane planes. These are probably α-helices. (b) A 3.5-Å electron diffraction-derived density map by Hayward and Stroud (17) showing the seven numbered helices found in bacteriorhodopsin. In this case, 21 helices surround a central lipid domain. The AChR channel structure postulated has four more amphipathic and charged helices within the lipid center that maintain a water-filled channel across the membrane.

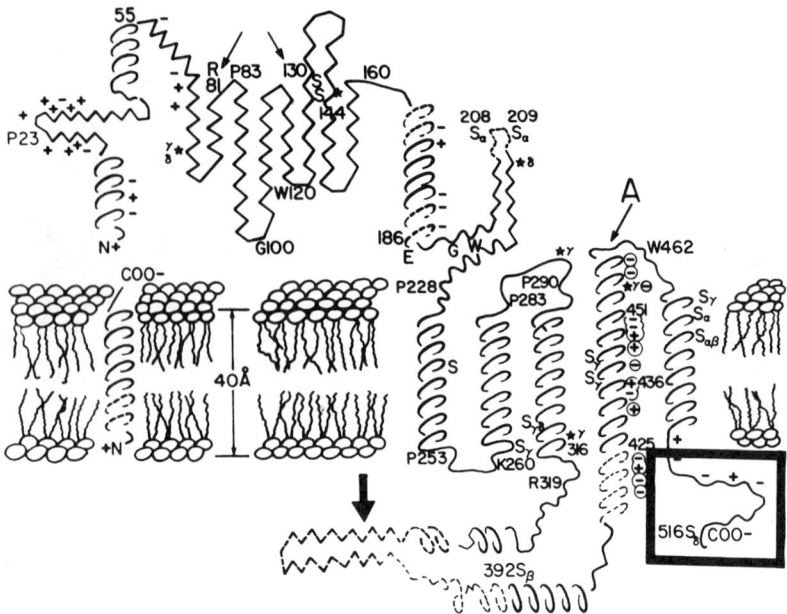

Fig. 3. Topology of the sequence in AChR subunits deduced from antibody labeling and electron microscopy. Stars indicate sites of N-glycosylation. Numbers and letters refer to the consensus aligned sequence numbers of Stroud and Finer-Moore (5).

expected in a particular secondary structure and provide a powerful tool for sequence analysis (5,20,21). Applied to chains of the AChR, this method shows that the N-terminus domain includes a sequence of 100 amino acids whose hydrophobicity alternates with the frequency expected for β-sheet. From the locations of turn-forming residues, we expect this sequence to form one eight-stranded, antiparallel β-pleated sheet structure per subunit. The analysis also identifies the strongly amphipathic helix "A" immediately preceding the final hydrophobic sequence, close to the carboxyl terminus of each chain. The implication of topological studies is that the amphipathic amino acid sequence in each subunit crosses the membrane (Fig. 3) and may provide the lining for the water-filled ion-conducting channel, which we have shown is located in the center of the molecule (6). One may therefore ask if an ion channel, constructed between a fivefold symmetric arrangement of these amphipathic segments of the five subunits, could create the characteristics of specificity and flux observed by electrophysiological studies for AChR.

A MODEL OF THE IONIC CHANNEL THAT CAN BE TESTED

To evaluate this possibility, a model was constructed in which the amphipathic sequences were close-packed around the ion channel (Fig. 4). The model is built with only two parameters, rotation around the axis of one helix and the tilt angle with respect to the vertical of the helical stretch. The former was optimized to bring the hydrophilic chains in toward the center of the molecule and the latter determined for optimum packing of side chains located between adjacent helices. Electrostatic energy calculations in which a sodium ion is placed at closely spaced positions throughout the length of the ion-conducting channel permit an estimation of the energy barriers to conductivity of that ion (5). These studies show that the energy barriers to conduction of a hydrated monovalent cation such as Na^+, assuming a dielectric constant of $\varepsilon = 80$ within the channel, are on the order of 3–4 kilocalories per mole, close to those estimated by electrophysiologists from ion conductivity measurements (Fig. 5) (22). The size of the model

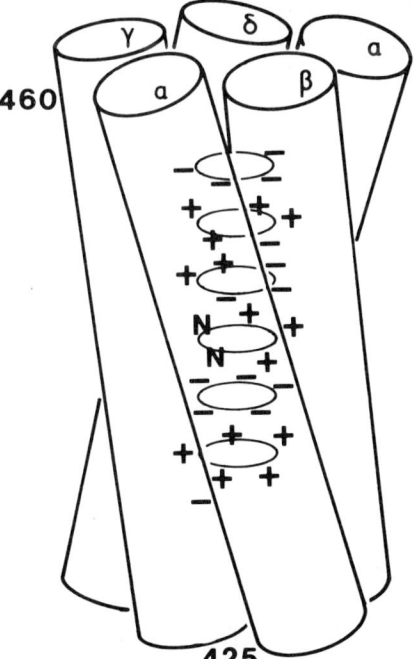

Fig. 4. Five-helix model of the ion channel in AChR where helices are represented by cylinders. One amphipathic helix, A in Fig. 3, derives from each subunit. Charged and hydrophilic groups in the channel lining are indicated.

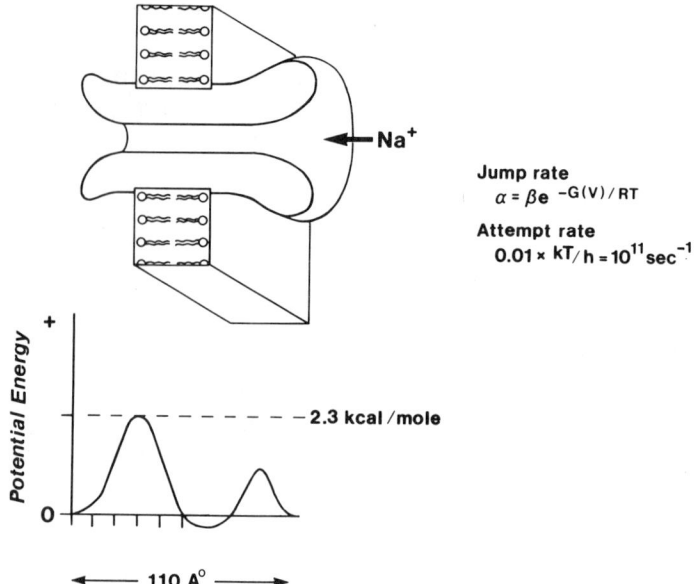

Fig. 5. Sketch of the energy profile for passage of a sodium ion through the AChR channel, deduced from electrophysiological experiments (22).

channel is sufficient to provide passage for larger ions like isopropyl ammonium, which are known to pass through the channel. The entry to the ion channel is negatively charged and provides a stable attractive space-charge force for cations and a repulsion for anions such as chloride. The net negative charge may explain the observed specificity for cations rather than anions in AChR, as revealed in the energy plots (see Fig. 6).

More recent calculations using a statistical mechanics method of steepest descent analysis, in which the entire channel model is permitted to librate and in which discrete water molecules fill the ionic channel, tend to show that the barriers to ion conductivity are much larger than estimated from the electrostatic calculations described above (P. Bash and R. M. Stroud, private communication). The difference is explicable most simply as being due to the fact that the water within the channel structure does not have a dielectric constant of 80, but one much closer to $\varepsilon \cong 8$. Negative charges within the ion channel model seem to provide a powerful attractive force for Na^+, thus impeding ion conductivity. Therefore, adjustments to the model are required to explain the rapid conductivity of cations observed (10^7–10^8

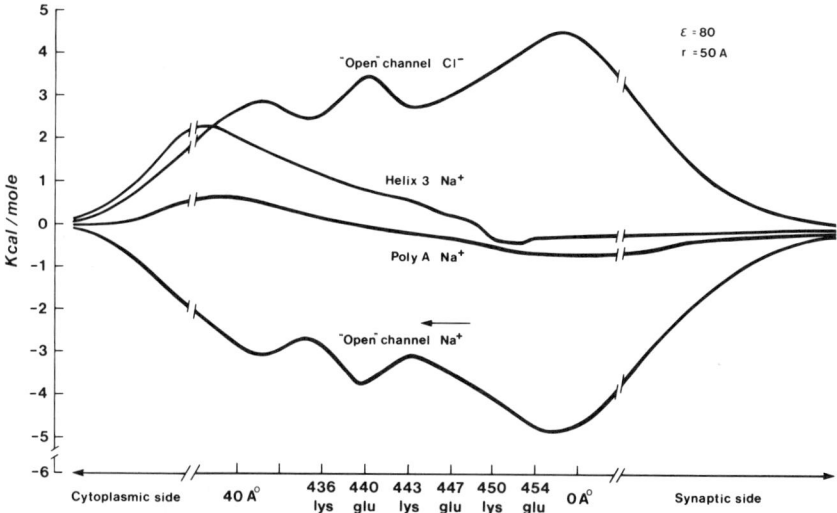

Fig. 6 Electrostatic energy profiles for a Na$^+$ and a Cl$^-$ ion as a function of distance from the extracellular side of the membrane are shown; they display cation selectivity. Calculations are for the model shown in Fig. 4 and for similar models constructed with hydrophobic helices M3 or for polyalanine helices. Conductivity for all models would be high, though for the latter two the channel could easily collapse without need for solvation. This calculation is based on the assumption of a dielectric constant of $\varepsilon = 80$ for water in the channel. Subsequent "Potential of Mean Force" calculations, in which discrete water molecules are included, show that $\varepsilon \sim 8$ is more appropriate. Thus we suspect that "rings of charge" in our original model will be disrupted and ions paired differently with each other to diminish the large energy barriers.

per second). Most likely, we propose, is that the five amphipathic helices may not all be at the same height with respect to each other. By relative vertical movements, charges within the channel would be better compensated by charges of opposite sign within the net neutral conducting zone. This may reduce the barriers and allow for the observed rates of conduction. Other model revisions are also plausible and may be tested. The statistical mechanical calculations do seem to verify that the basis for selectivity arises from the negative space-charge in the large open entrance to the ion-conducting channel.

The question may be raised as to why the central well in this model should carry charge at all. It is clear from electrostatic energy calculations in which the charged side chains are replaced by alanine, or by calculations in which the side chains are simply polar rather than charged, that the barriers to ion transduction by such models would be small (5). The (unproved) proposal for alamethecin action (23) depends on an arrangement of several copies of the short peptide helix that each contain only a single (negative) charge, around a central channel. Why then would the amphipathic helices in AchR carry charge? We propose that the high charge density within the channel would require hydration in the center of the molecular complex. Thus, it would stabilize the specific ion channel against collapse caused by repacking of the more hydrophobic helices.

In summary, this model suggests the following tentative conclusions: (i) The specificity for cations rather than anions arises from the predominantly negatively charged entry to the transmembranous region of the ion channel. (ii) The channel carries an almost equally balanced set of adjacent positive and negative charged side chains which preserve a water-filled volume. The arrangement of these side chains is crucial for ensuring that deep energy wells do not exist for a charged ion, since these energy wells would impede conductivity rather than enhance it. (iii) The size presented by the five-helix model, determined by the number of homologous subunits in any similar channel, is adequate to explain the size limitations for conductivity of organic and inorganic hydrated ions. (iv) The kind of structure based on amphipathic and charged helices we propose for the AChR channel will most likely be seen in other transmembrane channels such as those found in gap junctions or with colicins.

HOW IS THE IONIC CHANNEL GATED?

The question as to where the ion gating system might lie within the ion channel is addressed most directly by experiments of Fairclough *et al.* (12) in which terbium ions have been localized within the resting state of AChR. Terbium ions compete with calcium ions for sites on the acetylcholine receptor, and it has been shown that approximately 50% of these ion sites are displaced by the binding of ligands such as carbamyl choline. The Tb^{3+} binding sites were located perpendicular to the membrane by anomalous X-ray diffraction analysis. This technique is especially sensitive to lanthanide ions such as terbium with unfilled d-orbitals in which the wavelength-dependent dispersion is particularly large. One-dimensional diffraction data perpendicular to the membrane plane were recorded at five different wavelengths. The experiment is extremely well controlled and allows determination of the terbium ion positions, or rather the probability of Tb^{3+} occupation, relative to the membrane plane. In this case three distinct terbium ion locations were found in the transmembranous region of the protein. If it is assumed that these ion positions must be located in the most ionophilic region of the transmembrane complex, then it implies that terbium ions are bound inside the ion-conducting channel in the *resting* state of AChR, at positions which are approximately 11 Å apart (12). This distribution is consistent with the notion that there are stable positions for ions in the channel at each second turn of α-helix, surrounded by rings of negatively charged side chains, and it suggests the possibility that the resting state of acetylcholine receptor may contain bound ions in these locations *in vivo*.

Since the Tb^{3+} or Ca^{2+} sites displaced by Ach binding have not yet been assigned, no direct conclusion about mechanisms of channel gating can be inferred from this distribution. However, the minimum probability density of finding terbium ions lies 15 Å from the cytoplasmic surface of the protein, suggesting that this region is a possible focus for attention in the search for a gating mechanism.

LOCATION OF THE BINDING SITE FOR LIGANDS

Kao *et al.* (13) have localized the region of Cys 208/209 as being at the binding site for activating ligands. The neurotoxin binding site, which is most probably the Ach binding site, is located on the top crest of AchR (14). Thus, the effect of ligand binding must be relayed from the extracellular portion of the molecule to the channel. While it once seemed that a large-scale twisting of all subunits could be involved in gating, this kind of scheme does not answer the question at an atomic level, that is, it does not indicate which residues generate the "gate." There are still a variety of possible gating mechanisms being tested, in part by the elegant experiments of Mishina and co-workers (24). They have demonstrated that expression of protein followed by functional reconstitution in frog oocytes can permit the trial of different deletions or substituents in given regions of the protein structure using site-directed deletion and mutagenesis. Their studies pinpoint the functional domains in AchR and consequences for structure and/or function of microscopic changes of side chains in the sequence of this complex protein. These experiments, and those of Kao *et al.* (13), have illuminated the contribution of specific regions of the sequence to the mechanisms of the acetylcholine receptor.

ACKNOWLEDGMENTS

Work reported in these studies was carried out with support from National Institute of Health Grants GM24485 and GM32079 and from NSF Grant PCM83-16401.

REFERENCES

1. Karlin, A. (1983). *Neurosci. Comment.* **1**, 111–123.
2. Lindstrom, J. (1983). *Neurosci. Comment.* **1**, 139–156.
3. Popot, J.-L., and Changeux, J.-P. (1984). *Physiol. Rev.* **64**, 1162–1239.
4. McCarthy, M P., Earnest, J. P., Young, E. F., Choe, S., and Stroud, R. M. (1986). *Annu. Rev. Neurosci.* **9**, 383–413.
5. Stroud, R. M., and Finer-Moore, J. (1985). *Annu. Rev. Cell Biol.* **1**, 369–401.
6. Kistler, J., Stroud, R. M., Klymkowsky, M. W., Lalancette, R. A., and Fairclough, R. H. (1982). *Biophys. J.* **37**, 371–383.
7. Kistler, J., and Stroud, R. M. (1981). *Proc. Natl. Acad. Sci. U.S.A.* **78**, 3678–3682.
8. Ross, M. J., Klymkowsky, M. W., Agard, D. A., and Stroud, R. M. (1977). *J. Mol. Biol.* **116**, 635–659.
9. Brisson, A., and Unwin, P. N. T. (1984). *J. Cell Biol.* **99**, 1202–1211.
10. Brisson, A., and Unwin, P. N. T. (1985). *Nature (London)* **315**, 474–477.
11. Heuser, J. E., and Salpeter, S. R. (1979). *J. Cell Biol.* **82**, 150–173.
12. Fairclough, R. H., Miake-Lye, R. C., Stroud, R. M., Hodgson, K. O., and Doniach, S. (1986). *J. Mol. Biol.* **189**, 673–680.
13. Kao, P. N., Dwork, A. J., Kaldany, R.-R. J., Silver, M. L., Wideman, J., Stein, S., and Karlin, A. (1984). *J. Biol. Chem.* **259**, 11662–11665.
14. Fairclough, R. H., Finer-Moore, J., Love, R. A., Kristofferson, D., Desmeules, P. J., and Stroud, R. M. (1983). *Cold Spring Harbor Symp. Quant. Biol.* **48**, 9–20.
15. Deisenhofer, J., Epp, O., Miki, K., Huber, R., and Michel, H. (1985). *Nature (London)* **318**, 618–624.
16. Agard, D. A., and Stroud, R. M. (1982). *Biophys. J.* **37**, 589–602.
17. Hayward, S. B., and Stroud, R. M. (1981). *J. Mol. Biol.* **151**, 491–517.
18. Noda, M., Takahashi, H., Tanabe, T., Toyosato, M., Kikyotani, S., Furutani, Y., Hirose, T., Takashima, H., Inayama, S., Miyata, T., and Numa, S. (1983). *Nature (London)* **302**. 528–532.
19. Young, E. F., Ralston, E., Blake, J., Ramachandran, J., Hall, Z. W., and Stroud, R. M. (1985). *Proc. Natl. Acad. Sci. U.S.A.* **82**, 626–630.
20. Finer-Moore, J., and Stroud, R. M. (1984). *Proc. Natl. Acad. Sci. U.S.A.* **81**, 155–159.
21. Liscum, L., Finer-Moore, J., Stroud, R. M., Luskey, K. L., Brown, M. S., and Goldstein, J. L. (1985). *J. Biol. Chem.* **260**, 522–530.
22. Horn, R., and Stevens, C. F. (1980). *Comments Mol. Cell. Biophys.* **1**, 57–68.
23. Fox, R. O., and Richards, F. M. (1982). *Nature (London)* **300**, 325–330.
24. Mishina, M., Tobimatsu, T., Imoto, K., Tanaka, K., Fujita, Y., Fukuda, K., Kurasaki, M., Takahashi, H., Morimoto, Y., Hirose, T., Inayama, S., Takahashi, T., Kuno, M., and Numa, S. (1985). *Nature (London)* **313**, 364–369.

22
Membrane Protein Folding Motifs: An Examination of Empirical Predictions of Secondary Structure

D. L. MIELKE, M. CASCIO, AND B. A. WALLACE[1]
Department of Biochemistry and Molecular Biophysics
Columbia University
New York, New York 10032

INTRODUCTION

Two major goals of protein structural analyses are to determine the relationship between protein structure and function and to investigate the mechanism by which proteins fold into their native conformations. Since the secondary structures of soluble proteins have been shown to be determined predominantly by their amino acid sequences (Afinsen *et al.*, 1961), it may be possible to predict the folding of proteins from their sequences. Many empirical prediction schemes for secondary structures have been developed based on motifs found in soluble proteins of known three-dimensional structure and sequence (Lim, 1974; Garnier *et al.*, 1978; Argos *et al.*, 1982). The most widely used of these is the method of Chou and Fasman (1974a). Using a data base composed of 15 water-soluble proteins whose conformation had been determined by X-ray crystallographic analyses, this method calculated the probability of each of the 20 commonly found amino acids adopting α-helical or β-sheet conformations. These probabilities were then

---

[1] Present address: Department of Chemistry, Rensselaer Polytechnic Institute, Troy, New York 12180-3590.

applied to predict the secondary structures of proteins of known sequence but undetermined conformation. This scheme has been reported to predict the helix, sheet, and coil regions of soluble proteins with approximately 80% accuracy (Chou and Fasman, 1974b). Later studies extended the predictive power to include β-turns (Chou and Fasman, 1977, 1979). This paper examines the applicability of the Chou and Fasman parameters for structural prediction of membrane proteins.

Membrane proteins generally have one of three functions: the transport of molecules and ions (e.g., channels), information transfer (e.g., receptors), or anchorage and concentration of functional groups which act primarily in the aqueous phase. The structures of membrane proteins are influenced by the hydrophobic lipid environment of the membrane which, unlike an aqueous milieu, solvates neither peptide amino and carbonyl groups nor exposed polar side chains. Since the lipid "solvent" in a membrane cannot form hydrogen bonds with the peptide backbone, secondary structures which satisfy the hydrogen-bonding potentials of the amino and carbonyl groups are anticipated to be even more energetically favored in the hydrophobic environment of the membrane than in aqueous solution. The α-helical and β-sheet type structures, which permit hydrogen bond formation between all backbone groups, achieve this end, so these types of secondary structure might be expected to be formed with different probabilities than they are in soluble proteins. The additional energy required to bury unneutralized charges in the lipid hydrocarbon region, along with potential hydrophobic interactions and steric constraints between the lipid fatty acid chains and amino acid side chains, may also result in folding motifs for membrane proteins which differ substantially from those in soluble proteins. Additionally, functional properties of the membrane, such as asymmetry and the directional orientation of activity and insertion, may impose certain structural constraints.

Several polypeptides have, indeed, been found to adopt different conformations in their membrane-bound and soluble forms. For example, two channel-forming molecules, alamethicin (Cascio and Wallace, 1984) and gramicidin (Wallace *et al.*, 1983), exhibit this sensitivity to environment. Additionally, the helical content of the thermolysin fragment of colicin E1 (see below) is decreased dramatically from that in aqueous solution upon association of the molecule into dimyristoyl phosphatidylcholine (DMPC) vesicles.

Empirical prediction methods may not accurately predict secondary structures for proteins in membranes since their data bases are derived solely from soluble proteins and thus may not account for the energetic and steric constraints imposed on membrane proteins. With these considerations in mind, we compare the secondary structures predicted by the Chou and Fasman method with available experimental secondary structural data on a number of membrane proteins to determine the applicability of this predictive scheme.

MATERIALS AND METHODS

Twelve membrane proteins (or transmembrane fragments thereof) of known secondary structure and sequence were examined. In cases where the membrane-spanning portions of a protein had been identified and experimental data on their secondary structures were available, these fragments were analyzed rather than the intact protein since they may be more representative of a protein in contact with a hydrophobic environment. The proteins examined were the acetylcholine receptor, alpha-toxin from *Staphylococcus aureus*, bacteriorhodopsin, the C-terminal thermolysin fragment of colicin E1, crambin, the membrane-embedded fragment of cytochrome b_5, fd coat protein, the membrane-spanning region of glycophorin, lactose carrier, porin, proton-ATPase, and rhodopsin. The analysis employed a computerized version of the method of Chou and Fasman written by E. Elinopolis of Leeds University. The predicted secondary structures were compared to experimental secondary structural data as determined by X-ray diffraction, circular dichroism (CD), Raman spectroscopy, or infrared (IR) spectroscopy either taken from the literature or examined in this laboratory. Some structural studies were not conducted in membranes; in these cases, data reported are for the conformation of the protein in a detergent environment which may mimic the membrane. Whenever possible, experimental results from more than one laboratory and/or determined by more than one method are described and compared so that limits of reliability of the measurements may be evaluated.

For consistency, and because many published studies provide spectra without any secondary structural analysis or only a qualitative description of secondary structure, all CD data not originally from this laboratory were reanalyzed. The published spectra were digitized and the secondary structures were calculated by analysis of data in the wavelength range from 240 to 190 nm as previously described (Mao *et al.*, 1982). In general, the results thus obtained corresponded very closely to values given in the original reports.

Some ambiguity is inherent in the interpretation of the Chou and Fasman analysis because of overlapping stretches which exhibit high probabilities to form both α-helix and β-sheet structures. The lower end of the reported ranges in this paper include only those residues which are predicted as solely α-helix or β-sheet, but not both. The higher reported fractions include those residues predicted to be in one of these two conformations (i.e., overlapping regions), with the α-helical residues also including those helical regions which arise from aspartate, proline, or arginine promotion. The results are compiled in Table I, where the proteins are listed in order of increasing experimentally determined α-helix content.

As a control, two soluble proteins included in the Chou and Fasman reference data set—the primarily α-helical myoglobin (Kendrew et al., 1961) and ribonuclease, a protein relatively rich in β-structure (Wyckoff et al., 1970)—were subjected to analysis by this algorithm. The agreement between the predicted and the experimentally determined secondary structures of each test protein was reasonably good and agreed with the value reported by Chou and Fasman.

RESULTS

Each protein that was examined is described in terms of its disposition in the membrane, the experimental data available, and the envi-

TABLE I
Experimental and Predicted Secondary Structures

Protein	Experimental α-helix	Predicted α-helix	Experimental β-sheet	Predicted β-sheet
Alpha-toxin	5–10	24–31	55–62	25–28
Colicin E1[a]	9	40–65	33	24–30
Cytochrome b_5[b]	12–20	35–48	39–50	22–24
fd coat protein	17	48–54	83	38
Acetylcholine receptor	17–25	26–42	34–51	27–37
Porin	27	22–29	40	19–22
Crambin	36	0–15	22	17–20
Rhodopsin	66	28–34	0	26–32
Proton-ATPase proteolipid	75	57	0	0
Lactose carrier	70–81	29–41	12–19	31–39
Bacteriorhodopsin	74–83	39–53	0–9	22–28
Glycophorin[c]	76–100	17–34	0	60–69

[a] The C-terminal thermolysin fragment (residues 336–512).
[b] Membrane fragment
[c] A fragment encompassing the membrane-spanning region.

ronment and method from which such data are derived. Additionally, the correspondences between experimental and calculated values for α-helix and β-sheet are reported.

ACETYLCHOLINE RECEPTOR

The acetylcholine receptor (AChR) (MW = 285,000) forms pentameric membrane channels composed of four homologous subunits (Noda et al., 1982, 1983a,b; Claudio et al., 1983) with the stoichiometry $\alpha_2\beta\gamma\delta$ (Reynolds and Karlin, 1978; Lindstrom et al., 1979; Raftery et al., 1980). All subunits span the membrane (Flynn et al., 1982). Approximately 70% of this protein is extramembranous (Karlin et al., 1984); it protrudes from the bilayer by 55 Å on the extracellular side and 15 Å on the cytoplasmic side of the neuromuscular junction end plate (Kistler et al., 1982). On binding of two acetylcholine molecules, the channel opens to allow a Na^+ influx, resulting in membrane depolarization and subsequent nerve signal conductance.

Raman spectroscopy (Yager et al., 1984) of the AChR reconstituted in dielaidyl phosphatidylcholine vesicles indicates 25% α-helix and 39% β-sheet. CD studies (Mielke et al., 1986) of this protein solubilized in cholate or reconstituted in asolectin vesicles indicate 17–25% α-helix and 38–51% β-sheet, the exact value depending on the reconstitution condition. Given the error limits associated with both of these measurements, the calculated secondary structures derived from these studies are in reasonable agreement.

Analysis of the sequence (Noda et al., 1982, 1983a,b; Claudio et al., 1983) by the Chou and Fasman method predicts 26–42% α-helix and 27–37% β-sheet for this protein. Comparison of these predictions with the experimental data (Table I) shows a reasonably good agreement, with a small overprediction of α-helical structure and an underprediction of β-sheet.

ALPHA-TOXIN

The alpha-toxin of Staphylococcus aureus is secreted as a 33,000-Da monomer (Gray and Kehoe, 1984). On contact with susceptible cells, this amphiphilic polypeptide assembles into a hexameric membrane channel with a pore diameter of about 30 Å (for review, see Freer and Arbuthnott, 1982).

Both the monomeric and hexameric forms of alpha-toxin have been examined by CD spectroscopy (Tobkes et al., 1985) in 10 mM NaP_i–150 mM NaCl, pH 8.0, with and without 1.25 mM deoxycholate (DOC). The secondary structure composition of the toxin appears to be 5–10% α-helix and 55–62% β-sheet under all these conditions.

Chou and Fasman analysis of the sequence (Gray and Kehoe, 1984) predicts 24–31% α-helix and 25–28% β-sheet, significantly overestimating α-helix and underestimating β-sheet for the protein both in the membrane-mimetic DOC and in an aqueous solution. So, in this case, Chou and Fasman analysis also fails to accurately predict the structure of the solubilized protein.

BACTERIORHODOPSIN

Bacteriorhodopsin (BR) is the sole protein component of the purple membrane of *Halobacterium halobium* (Oesterhelt and Stoeckenius, 1971). This protein (for review, see Wallace, 1982) is composed of a single polypeptide chain of 26,500 Da (Ovchinnikov et al., 1979), most of which is localized in the transmembrane region as evidenced by protection from proteolysis (Khorana et al., 1979). As a light-driven proton pump, BR creates an electrochemical gradient across the plasma membrane, the energy from which is used by the cell for ATP synthesis (Oesterhelt and Stoeckenius, 1973).

Several CD studies have been done on BR in DMPC vesicles (Mao and Wallace, 1984; Nabedryk et al., 1985; Wallace and Teeters, 1986; Popot et al., submitted for publication) that indicate a secondary structure composed of ~80% α-helix and ~5% β-sheet. The data correlate well with the three-dimensional structure determined by image reconstruction (Henderson and Unwin, 1975; Leifer and Henderson, 1983; Engelman et al., 1980), which includes approximately 80% α-helix and no detectable β-sheet.

Chou and Fasman analysis of the sequence (Ovchinnikov et al., 1979; Khorana et al., 1979) of BR predicts only 39–53% α-helix whereas β-sheet content is substantially overpredicted at 22–28%.

COLICIN E1

Colicin E1 is a 60,000-Da antibacterial toxin which acts by depolarizing the inner membrane of *Escherichia coli*. This protein forms a trans-negative, voltage-dependent ion channel which is monomeric, nonspecific, and dissipates the cell membrane potential (Davidson et al., 1984). The channel is 6–8 Å in diameter and its activity lies in the hydrophobic C-terminal fragment (residues 370–522) created by cyanogen bromide (CNBr) cleavage (Cleveland et al., 1983). Similar, slightly larger channel-forming fragments can be prepared by digestion with thermolysin and trypsin. The receptor binding activity is found in the 231–370 residue fragment produced by CNBr cleavage, suggesting that much of the molecule is extramembranous.

CD studies indicate that the C-terminal thermolysin fragment (residues 336–512) is approximately 9% α-helical with 33% β-sheet content (D. Mao., M. Cleveland, C. Levinthal, and B. A. Wallace, unpublished data) in small, unilamellar DMPC vesicles, whereas the slightly larger C-terminal tryptic peptide (residues 336–522) is 41% α-helical in aqueous solution (Brunden et al., 1984), with a 16% β-sheet content.

The C-terminal fragment of colicin E1 (Yamada et al., 1982) is predicted to be 40–65% α-helical, much greater than that determined experimentally in membranes. Additionally, the 24–30% β-sheet predicted is slightly less than the fraction found in membranes. However, these predictions are closer to the secondary structure of the molecule in an aqueous environment.

CRAMBIN

Crambin is a small, hydrophobic protein isolated from aqueous acetone extracts of delipidated embryonic tissue in seeds of the plant *Crambe abyssinica* (Van Etten et al., 1965). This 46-amino acid protein was found to have extensive sequence homology with a family of membrane-active plant toxins, e.g., viscotoxin and purothionin (Teeter et al., 1981), although it does not exhibit similar lytic activity. Crambin has been shown to be capable of being solubilized in and associating with lipid bilayers (Wallace et al., 1984), although its function is unknown.

Crambin has been crystallized from 60% ethanol and its structure determined to 1.5 Å resolution (Hendrickson and Teeter, 1981). These studies indicate that the molecule is composed of 37% α-helix and 17% β-sheet structure. Comparative CD studies of this protein solubilized in the same solvent (60% ethanol) and reconstituted into small unilamellar DMPC vesicles show the protein to have identical secondary structures—36% α-helix and 22% β-sheet (Wallace et al., 1984).

Chou and Fasman analysis on the sequence of crambin predicts 0–15% α-helix and 17–20% β-sheet structure. Comparison of these estimates with the experimentally determined secondary structures shows a pronounced underprediction of α-helical structure in the protein.

CYTOCHROME b_5

Cytochrome b_5 is a 16,000-Da membrane-bound heme protein isolated from hepatic endoplasmic reticulum. The single polypeptide chain is folded into two functionally distinct and independent segments (Spatz and Strittmatter, 1971): the 11,000-Da N-terminal catalytic subunit of polar residues with the noncovalently bound heme group, and the 5000-Da hydrophobic segment which anchors the protein in the endoplasmic reticulum. This cytochrome plays a key role in electron transport, where it is required for the function of the stearyl-CoA desaturase system (Spatz and Strittmatter, 1971; Strittmatter et al., 1972).

CD studies of the nonpolar fragment of cytochrome b_5 in a variety of environments (Dailey and Strittmatter, 1978), including DMPC vesicles and sodium dodecyl sulfate (SDS) and DOC detergents, show 12–20% α-helix and 39–50% β-sheet.

Chou and Fasman analysis of this nonpolar segment sequence (Fleming et al., 1978) predicts 35–48% α-helix and 22–24% β-sheet, so the α-helical content is overpredicted whereas the β-sheet is underpredicted.

fd COAT PROTEIN

The fd coat protein is the major coat protein of class 1 filamentous bacteriophages. This 5200-Da polypeptide has a known sequence of 50 amino acids which contains a central core of 19 hydrophobic residues flanked by stretches of charged residues (Nakashima and Konigsberg, 1974). During biosynthesis, the protein inserts into and spans the cell membrane (Wickner, 1976). Subsequently, the phage coat assembles to form the virus as the single strand of viral DNA is extruded through the host membrane (Smilowitz, 1974).

Early structural studies using CD indicated that the conformation of this protein is strongly dependent on environment—the secondary structure of the membrane-bound state differs from that in the intact phage (Nozaki et al., 1976). Additional studies have shown that the fd coat protein is entirely α-helical in intact phage (Williams and Dunker, 1977), but in reconstituted vesicles its structure is approximately 17% α-helical with 83% β-sheet content (B. A. Wallace and M. R. Kimball, unpublished results).

Chou and Fasman analysis on the coat protein sequence estimated the α-helical content to be 48–54% and 38% of the molecule to be in a β-sheet conformation. These predictions do not correlate with the experimentally determined structure of either membrane-bound or virus-associated coat protein.

GLYCOPHORIN

Glycophorin is the major sialoglycoprotein of the human erythrocyte plasma membrane. Its 131-amino acid sequence (Tomita and Marchesi, 1975; Furthmayr et al., 1978) contains three distinct domains: an intracellular C-terminus domain, a hydrophobic 23-residue membrane-spanning domain which is responsible for the self-association of glycophorin, and a heavily glycosylated N-terminus extracellular domain (Silverberg et al., 1976; Tomita et al., 1978). The 16 oligosaccharide chains on the N-terminus domain account for 60% of the protein's 31,000-Da mass. Glycophorin's only known functional role is to carry the M and N blood group antigen activities and the influenza virus binding sites (Winzler et al., 1969).

The hydrophobic, proteolytically derived, membrane-spanning segment was examined by CD spectroscopy in SDS detergent (Schulte and Marchesi, 1979) and found to be entirely α-helical. Additionally, a similar synthetic segment of the membrane-spanning fragment of glycophorin (residues 73–94) was examined in SDS/trifluoroethanol (Goldstein et al., 1986). CD studies indicate its secondary structure to be 76% α-helical and to contain no β-sheet structure.

Analysis of the sequence of the glycophorin transmembrane residues predicts this fragment to be 17–34% α-helix and 60–69% β-sheet. This prediction is in direct contradiction with the experimentally determined secondary structure, which indicates the molecule to be entirely helical.

LACTOSE CARRIER

The lactose carrier is a 417-amino acid protein encoded by the *lac y* gene. This 46,500-Da protein is very hydrophobic, spans the membrane, and acts to translocate lactose in symport with protons. This translocation results from the coupling of the uphill accumulation of solute with the downhill movement of protons in response to a transmembrane electrochemical proton gradient ($\Delta\mu$). The lack of antigenic sites indicates that most of the protein is buried in the bilayer (for review, see Kaback, 1983).

The secondary structures of the lac carrier protein in either octylglucoside or reconstituted into proteolipid vesicles were found to be identical by CD spectroscopic studies and had spectra compatible with a protein that was 81% α-helical and 19% β-sheet (Foster et al., 1983).

Analysis of the protein sequence, which was derived from the nucleotide sequence (Buchel et al., 1980), predicts the molecule to be 29–41% α-helical. This value is significantly less than that determined by the structural studies. Additionally, 31–39% of the molecule is predicted to be in a β-sheet conformation, much more than that fraction found experimentally.

PORIN

Porin is the major envelope protein of *Escherichia coli*. Although it has a high proportion of charged residues (polar:apolar = 1.8), this 36,500-Da protein is located mostly within the outer membrane (Rosenbusch, 1974). It provides mechanical and osmotic stability for the cell by forming a hexagonal network (Garavito et al., 1983) of trimeric channels (Steven et al., 1977) on the peptidoglycan layer.

IR spectroscopic studies (Rosenbusch, 1974) suggest substantial β-sheet conformation, and the X-ray diffraction pattern of the hexagonal crystal form of porin (Garavito et al., 1983) indicates the presence of a significant amount of β-sheet perpendicular to the membrane plane. CD studies of porin in 1% SDS (Schindler and Rosenbusch, 1984) indicate that the molecule is 27% α-helix and 40% β-sheet.

Chou and Fasman analysis of the porin sequence (Overbeeke et al., 1983) predicts 22–29% α-helix and 19–22% β-sheet. Although the α-helix prediction correlates well with experimental data, the extent of β-structure is greatly underestimated.

PROTON-ATPASE PROTEOLIPID

The DCCD-binding (dicyclohexylcarbodiimide) proteolipid is an 8000-Da mitochondrial subunit of the proton-ATPase (H^+-ATPase) (Sebald et al., 1979) which reversibly couples oxidative phosphorylation of ADP to the translocation of protons across the membrane (for reviews, see Kagawa, 1972; Racker and Stoeckenius, 1974). The ATPase enzyme complex consists of two components, F_0 and F_1. F_1 is a water-soluble protein responsible for the ATPase activity of this complex, whereas F_0 is an integral membrane protein which acts as a channel facilitating proton translocation across the membrane (Foster and Fillingame, 1979). Despite the absence of any covalently bound lipid, one subunit of the F_0 component has been designated "proteolipid" because of its solubility in $CHCl_3$/MeOH. Seventy-five percent of the residues in this subunit are hydrophobic (Sebald et al., 1979), and in the assembled complex essentially all the proteolipid is buried in the membrane.

CD studies (Mao et al., 1982) on the proteolipid indicate that in DMPC vesicles its structure is 75% α-helical and contains no β-sheet structure.

Chou and Fasman analysis of the sequence (Sebald et al., 1979) predicts 57% α-helix and 0% β-sheet. Thus, the helix is underpredicted, but the low β-sheet content is well estimated.

RHODOPSIN

Rhodopsin, a 28,600-Da protein, is the photoreceptor pigment in vertebrate retinal rod cells. Rhodopsin functions as a transducer: it converts the energy of an absorbed photon into a transient change in the cell membrane electrical potential, which is subsequently transmitted to the nervous system. The molecule consists of a chromophore, the 11-cis isomer of retinal (an aldehyde of vitamin A) covalently bound to the apoprotein opsin (for review, see Hargrave, 1984). The inaccessibility of peptide N–H groups to proton exchange (except under severe denaturation conditions) in H–D exchange measurements by IR spectroscopy indicates that most of this molecule is embedded in the bilayer (Osborne and Nabedryk-Viala, 1977).

The secondary structure of rhodopsin was examined by CD spectroscopic studies in 0.05 M octylglucoside (Stubbs et al., 1976). Analysis of the spectra shows that 66% of the molecule is α-helical and that it lacks β-sheet.

The 348-amino acid sequence of rhodopsin is known (Ovchinnikov, 1982), and Chou and Fasman analysis predicts 28–34% of the molecule to be α-helical and 26–32% to be in a β-sheet conformation. Both of these values vary considerably from those determined experimentally.

DISCUSSION

The Chou and Fasman method, which utilizes a data base derived from a set of 15 soluble proteins, was used to predict the secondary structural characteristics of 12 membrane proteins of known sequence and secondary structure. These predictions were compared with the structural composition as determined experimentally (Table I). A chi-square test indicated that there is less than 0.5% correlation between the two data sets, i.e., the predicted secondary structures and the experimental results are independent. The 12 membrane proteins were selected to represent a wide range of structural motifs; their composition ranged from the predominantly helical membrane-spanning fragment of glycophorin to the extensively β-sheet-containing alpha-toxin. Assuming that the examined proteins are a representative subset of membrane proteins in general, it follows that the present Chou and Fasman predictive scheme does not accurately predict membrane protein secondary structure. It is also reasonable that other empirical prediction methods which, like the Chou and Fasman method, use a data base derived from soluble proteins, may also prove to be inapplicable for membrane proteins (Wallace et al., 1986).

The failure of the Chou and Fasman method in predicting membrane protein folding is understandable in light of the hydrophobic and anisotropic nature of the bilayer. This environment cannot solvate exposed polar groups, therefore α-helices and β-sheets, which satisfy the hydrogen-bonding potentials of the peptide backbone, are differentially favored relative to an aqueous environment.

It follows that membrane proteins with large, solvent-accessible, extramembranous domains may be more amenable to analysis by the Chou and Fasman predictive scheme. This seems to be the case for the large, pentameric acetylcholine receptor (see Table I), where the predicted and experimental secondary structures more closely correspond. The predictive power of this scheme for intra- and extracellular domains is additionally evidenced in examination of the structure of the extramembranous domains of glycophorin. When the secondary structures of these isolated regions were determined by CD spectroscopy, they were found to be 10% α-helical and 10% β-sheet (Schulte and Marchesi, 1979). Chou and Fasman analysis of these domains predicts 13–26% α-helix and 10–15% β-sheet, both of which are considerably closer to the experimentally determined secondary structures than are the predicted values for the transmembrane domain.

A number of membrane proteins are amphimorphic: they can exist in both aqueous-soluble and membrane environments. The Chou and Fasman method was expected to more accurately reflect the secondary structure of aqueous-solubilized forms of these proteins. This is the case for the C-terminal peptide of colicin E1. CD studies of this molecule in aqueous solution at pH 6.5 indicate 41% helix and 16% β-sheet content. These results are close to the predicted values of 40–65% and 24–30%, respectively, and much different than the experimental values for the thermolysin fragment in membranes. This result seems reasonable, as the aqueous-solubilized form should be hydrated in the manner of other soluble proteins. However, alpha-toxin, another amphimorphic protein which adopts similar secondary structures but different tertiary structures in aqueous and membrane environments (Tobkes et al. 1985), is not well predicted in either environment.

In summary, though the Chou and Fasman method may be a valuable tool for predicting secondary structures of soluble proteins, it fails to accurately predict secondary structures when applied to membrane proteins. It is anticipated that progress in the field of membrane protein studies will eventually provide sufficient structural and sequence information to allow compilation of a data base for the prediction of membrane protein folding. Until that time, however, it is clear that this method should not be used to *ab initio* predict secondary structures of membrane proteins and that information from experimental methods must serve as the basis for model building of membrane-embedded proteins.

ACKNOWLEDGMENTS

This work was supported by NIH Grant GM27292. D.L.M. is the recipient of National Institutes of Health Traineeship NSO7258. B.A.W. was the recipient of a Hirschl Career Scientist Award and a Camille and Henry Dreyfus Teacher–Scholar Award during the course of these investigations.

REFERENCES

Anfinsen, C. B., Haber, E., Sela, M., and White, F. H., Jr. (1961). *Proc. Natl. Acad. Sci. U.S.A.* **47**, 1309–1314.
Argos, P., Rao, J. K. M., and Hargrave, P. A. (1982). *Eur. J. Biochem.* **128**, 565–575.
Brunden, K. R., Uratani, Y., and Cramer, W. A. (1984). *J. Biol. Chem.* **259**, 7682–7687.
Buchel, D. E., Groneborn, B., and Muller-Hill, B. (1980). *Nature (London)* **283**, 541–545.
Cascio, M., and Wallace, B. A. (1984). *Ann. N. Y. Acad. Sci.* **435**, 527–529.
Chou, P. Y., and Fasman, G. D. (1974a). *Biochemistry* **13**, 211–222.
Chou, P. Y., and Fasman, G. D. (1974b). *Biochemistry* **13**, 222–245.
Chou, P. Y., and Fasman, G. D. (1977). *J. Mol. Biol.* **115**, 135–175.
Chou, P. Y., and Fasman, G. D. (1979). *Biophys. J.* **26**, 367–384.
Claudio, T., Ballivet, M., Patride, J., and Heinemann, S. (1983). *Proc. Natl. Acad. Sci. U.S.A.* **80**, 1111–1115.
Cleveland, M., Slatin, S., Finkelstein, A., and Levinthal, C. (1983). *Proc. Natl. Acad. Sci. U.S.A.* **80**, 3706–3710.
Dailey, H. A., and Strittmatter, P. (1978). *J. Biol. Chem.* **253**, 8203–8209.
Davidson, V. L., Brunden, K. R., Cramer, W. A., and Cohen, F. S. (1984). *J. Membr. Biol.* **79**, 105–118.

Engelman, D. M., Henderson, R., McLachlan, A., and Wallace, B. A. (1980). *Proc. Natl. Acad. Sci. U.S.A.* **77**, 2023–2027.

Fleming, P. J., Dailey, H. A., Corcoran, D., and Strittmatter, P. (1978). *J. Biol. Chem.* **253**, 5369–5372.

Flynn, D. D., Kloog, Y., Potter, L. T., and Axelrod, J. (1982). *J. Biol. Chem.* **257**, 9513–9517.

Foster, D. L., and Fillingame, R. H. (1979). *J. Biol. Chem.* **254**, 8230–8236.

Foster, D. L., Boublik, M., and Kaback, H. R. (1983). *J. Biol. Chem.* **258**, 31–34.

Freer, J. H., and Arbuthnott, J. P. (1982). *Pharmocol. Ther.* **19**, 55–106.

Furthmayr, H., Galardy, R. E., Tomita, M., and Marchesi, V. T. (1978). *Arch. Biochem. Biophys.* **185**, 21–29.

Garavito, R. M., Jenkins, J., Jansonius, J. N., Karlsson, R., and Rosenbusch, J. P. (1983). *J. Mol. Biol.* **164**, 313–327.

Garnier, J., Osguthorpe, D. J., and Robson, B. (1978). *J. Mol. Biol.* **120**, 97–120.

Goldstein, N., Kortylewicz, Z., Galardy, R., and Wallace, B. A. (1986). *Ann. N. Y. Acad. Sci.* **463**, 384–388.

Gray, G., and Kehoe, M. (1984). *Infect. Immun.* **46**, 615–618.

Hargrave, P. A. (1984). *Vision Res.* **24**, 1487–1499.

Henderson, R., and Unwin, P. N. T. (1975). *Nature (London)* **257**, 28–33.

Hendrickson, W. A., and Teeter, M. M. (1981). *Nature (London)* **290**, 107–113.

Kaback, H. R. (1983). *J. Membr. Biol.* **76**, 95–112.

Kagawa, Y. (1972). *Biochim. Biophys. Acta* **265**, 297–338.

Karlin, A., Cox, R., Kaldany, R.-R., Lobel, P., and Holtzman, E. (1984). *Cold Spring Harbor Symp. Quant. Biol.* **48**, 1–8.

Kendrew, J. C., Watson, H. C., Strandberg, B. E., Dickerson, R. E., Phillips, D. C., and Shore, V. C. (1961). *Nature (London)* **190**, 666–670.

Khorana, H. G., Gerber, G. E., Herlihy, W. C., Gray, C. P., Anderegg, R. J., Nehei, K., and Biemann, K. (1979). *Proc. Natl. Acad. Sci. U.S.A.* **76**, 5047–5050.

Kistler, J., Stroud, R. M., Klymkowsky, M. W., Lalancette, R. A., and Fairclough, R. H. (1982). *Biophys. J.* **37**, 371–383.

Leifer, D., and Henderson, R. (1983). *J. Mol. Biol.* **163**, 451–466.

Lim, V. I. (1974). *J. Mol. Biol.* **88**, 857–872.

Lindstrom, J., Merlie, J., and Yogeeswaran, G. (1979). *Biochemistry* **18**, 4465–4470.

Mao, D., and Wallace, B. A. (1984). *Biochemistry* **23**, 2667–2673.

Mao, D., Wachter, E., and Wallace, B. A. (1982). *Biochemistry* **21**, 4960–4968.

Mielke, D. L., Kaldany, R.-R., Karlin, A., and Wallace, B. A. (1986). *Ann. N. Y. Acad. Sci.* **463**, 392–395.

Nabedryk, E., Bardin, A. M., and Breton, J. (1985). *Biophys. J.* **48**, 873–876.

Nakashima, Y., and Konigsberg, W. (1974). *J. Mol. Biol.* **88**, 598–600.

Noda, M., Takahashi, H., Tanabe, T., Toyosato, M., Furutani, Y., Hirose, T., Asai, M., Inayama, S., Miyata, T., and Numa, S. (1982). *Nature (London)* **299**, 793–797.

Noda, M., Takahashi, H., Tanabe, T., Toyosato, M., Kikyotani, S., Hirose, T., Asai, M., Takashima, H., Inayama, S., Miyata, T., and Numa, S. (1983a). *Nature (London)* **301**, 251–255.

Noda, M., Takahashi, H., Tanabe, T., Toyosato, M., Kikyotani, S., Furutani, Y., Hirose, T., Takashima, H., Inayama, S., Miyata, T., and Numa, S. (1983b). *Nature (London)* **302**, 528–532.

Nozaki, Y., Chamberlain, B. K., Webster, R. E., and Tanford, C. (1976). *Nature (London)* **259**, 335–337.

Oesterhelt, D., and Stoeckenius, W. (1971). *Nature (London), New Biol.* **233**, 149–152.

Oesterhelt, D., and Stoeckenius, W. (1973). *Proc. Natl. Acad. Sci. U.S.A.* **70**, 2853–2857.
Osborne, H. B., and Nabedryk-Viala, E. (1977). *FEBS Lett.* **84**, 217–220.
Ovchinnikov, Y. A. (1982). *FEBS Lett.* **148**, 179–191.
Ovchinnikov, Y. A., Abdulaev, N. G., Feigina, M. Y., Kiselev, A. V., and Lobanov, N. A. (1979). *FEBS Lett.* **100**, 219–224.
Overbeeke, N., Bergmans, H., van Mansfield, F., and Lugtenberg, B. (1983). *J. Mol. Biol.* **163**, 513–532.
Racker, E., and Stoeckenius, W. (1974). *J. Biol. Chem.* **249**, 662–664.
Raftery, M. A., Hunkapiller, M. W., Strader, C. D., and Hood, L. E. (1980). *Science* **208**, 1454–1457.
Reynolds, J. A., and Karlin, A. (1978). *Biochemistry* **17**, 2035–2038.
Rosenbusch, J. P. (1974). *J. Biol. Chem.* **249**, 8019–8029.
Schindler, M., and Rosenbusch, J. P. (1984). *FEBS Lett.* **173**, 85–89.
Schulte, T. H., and Marchesi, V. T. (1979). *Biochemistry* **18**, 275–280.
Sebald, W., Hoppe, J., and Wachter, E. (1979). In "Function and Molecular Aspects of Biomembrane Transport" (E. Quagliariello, F. Palmieri, S. Papa, and M. Klingenberg, eds.), p. 63. Elsevier/North-Holland Biomedical Press, New York.
Silverberg, M., Furthmayr, H., and Marchesi, V. R. (1976). *Biochemistry* **15**, 1448–1451.
Smilowitz, H. (1974). *J. Virol.* **13**, 94–99.
Spatz, L., and Strittmatter, P. (1971). *Proc. Natl. Acad. Sci. U.S.A.* **68**, 1042–1046.
Steven, A. C., von Heggeler, B., Muller, R., Kistler, J., and Rosenbusch, J. P. (1977). *J. Cell Biol.* **72**, 292–301.
Strittmatter, P., Rogers, M. J., and Spatz, L. (1972). *J. Biol. Chem.* **247**, 7188–7194.
Stubbs, G. W., Smith, H. G., Jr., and Litman, B. J. (1976). *Biochim. Biophys. Acta* **425**, 46–56.
Teeter, M. M., Mazer, S. A., and L'Italien, J. J. (1981). *Biochemistry* **20**, 5437–5443.
Tobkes, N., Wallace, B. A., and Bayley, H. (1985). *Biochemistry* **24**, 1915–1920.
Tomita, M., and Marchesi, V. T. (1975). *Proc. Natl. Acad. Sci. U.S.A.* **72**, 2964–2968.
Tomita, M., Furthmayr, H., and Marchesi, V. R. (1978). *Biochemistry* **17**, 4756–4770.
Van Etten, C. H., Nielsen, H. C., and Peters, J. E. (1965). *Phytochemistry* **4**, 467–473.
Wallace, B. A. (1982). In "Methods in Enzymology" (L. Packer, ed.), Vol 88, pp. 447–462. Academic Press, New York.
Wallace, B. A. (1983). *Biopolymers* **22**, 397–402.
Wallace, B. A., Kohl, N., and Teeter, M. M. (1984). *Proc. Natl. Acad. Sci. U.S.A.* **81**, 1406–1410.
Wallace, B. A., Cascio, M., and Mielke, D. L. (1986). *Proc. Natl. Acad. Sci. U.S.A.* **83**, 9423–9427.
Wallace, B. A., and Teeters, C. L. (1987). *Biochemistry* **26**, (in press).
Wickner, W. (1976). *Proc. Natl. Acad. Sci. U.S.A.* **73**, 1159–1163.
Williams, R. W., and Dunker, A. K. (1977). *J. Biol. Chem.* **252**, 6253–6255.
Winzler, R. J., Jamison, G. A., and Greenwalt, T. J., eds. (1969). "The Red Cell Membrane, Structure, and Function," pp. 157–171. Lippincott, Philadelphia, Pennsylvania.
Wyckoff, H. W., Tsernoglou, D., Hanson, A. W., Knox, J. P., Lee, B., and Richards, F. M. (1970). *J. Biol. Chem.* **245**, 30–38.
Yager, P., Chang, E. L., Williams, R. W., and Dalziel, A. W. (1984). *Biophys. J.* **45**, 26–28.
Yamada, M., Efira, Y., Miyata, T., Nakazawa, T., and Nakazawa, A. (1982). *Proc. Natl. Acad. Sci. U.S.A.* **79**, 2827–2831.

PART VI

CONCLUSION

23
Structure and Function of the LDL Receptor[1]

JOSEPH L. GOLDSTEIN AND MICHAEL S. BROWN
Department of Molecular Genetics
University of Texas Health Science Center at Dallas
Dallas, Texas 75235

INTRODUCTION

The low-density lipoprotein (LDL) receptor is a cell surface glycoprotein that binds and internalizes plasma LDL, the major cholesterol transport protein in human plasma. Once inside the cell, the LDL is delivered to lysosomes, where the entire lipoprotein is degraded and its cholesterol becomes available to cells for use in the synthesis of plasma membranes (most cell types), steroid hormones (adrenal cortex and ovarian corpus luteum), and bile acids (hepatocytes). The binding, uptake, and lysosomal degradation of LDL occur by receptor-mediated endocytosis, a general biological process by which a variety of nutritional and regulatory molecules are internalized by cells. Such receptor-mediated endocytosis requires that the cell surface receptors have multiple domains that allow them to bind their specific ligands, to cluster within clathrin-coated pits on the plasma membrane, and in many cases to recycle to the cell surface after ligand delivery. New insights into the structure of the LDL receptor have recently begun to shed light on the multidomain nature of coated pit receptors. For a comprehensive review of the structure and function of the LDL receptor, the reader is referred to Goldstein *et al.* (1).

---

[1] Based on the Concluding Address delivered at the symposium, "Biological Organization: Macromolecular Interactions at High Resolution," held at Arden House on the Harriman Campus of Columbia University, May 31 through June 2, 1985.

ITINERARY OF THE LDL RECEPTOR

The mature LDL receptor is a cell surface glycoprotein that contains 2 N-linked and 18 O-linked sugar chains. Figure 1 illustrates the itinerary followed by the LDL receptor from its site of synthesis to its site of internalization in coated pits. The receptor is synthesized in the rough endoplasmic reticulum as a precursor molecule to which high-mannose N-linked carbohydrate chains and the core sugar (i.e., N-acetylgalactosamine) of each O-linked tetrasaccharide chain are attached (2,3). These O-linked core sugars are added before the mannose residues of the N-linked chains are trimmed, i.e., while the receptor is still in the endoglycosidase H-sensitive stage. Thus, the O-linked core sugars must be added either in the endoplasmic reticulum or in the transitional zone between the endoplasmic reticulum and the Golgi apparatus.

Within 30 min after its synthesis, the LDL receptor precursor de-

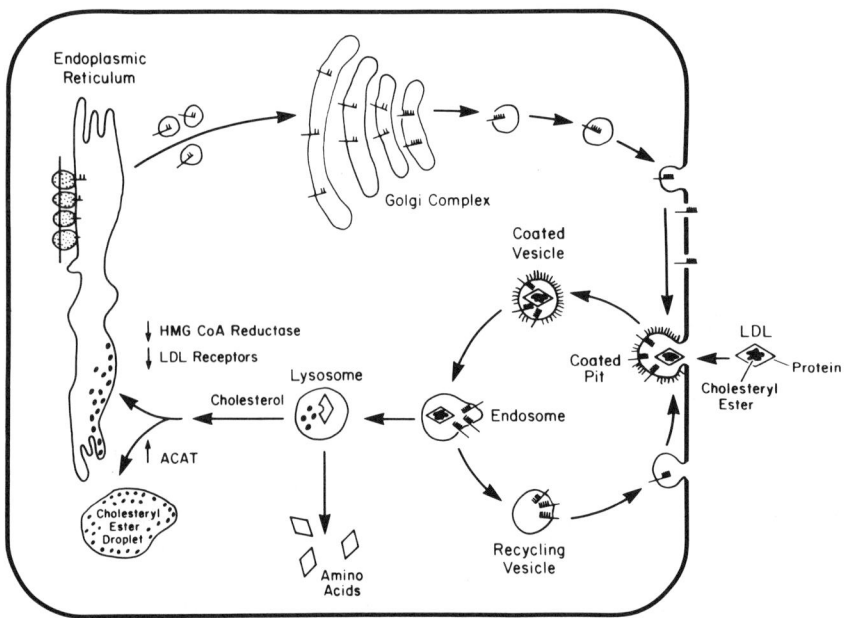

Fig. 1. The LDL receptor pathway in mammalian cells. The receptor undergoes a circuitous itinerary, beginning life as a newly synthesized protein on the rough endoplasmic reticulum. The remainder of the itinerary is discussed in the text. HMG-CoA reductase denotes 3-hydroxy-3-methylglutaryl-CoA reductase; ACAT denotes acyl-CoA : Cholesterol acyltransferase. Vertical arrows indicate regulatory effects.

creases in mobility on NaDodSO$_4$ gels from a mobility corresponding to 120,000 Da to a mobility corresponding to 160,000 Da (2). This change is coincident with the enzymatic reactions that modify the high-mannose N-linked oligosaccharide chains to the complex, sialic acid-containing, endoglycosidase H-resistant type. At the same time, the core N-acetylgalactosamine of each O-linked chain is elongated by the addition of one galactose and two sialic acid residues (3). The net amount of carbohydrate that is added is not sufficient to account for a true increase of 40,000 Da; the bulk of the change is due to conformational alterations in the protein and changes in the amount of NaDodSO$_4$ binding that retard its mobility on NaDodSO$_4$ gels (3).

About 45 min after its synthesis, the LDL receptor appears on the surface, where it clusters together with other receptors in indented regions of the plasma membrane that are coated on the cytoplasmic surface with a protein called clathrin. These are the so-called "coated pits" that are responsible for the receptor-mediated endocytosis of a variety of receptor-bound molecules (1). While in the coated pit, the receptor binds LDL by attaching to the protein component of the lipoprotein. Within 3 to 5 min of their formation, the coated pits invaginate to form coated endocytic vesicles that pinch off from the plasma membrane and exist for a brief moment (less than 2 min) as free structures in the cytoplasm. Very quickly, the clathrin coat dissociates from the surface of the coated vesicle, which now appears as a smooth-surfaced structure. Multiple endocytic vesicles fuse with each other to create larger membrane-enclosed sacs of irregular contour called endosomes. The pH of the endosomes is lower than that of the surrounding cytoplasm, owing to the presence of proton pumps in the membrane of the endosome that acidify its contents (4). Under the influence of the acid pH, the LDL dissociates from the receptor. The receptor can then return to the surface, apparently by clustering together with other receptors in a segment of the endosome membrane that pinches off to form a recycling vesicle that carries the receptors back to the surface (1). Once it reaches the surface the receptor binds another LDL particle and initiates another cycle of endocytosis and recycling. Each receptor makes one round-trip every 10 min in a continuous fashion.

The LDL that dissociates from the receptor remains within the lumen of the endosome and is eventually delivered to a lysosome when the membranes of the endosome and lysosome fuse. The LDL is now exposed to a variety of acid hydrolases. Its protein component is hydrolyzed to amino acids and its cholesteryl esters are hydrolyzed by an acid lipase, liberating cholesterol for use in the synthesis of new membranes, steroid hormones, and bile acids (1). After leaving the lysosome, the cholesterol liberated from LDL also regulates the cell's cholesterol metabolism, thereby assuring a steady level of cholesterol within the cell (5).

STRUCTURE OF THE LDL RECEPTOR

The LDL receptor was purified from bovine adrenal cortex by Wolfgang Schneider (6), a partial amino acid sequence was obtained, and this sequence was used by David Russell and Tokuo Yamamoto to isolate a full-length cDNA for the human LDL receptor (7,8). Studies of the receptor protein, coupled with the amino acid sequence that was deduced from the nucleotide sequence of the cDNA, have provided insight into the structure of the LDL receptor (8,9). A model of the domain structure is shown in Figs. 2 and 3.

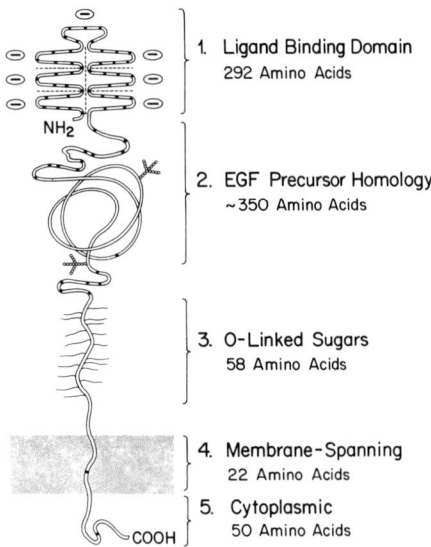

Fig. 2. Model is showing five domains of the human LDL receptor. The 839-amino acid sequence of the mature receptor is drawn to scale. The first domain is divided into seven repeat units each of which has a net negative charge, indicated by the symbol ⊖. Cysteine residues are indicated by solid dots. Two potential sites of attachment of N-linked oligosaccharide chains in the second domain are indicated; three other potential N-linked glycosylation sites in the first domain are not shown. Potential sites of attachment of 18 O-linked oligosaccharide chains are indicated in the third domain.

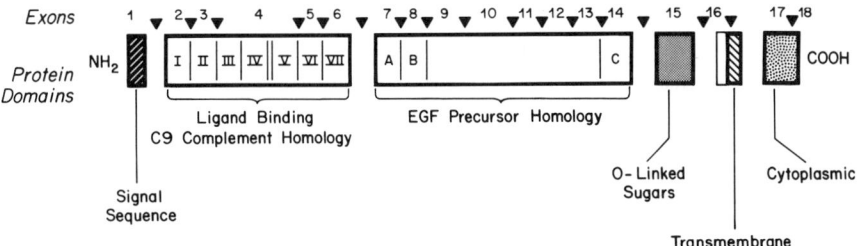

Fig. 3. Exon organization and protein domains in the human LDL receptor. The domains of the protein are delimited by thick black lines and are labeled in the lower portion of the figure. The seven cysteine-rich, 40-amino acid repeats in the LDL binding domain are assigned roman numerals, I–VII. Repeats IV and V are separated by eight amino acids. The three cysteine-rich repeats in the EGF precursor homology domain are lettered A–C. The positions at which introns interrupt the coding region are indicated by arrow heads. Exon numbers are shown between the arrow heads. Reproduced from ref. 10 with permission.

FIRST DOMAIN: LIGAND BINDING

The extreme NH$_2$-terminus of the LDL receptor consists of a hydrophobic sequence of 21 amino acids that is cleaved from the receptor and is not present in the mature protein. This segment presumably functions as a classic signal sequence to direct the receptor-synthesizing ribosomes to the ER membrane. Because it does not appear in the mature receptor, the signal sequence is omitted from the numbering system that is described below. The mature receptor (without the signal sequence) consists of 839 amino acids (8).

The first domain of the mature LDL receptor consists of the NH$_2$-terminus 292 amino acids, which is composed of a sequence of 40 amino acids that is repeated seven times (8,10). Studies of anti-peptide antibody binding to intact cells revealed that this domain is located on the external surface of the plasma membrane (11). Each of the seven 40-amino acid repeats contains six cysteine residues, which are essentially in register for all of the repeats. The receptor cannot be labeled with [^{3}H]iodoacetamide without prior reduction, suggesting that all these cysteines are involved in disulfide bonds. This region of the receptor must therefore exist in a tightly cross-linked, convoluted state.

A striking feature of each repeat sequence is a cluster of negatively charged amino acids near the COOH-terminus (8,10). These sequences are complementary to positively charged sequences in the best-characterized ligand for the LDL receptor, apolipoprotein E (apo

E), a 33-kDa protein component of the plasma lipid transport system. As first noted by Innerarity and Mahley, apo E contains a cluster of positively charged residues which is believed to face one side of a single α-helix (12,13). Studies with mutant and proteolyzed forms of apo E and with monoclonal antibodies against different regions of apo E showed that the positively charged region contains the site whereby this protein binds to the LDL receptor (12). It is therefore tempting to speculate that the negatively charged clusters of amino acids within the cysteine-rich repeat sequence of the LDL receptor constitute multiple binding sites, each of which binds a single apo E molecule by attaching to its positively charged α-helix.

Each of the seven 40-amino acid repeats in the LDL receptor is strongly homologous to a single 40-residue sequence that occurs within the cysteine-rich region of human complement component C9, a plasma protein of 537 amino acids (14,15). Of the 19 conserved amino acids in the LDL receptor repeats, 14 are found in the C9 sequence, including the highly conserved negatively charged cluster:

LDL Receptor Consensus (ref. 10)

```
x x T [C] x x x E [F] x [C] x x [G] x [C I]
E D D [C] - G N D [F] Q [C] S T [G] R [C I]

x x x W x [C] D x x x [D C] x [D] G [S D E] x x [C]
K M R L R [C] N G D N [D C] G [D] F [S D E] D D [C]
```

Complement Factor C9 (residues 77–113)

This finding raises the possibility that C9 might have measurable binding affinity for lipoproteins containing apo B or E, the two ligands from the LDL receptor.

SECOND DOMAIN: HOMOLOGY WITH THE EPIDERMAL GROWTH FACTOR PRECURSOR

Epidermal growth factor is a peptide of 53 amino acids that is synthesized as a large precursor of 1217 amino acids (16,17). Analysis of the amino acid sequence, as revealed from the sequence of the cloned cDNA, suggested that the EGF precursor is synthesized as a membrane-bound molecule (18). During synthesis, the first 1038 amino acids of the precursor penetrate into the lumen of the ER, whereupon a stretch of 22 hydrophobic amino acids is encountered. According to the current view of protein synthesis in the ER, such a hydrophobic stretch should become anchored in the membrane (19). On completion of translation, a short tail of 158 amino acids would face the cytoplasm, constituting the cytoplasmic domain of the precursor. The 53-amino acid EGF sequence lies just outside the membrane-spanning region in the extracellular domain of 1038 amino acids. This extracellular sequence also contains multiple repeats of the EGF sequence that have diverged during evolution as well as spacer sequences that are not related to EGF. EGF is presumably liberated from this putative membrane-bound precursor by proteolysis, and the peptide is then transported to receptors on epithelial cells, which it stimulates to divide.

The second domain of the LDL receptor, consisting of ~400 amino acids (8,10), is homologous to a portion of the extracellular domain of the EGF precursor (9,20). Within this region approximately 35% of the amino acids are identical with a few short gaps. At the ends of this domain there are three short, repetitive sequences of ~40 amino acids each that are designated A, B, and C in Fig. 3. Each of these repeats contains six cysteine residues spaced at similar intervals. The A, B, and C sequences are homologous to four repeat sequences in the EGF precursor (18,20). Repeats A, B, and C in the LDL receptor are also homologous to certain proteins of the blood clotting system, including factor IX, factor X, and protein C (18,20).

THIRD DOMAIN: O-LINKED SUGARS

Immediately external to the membrane-spanning domain of the human LDL receptor is a stretch of 58 amino acids that contains 18 serine or threonine residues (8). This domain is encoded within a single exon (see below). Proteolysis studies reveal that this region contains carbohydrate chains attached in O-linkage (9). Each O-linked sugar chain consists of a core N-acetylgalactosamine plus a single galactose and one or two sialic acids. In this respect, the LDL receptor resembles glycophorin, a red cell membrane protein that contains short O-linked sugar chains attached to clusters of serines and threonines (21). Another cell surface receptor, that for interleukin-2 (IL-2) on T lymphocytes, contains O-linked sugars and a cluster of serine and threonine residues immediately external to the membrane-spanning region (22,23).

FOURTH DOMAIN: MEMBRANE-SPANNING REGION

This domain consists of a stretch of 22 hydrophobic amino acids. Proteolysis experiments confirmed that this domain spans the membrane (9). Comparison of the amino acid sequences of the bovine and human LDL receptors reveals that the membrane-spanning region is relatively poorly conserved (1). Of the 22 amino acids in this region, 7 differ between human and cow, but all of the substitutions retain a hydrophobic character.

FIFTH DOMAIN: CYTOPLASMIC TAIL

The human and bovine LDL receptors both contain a COOH-terminus segment of 50 amino acids that projects into the cytoplasm. This sequence is strongly conserved; only 4 of the 50 amino acids differ between the two species, and each of these substitutions is conservative with respect to the charge of the amino acid (1). Localization of this domain to the cytoplasmic side of the membrane was determined through use of an anti-peptide antibody directed against the COOH-terminus sequence (9). When inside-out membrane vesicles containing receptor were digested with pronase, the antibody-reactive material was removed, and the molecular weight of the receptor was reduced by approximately 5000.

Immediately internal to the membrane, the cytoplasmic tail contains a cluster of positively charged amino acids (three of the first six residues are lysines or arginines). This is a frequent feature of plasma membrane proteins (19). Near the COOH-terminus end of the receptor lies a cluster of negatively charged residues (glutamic–aspartic–aspartic) (9). The cytoplasmic segment also contains several serine and threonine residues and two tyrosines, which might be potential sites for phosphorylation. This domain also contains a single cysteine, which might be a site for disulfide bond formation or for fatty acylation. None of these modifications have been detected as yet.

The cytoplasmic domain of the LDL receptor plays an important role in clustering in coated pits, either through interaction with clathrin itself or with some protein that is associated with clathrin on the cytoplasmic side of the membrane (24–26). This conclusion is based on a molecular analysis of three naturally occurring mutations at the LDL receptor locus that produce receptors that bind LDL normally but fail to cluster in clathrin-coated pits and therefore cannot transport LDL into cells. We prepared genomic DNA libraries from cells of three unrelated individuals with this phenotype (internalization-defective form of familial hypercholesterolemia) and isolated the segments of the gene encoding the COOH-terminus cytoplasmic domain of the receptor (24). In all three mutants, a nucleotide abnormality (one missense mutation, one nonsense mutation, and one frameshift mutation) was found in the one of the two exons encoding the cytoplasmic domain. These results strongly support the proposal that this domain is crucial in directing the LDL receptor to coated pits (1,24).

EXON ORGANIZATION AND PROTEIN DOMAINS OF THE LDL RECEPTOR

Southern blotting of genomic DNA demonstrated that the haploid human genome contains a single copy of the receptor gene (10). This gene resides on chromosome 19 as determined by somatic cell genetic techniques (27). The gene spans more than 45 kb. Sequences representing almost the entire gene have been isolated from bacteriophage lambda and cosmid libraries (10). The position of each intron within the gene has been mapped and the sequence of each exon–intron junction has been determined.

These studies reveal that the receptor gene is made up of 18 exons. The sites of the introns in relation to the protein sequence are indicated in Fig. 3. Most of the introns separate regions of the protein that correspond to domains that were identified through the protein chemistry studies described above (10). The first intron is located just at the end of the DNA encoding the cleaved signal sequence of the protein. Within the binding domain of the receptor (which contains the seven cysteine-rich repeats), introns occur precisely between repeats I and II, II and III, V and VI, and VI and VII (Fig. 3). Repeats III, IV, and V are included in one exon. The binding domain is terminated by an intron at amino acid 292, the last residue in the seventh repeat.

The next domain, the region of homology with the EGF precursor, is encoded in eight contiguous exons. Within this 400-amino acid region of homology are located three copies of a repeated sequence (repeats A,B, and C in Fig. 3), each of which is encoded by a single exon (10,20). The striking similarity in the exon–intron organization of this region of the LDL receptor gene and the EGF precursor gene is discussed above.

The O-linked sugar domain is also demarcated neatly by two introns (Fig. 3). However, not all domains of the protein are encoded by single exons. Thus, the membrane-spanning region is interrupted by an intron. Another intron interrupts the coding region for the cytoplasmic tail 11 amino acids from the COOH-terminus.

The placement of the introns is consistent with the notion that the human LDL receptor gene was constructed by the stepwise assembly of exons that encode useful protein sequences (10). Thirteen of the 18 exons comprising the LDL receptor gene encode protein sequences that are homologous to sequences in other proteins: five of these exons encode a sequence similar to one in the C9 component of complement; three exons encode a sequence similar to a repeat sequence in the EGF precursor and in three proteins of the blood clotting system; and five other exons encode nonrepeated sequences that are shared only with the EGF precursor. The LDL receptor thus appears to be a mosaic protein built up of exons shared with different proteins (10).

The occurrence of shared exons in the LDL receptor gene provides strong evidence to support Walter Gilbert's hypothesis concerning the nature and function of exons and introns (28). As originally proposed by Gilbert, the existence of introns permits functional domains encoded by discrete exons to shuffle between different proteins, thus allowing proteins to evolve as new combinations of preexisting functional units. The LDL receptor is a vivid example of such a mosaic protein.

REFERENCES

1. Goldstein, J. L., Brown, M. S., Anderson, R. G. W., Russell, D. W., and Schneider, W. J. (1985). *Annu. Rev. Cell Biol.* **1**, 1–39.
2. Tolleshaug, H., Goldstein, J. L., Schneider, W. J., and Brown, M. S. (1982). *Cell (Cambridge, Mass.)* **30**, 715–724.
3. Cummings, R. D., Kornfeld, S., Schneider, W. J., Hobgood, K. K., Tolleshaug, H., Brown, M. S., and Goldstein, J. L. (1983). *J. Biol. Chem.* **258**, 15261–15273.
4. Helenius, A., Mellman, I., Wall, D., and Hubbard, A. (1983). *Trends Biochem. Sci.* **8**, 245–250.
5. Goldstein, J. L., and Brown, M. S. (1977). *Annu. Rev. Biochem.* **46**, 897–930.
6. Schneider, W. J., Beisiegel, U., Goldstein, J. L., and Brown, M. S. (1982). *J. Biol. Chem.* **257**, 2664–2673.
7. Russell, D. W., Yamamoto, T., Schneider, W. J., Slaughter, C. J., Brown, M. S., and Goldstein, J. L. (1983). *Proc. Natl. Acad. Sci. U.S.A.* **80**, 7501–7505.
8. Yamamoto, T., Davis, C. G., Brown, M. S., Schneider, W. J., Casey, M. L., Goldstein, J. L., and Russell, D. W. (1984). *Cell (Cambridge, Mass.)* **39**, 27–38.
9. Russell, D. W., Schneider, W. J., Yamamoto, T., Luskey, K. L., Brown, M. S., and Goldstein, J. L. (1984). *Cell (Cambridge, Mass.)* **37**, 577–585.
10. Südhof, T. C., Goldstein, J. L., Brown, M. S., and Russell, D. W. (1985). *Science* **228**, 815–822.

11. Schneider, W. J., Slaughter, C. J., Goldstein, J. L., Anderson, R. G. W., Capra, D. J., and Brown, M. S. (1983). *J. Cell Biol.* **97**, 1635–1640.
12. Innerarity, T. L., Weisgraber, K. H., Arnold, K. S., Rall, S. C., Jr., and Mahley, R. W. (1984). *J. Biol. Chem.* **259**, 7261–7267.
13. Mahley, R. W., and Innerarity, T. L. (1983). *Biochim. Biophys. Acta* **737**, 197–222.
14. Stanley, K. K., Kocher, H.-P., Luzio, J. P., Jackson, P., and Tschopp, J. (1985). *EMBO J.* **4**, 375–382.
15. DiScipio, R. G., Gehring, M. R., Podack, E. R., Kan, C. C., Hugli, T. E., and Fey, G. H. (1984). *Proc. Natl. Acad. Sci. U.S.A.* **81**, 7298–7302.
16. Scott, J., Urdea, M., Quiroga, M., Sanchez-Pescador, R., Fong, N., Selby, M., Rutter, W. J., and Bell, G. I. (1983). *Science* **221**, 236–240.
17. Gray, A., Dull, T. J., and Ullrich, A. (1983). *Nature (London)* **303**, 722–725.
18. Doolittle, R. F., Feng, D.-F., and Johnson, M. S. (1984). *Nature (London)* **307**, 558–566.
19. Sabatini, D. D., Kriebich, G., Morimoto, T., and Adesnik, M. (1982). *J. Cell Biol.* **92**, 1–21.
20. Südhof, T. C., Russell, D. W., Goldstein, J. L., Brown, M. S., Sanchez-Pescador, R., and Bell, G. I. (1985). *Science* **228**, 893–895.
21. Marchesi, V. T., Furthmayr, H., and Tomita, M. (1976). *Annu. Rev. Biochem.* **45**, 667–698.
22. Leonard, W. J., Depper, J. M., Crabtree, G. R., Rudikoff, S., Pumphrey, J., Robb, R. J., Kronke, M., Svetlik, P. B., Peffer, N. J., Waldmann, T. A., and Greene, W. D. (1984). *Nature (London)* **311**, 626–631.
23. Nikaido, T., Shimizu, A., Ishida, N., Sabe, H., Teshigawara, K., Maeda, M., Uchiyama, T., Yodoi, J., and Honjo, T. (1984). *Nature (London)* **311**, 631–635.
24. Lehrman, M. A., Goldstein, J. L., Brown, M. S., Russell, D. W., and Schneider, W. J. (1985). *Cell (Cambridge, Mass)* **41**, 735–743.
25. Brown, M. S., and Goldstein, J. L. (1976). *Cell (Cambridge, Mass)* **9**, 663–674.
26. Goldstein, J. L., Brown, M. S., and Stone, N. J. (1977). *Cell (Cambridge, Mass.)* **12**, 629–641.
27. Francke, U., Brown, M. S., and Goldstein, J. L. (1984). *Proc. Natl. Acad. Sci. U.S.A.* **81**, 2826–2830.
28. Gilbert, W. (1978). *Nature (London)* **271**, 501.

Index

A

Acetylcholine receptor, 307–318
  amino acid sequences, 310, 312
  amphipathic membrane crossing, 310–317
  α-helix, 310, 311, 312, 322, 323
  β-sheet, 322, 323
  ion conductance, 307–318
  ionic channel gating, 316–317
  ionic channel location, 309–310, 311
  ionic channel model, 313–316
  ligand-binding site, 317
  quasi-equivalent subunits, 309–310, 311
  secondary structure prediction, 322, 323
  structure, 307–312
Actomyosin complex, 236
Adenosine triphosphatase, proton proteolipid, 322, 328–329
Adenovirus
  hexon, 113–124
    β-barrels, 114, 116–123
    characteristics, 113–114
    envelope model, 114, 115
    polypeptide chain interweaving, 114, 116–117
    structure, 114–124
    structural proteins, 113
    virion, 113–114
Agmenellum quadruplicatum, chromophore, 285
Alamethicin, 320
Alpha-toxin, secondary structure prediction, 322, 323–324
  α-helix, 322, 323–324
  β-sheet, 322, 323–324
Alphavirus, budding, 141–145

Amino acid sequences
  acetylcholine receptor, 310, 312
  antigenic specificity recognition effects, 221
  bacteriophage T4 lysozyme mutant, 245–258
  $C_H 1 : C_L$ interface residues, 194–203
  catabolic gene activator protein, 61, 62
  Cro protein, 61, 62
  DNA polymerase I Klenow fragment, 47–48, 151
  DNA-binding protein, 61–62
  epidermal growth factor, 342–343
  immunoglobulin E, 203, 205–213
  immunoglobulin G, 205–213
  lac repressor, 61
  λ repressor, 61, 62
  low-density lipoprotein receptor, 340, 341–343, 344, 345, 346
  lysozyme, 220–221, 222, 223
  in protein secondary structure prediction, 319–333
  trp repressor, 61
Androctonus australis, hemocyanin, 155, 157–169, 176–188
  epitope monoclonal antibody labeling, 180–188
  oligomer autoassembly, 176–177, 178
  quaternary structure, 164–176
  subunit heterogeneity, 160–164
  subunit localization, 164–169, 172, 176–180
Antibody, *see also* names of specific antibodies
  constant domain structure, 193–214
  $C_H 1 : C_L$ interface residues, 194–203
  Fc fragment, 194–214
  functions, 193

349

immunoglobulin E Fc fragment, 203, 205–213
$V_H : V_L$ interface residues, 194
Antibody–antigen complex
  lysozyme–Fab, 215–225
  order–disorder phenomenon, 227–234
  protein interface, 236
Antigenic determinants, molecular immunoelectron microscopic localization, 153–191
  Androctonus australis, 157–169, 172, 176–188
  arachnid, 163
  Cupiennius salei, 160
  epitope monoclonal antibody labeling, 180–188
  intramolecular localization, 164–189
  quaternary structure, 164–176
  of subunits with polyclonal antibodies, 164–180
  Limulus polyphemus, 155, 157–164, 170–177, 178
  merostome, 163
  Panulirus interruptus, 155–157, 183
  structure, 155–164
    dodecameric, 157–160
    FLIP/FLOP effect, 158–159
    hexameric, 156, 157
    oligomer autoassembly, 170–177, 178
    rocking effect, 158
    75-kDa subunit, 155–157
    subunit heterogeneity, 160–164
  Uca pugnax, 177, 179
Antigenic recognition, amino acid sequence effects, 221
Antigenic sites
  foot-and-mouth disease virus, 88, 102–104
  poliovirus, 88, 102–104
Antigenic specificity, amino acid sequence effects, 221
Anti-peptide antibody, as immunogenic peptide probe, 227–234
  influenza virus hemagglutinin, 228–231, 233
  myohemerythrin, 231–232, 233
Aphthovirus, classification, 86
Apolipoprotein E, 341–342
Arachnid, hemocyanin, 163

Association constant, protein interfaces, 238–239
Autoimmune disease, Z-DNA antibodies in, 4

B

Bacillus stearothermophilus, DNA-binding protein II, 13
Bacteria, *see also* names of specific bacteria
  photosynthetic light-harvesting complex, 283, 284
  photosynthetic reaction centers, 271–286
Bacteriochlorophyll, 271–272, 273, 284
Bacteriophage
  cII gene activator protein, 57, 64
  repressor, *see* λ repressor
Bacteriophage T4 mutant, lysozyme, amino acid substitutions, 245–258
  environmental effects, 253
  global thermal stability, 253
  hydrogen bonding, 249, 253, 256, 257
  interactional effects, 253–254
  protein evolution and, 257
  secondary structure stability, 255
  self-directed mutagenesis and, 248–253, 256–257
  structural change and, 254–255
  temperature-sensitive mutants, 247–248, 249, 253, 254–255, 256, 257
  wild-type mutants, 246–247, 248, 249, 254–255, 256–257
Bacteriopheophytin, 272, 273, 284
Bacteriorhodopsin
  photoreactions, 262, 263, 264, 265, 266
  structure
    helices, 311, 322, 324
    secondary structure prediction, 322, 324
Beta-structure
  acetylcholine receptor, 322, 323
  adenovirus, 114, 116–123
  alpha-toxin, 322, 323–324
  bacteriorhodopsin, 322, 324
  colicin E1, 322, 325
  crambin, 322, 325

cytochrome, b_5, 322, 326
EcoRI endonuclease, 11, 12, 23–25, 26–27, 40–41
Fab fragment, 121
glycophorin, 322, 327
hemocyanin, 121
lactose carrier, 322, 328
nonviral proteins, 121–122
porin, 322, 328
proton-ATPase proteolipid, 322, 329
rhodopsin, 322, 329
southern bean mosaic virus, 121
spherical RNA viruses, 120, 121
Budding, of enveloped viruses, 139–150
alphaviruses, 141–145
G protein and, 145, 146–147, 148
M protein and, 145–146, 148
nucleocapsid and, 140–141, 142–144, 145, 146, 147
thy-1 protein and, 147
vesicular stomatitis virus, 145–147, 148

C

$C_H1 : C_L$, amino acid sequences
interface cavity, 195, 203, 204
interface residue analysis, 194–203
kappa chain, 201–203
lambda chain, 201–203
cII gene activator protein, 57, 64
Capsid protein
human rhinovirus 14, 86–87, 94–100
picornavirus, 107
plant viruses, 95, 96, 97, 98, 99
Carboxyl charge group, tobacco mosaic virus, 71–72, 74–77, 81
Cardiovirus, classification, 86
Catabolic gene activator protein
amino acid sequences, 61, 62
DNA binding, 59, 61–62
helix-turn-helix unit, 12, 61, 62
domains, 58, 59
structural studies, 58, 59, 61–63
C-helix peptide, 231–232, 233
m-Chlorophenylhydrozone, 262
Cholesterol, low-density lipoprotein degradation and, 337, 339–340
Chromophore, of photosynthetic reaction centers, 271, 284–285, 286

Clathrin, 339
Coat protein, fd, secondary structure prediction, 322, 326–327
Coated pit, low-density lipoprotein
clustering, 337, 339, 344–345
internalization, 338, 339
multidomain nature, 337
Cold virus, see Human rhinovirus 14
Colicin E1, secondary structure prediction, 320, 322, 324–325
α-helix, 322, 325
β-sheet, 322, 325
Complementarity, in protein interfaces, 236–238
Concanavalin A
β-barrel, 121–122
receptor binding, 88, 105–106
Copper superoxide dismutase, 121
Coxsackie virus, 88
Crambin
hydrated interfaces, 241
secondary structure prediction, 322, 325
α-helix, 322, 325
β-sheet, 322, 325
Cro protein
amino acid sequences, 61, 62
DNA binding, 12, 58–59, 61–64
DNA structural complementarity, 60
helix-turn-helix unit, 61, 62, 63
hydrogen bonds, 59
domain, 58, 59
structural studies, 12, 58–59, 60, 61–64
α-helix, 58–59, 60
studies in progress, 63–64
Crystalline protein
hydrated interfaces, 241–242
lattice contacts, 236
Cupiennius salei, hemocyanin, 160
Cyanobacteria, photosynthetic reaction centers, 283–286
Cytochrome, of photosynthetic reaction centers, 284, 285, 286
Cytochrome b_5, secondary structure prediction, 322, 326
α-helix, 322, 326
β-sheet, 322, 326
Cytoplasmic tail, of low-density lipoprotein, 340, 344–345

D

Deaminooxytocin, 289–306
 conformational flexibility, 290–301
 crystal structure description, 294–298
 crystal structure determination, 291–294
 hydrogen bonds, 294, 295–298, 303–304
 hydrophilic regions, 294, 295–296, 303–304
 receptor binding, 302, 303–304
 structure in dimethyl sulfoxide, 290, 298–301
Deoxyguanosine, 3
Deoxyhemoglobin, 239
Deoxynucleoside monophosphate, 45–46, 48–49
Deoxynucleoside triphosphate, 45–46, 48, 52
Deoxyribonucleic acid
 adenoviruses, 113
 neokink, 27–31, 32, 34, 40
cDeoxyribonucleic acid, low-density lipoprotein, 340
B-Deoxyribonucleic acid, 1
 van der Waals model, 2
 Z-DNA conformational relationship, 1–2, 3, 4, 5
 Z-DNA interconversion, 3, 4, 5
Z-Deoxyribonucleic acid
 left-handed double helix, 1–7
  B-DNA interconversion, 3, 4, 5
  binding proteins, 5–7
  biological roles, 4, 6–7
  chemical characteristics, 2, 4–5
  deoxyguanosine conformation, 3
  immunogenicity, 4
  van der Waals model, 2
 right-handed double helix, bases location, 1–2, 4
 stabilization, 4–6
Deoxyribonucleic acid polymerase I
 binding sites, 45–46
 domains, 46
 enzymatic activities, 45, 46
 Klenow fragment structure, 12, 45–55
  amino acid sequences, 47–48, 51
  DNA binding, 49–50, 51, 54
  DNA co-crystallization, 50–51, 54
  DNA synthesis fidelity, 52–53
  domains, 46, 47–48
  3'-5' exonuclease activity, 48, 52, 54
  large domain function, 49–50
  processivity, 51–52
  relation to other polymerases, 51
  small domain function, 48–49
  structure, 47–48
 molecular weight, 45
Deoxyribonucleic acid protein recognition
 catabolic gene activator protein, 58, 59, 61–63
 Cro protein, 12, 58–59, 61–64
 DNA-binding models, 58–60
 *Eco*RI endonuclease
  α/β domain, 11, 25, 40, 41
  β-sheet, 11, 23–25, 26–27, 40–41
  cocrystalline recognition complex, 11–33
  conformational mobility, 36–40
  contact points, 35
  DNA structural features, 27–31
  DNA—encircling arms, 11, 26–27, 40, 41
  DNA-protein interactions, 31–35
  *Eco*RIendonuclease specificity, 36–40
  electron density maps, 15–22
  general features, 22–23
  α-helix, 23, 24, 25–26, 32–34
  hydrogen bonds, 36–39
  iterative single isomorphous replacement method, 15–22
  methods, 14–22
  neokink, 27–31, 32, 34, 40
  protein subunit, 23–26
 hydrogen bonds, 13
 indirect, 14
 λ repressor, 12, 58, 59, 61–63
 macromolecular energy differences, 13–14
 structural studies, 57–65
Deoxyribonucleic acid recombination, Z-DNA and, 6–7
Deoxyribonucleic acid synthesis, DNA polymerase I Klenow fragment and, 45–55
Deoxyribonucleic acid-binding protein, *see also* Deoxyribonucleic acid

INDEX 353

protein recognition; names of
specific proteins
amino acid sequences, 61–62
as DNA-binding model, 58–60
helix-turn-helix binding motif, 57,
61–63, 64
Deoxyribonucleic acid-binding protein
II, 13
Dimethyl sulfoxide, deaminooxytocin
structure determination in, 290,
298–301
Dissociation reaction, protein interfaces,
238

E

*Eco*RI endonuclease, cocrystalline
recognition complex, 11–33
α/β domain, 25, 40, 41
β-sheet, 11, 12, 23–25, 26–27, 40–41
conformational mobility, 36–40
contact points, 35
DNA conformation, 11–12
DNA neokink, 27–31, 32, 34, 40
DNA structural features, 27–31
DNA-catalyzing specificity, 13–14
DNA—encircling arm, 26–27, 36, 40, 41
DNA-protein interactions, 31–35
*Eco*RI endonuclease specificity, 36–40
electron density maps, 15–22
general features, 22–23
α-helix, 12, 23, 24, 25–26, 32–34
hydrogen bonds, 36–39
iterative single isomorphous
replacement method, 15–22
methods, 14–22
protein subunit, 23–26
Electron density map
antineuraminidase Fab fragment,
126–129
*Eco*RI endonuclease-deoxyribonucleic
acid complex, 15–22
lysozyme-Fab fragment complex, 219,
220, 222
Electron transfer, in photosynthetic
reaction centers, 271–282
Electron transfer complex, 236
Encephalomyocarditis virus, 87

Endocytosis, receptor-mediated, 337, 339
Endoplasmic reticulum, protein
localization, 139
Enterovirus, classification, 86
Entropy, protein interfaces, 238–239,
240–241, 242
Enzyme, DNA, catalyzing, 13, *see also*
*Eco*RI endonuclease
Enzyme–inhibition complex, 236
Enzyme–substrate complex, 236
Epidermal growth factor, low-density
lipoprotein receptor precursor
homology, 340, 342–343
Epitope, monoclonal antibody labeling,
180–188
Erabutaxin, 241
Exon, low-density lipoprotein, 341,
345–346
3′-5′ Exonuclease, 45, 46, 48, 52, 54

F

Fab fragment, *see also* Lysozyme-Fab
complex
antigen recognition site, 215
antineuraminidase monoclonal,
125–133
antigen interaction, 129–132
binding properties, to monoclonal
variants, 129–130
binding properties, to
neuraminidase field strains,
130–131
electron density map, 126–129
neuraminidase-Fab complex,
131–132
structure, 126–129
β-barrel, 121
as hemocyanin immunolabel, 165–189
McPC 603, 215
new, 215
S10/1, *see* Fab fragment,
antineuraminidase monoclonal
FC fragment
antibody constrant domains, 194–214
immunoglobulin E, 203, 205–213
Fiber diffraction, phase problem, 72–73
Foot-and-mouth disease virus
antigenic sites, 88, 102–104

gene order, 87
receptor binding site, 105
ribonucleic acid, 87
serotypes, 86

G

G protein, in viral budding, 145,
  146–147, 148
Genetic recombination, Z-DNA and, 6–7
Glycophorin
  O-linked sugar chains, 343
  secondary structure prediction, 322,
    327
  α-helix, 322, 327
  β-sheet, 322, 327
Golgi complex, protein localization, 139
Gramicidin, 320

H

Halobacterium halobium
  phototactic response, 265–267, 368
  rhodopsins, 261–269
    bacteriorhodopsins, 324
    halorhodopsin, 262–265
    sensory, 265–267
Halorhodopsin, 261, 262–265
  osmoregulation function, 265
  photoreaction cycle, 263–265
Heavy chain, $C_H 1$ domain amino acid
  sequence, 195–201
α-Helix
  acetylcholine receptor, 310, 311, 312,
    322, 323
  alpha-toxin, 322, 323–324
  bacteriorhodopsin, 311, 322, 324
  colicin E1, 322, 325
  crambin, 322, 325
  Cro protein, 58–59, 60
  cytochrome b_5, 322, 326
  *eco*RI endonuclease, 23, 24, 25–26,
    32–34
  gene expression-regulating proteins,
    57
  glycophorin, 322, 327
  lactose carrier, 322, 328
  λ repressor, 59

porin, 322, 328
proton-ATPase proteolipid, 322, 329
rhodopsin, 322, 329
Hemagglutinin, influenza virus, 135–138
  β-turn, 228–231
  membrane fusion, 135, 136–138
  receptor binding, 135–136
Hemerythrin, 243
Hemocyanin
  arthropod, antigenic determinants,
    153–191
  aggregation level, 154–155
  Androctonus australis, 155,
    157–169, 172, 176–188
  arachnid, 163
  β-barrel, 121
  Cupiennius salei, 160
  dodecameric structure, 157–160
  epitope monoclonal antibody
    labeling, 180–188
  FLIP/FLOP effect, 158–159
  hexameric structure, 156, 157
  intramolecular localization, with
    polyclonal antibodies, 164–180
  Limulus polyphemus, 155, 157–164,
    170–177, 178
  merostome, 163
  oligomer autoassembly, 176–177,
    178
  Panulirus interruptus, 155–157, 183
  quaternary structure, 164–176
  rocking effect, 158
  75-kDa subunit, 155–157
  structure, 155–164
  subunit heterogeneity, 160–164
  Uca pugnax, 177, 179
Hemoglobin, interfaces, 236, 237,
  239–240
  hydrated, 241
Hen egg white lysozyme–Fab complex,
  216–225
Hepatitis A virus, RNA, 87
Hexon, of adenoviruses, 113–124
  characteristics, 113–114
  structure, 114–124
    β-barrels, 114, 116–123
    envelope model, 114, 115
    polypeptide chain interweaving,
      114, 116–117
Histone, octamer, 13

Human rhinovirus 2
  host/tissue specificities, 88
  receptor binding site, 105
Human rhinovirus 14
  capsid protein cleavage, 86–87
  host/tissue specificities, 88
  immunogenic sites, 88, 100–102
  receptor binding site, 88, 104–106
  serotypes, 106
  structural proteins, 87
  structure, 85–112
    capsid proteins, 86–87, 94–100
    RNA, 94, 98, 100
    structure determination, 89–94
Hydration, in protein interfaces, 241–242
Hydrogen bonding
  bacteriophage T4 lysozyme, 249, 253, 256, 257
  Cro protein-deoxyribonucleic acid complex, 59
  deaminooxytocin, 294, 295–296, 303–304
  DNA protein recognition, 13
  EcoRIendonuclease, 36–39
  oxytocin, 290
  proteins, 238, 241, 242
Hydrophobic zones
  deaminooxytocin, 294, 295–296, 303–304
  protein interfaces, 238, 239–240, 241, 242
Hypercholesterolemia, familial, 345

I

Immunogenic sites, human rhinovirus 14, 88, 100–102
Immunoglobulin E
  amino acid sequences, 203, 205–213
  disulfide bridge linkage, 205, 213
  Fc model, 203, 205–213
Immunoglobulin G, amino acid sequences, 205–213
Immunoglobulin M, disulfide bridge linkage, 213
Influenza virus
  budding, 145
  hemagglutinin, 135–138
    β-turn, 228–231, 233
  in membrane fusion, 135, 136–318
  receptor binding, 135–136
  infection process, 135
  neuraminidase, 129–131
Ion conductance, acetylcholine receptor, 307–318
Interleukin-2, receptor, 343
Intron, low-density lipoprotein, 345, 346
Iterative single isomorphous replacement, EcoRIendonuclease, 15–22

L

lac repressor
  amino acid sequences, 61
  helix-turn-helix unit, 61, 63
Lactose carrier, secondary structure prediction, 322, 327–328
  α-helix, 322, 328
  β-sheet, 322, 328
λ repressor
  DNA binding, 59, 61
  helix-turn-helix unit, 61, 62
  structural studies, 12, 58, 59, 61–63
    amino acid sequences, 61, 62
    domains, 58, 59
    α-helix, 59
Ligand-binding site
  acetylcholine receptor, 317
  low-density lipoprotein, 340, 341–342
Ligand–receptor complex, 236
Light chain, $C_H 1$ domain amino acid sequences, 201–203
Limulus polyphemus, hemocyanin, 155, 157–164, 170–177, 178
  oligomer autoassembly, 176, 177, 178
  quaternary structure, 170–176
  subunit heterogeneity, 160–164
  subunit localization, 170–177, 178
Low-density lipoprotein, cholesterol release, 337, 339–340
Low-density lipoprotein receptor
  coated pit
    clustering, 337, 339, 344–345
    internalization, 338, 339
    multidomain nature, 337
  cDNA, 340
  function, 337

gene, 345–346
low-density lipoprotein dissociation, 339
as mosaic protein, 346
pathway, 338–340
structure, 340–346
  cytoplasmic tail, 340, 344–345
  domains, 340–346
  epidermal growth factor precursor homology domain, 340, 342–343, 345, 346
  exon organization, 341, 345–346
  intron organization, 345, 346
  ligand-binding domain, 340, 341–342
  membrane-spanning domain, 340, 344, 345–346
  O-linked sugars, 338, 339, 340, 343, 345
synthesis, 338
Lysozyme, of mutant bacteriophage T4, amino acid sequences, 245–258
  environmental effects, 253
  global thermal stability, 253
  hydrogen bonding, 249, 253, 256, 257
  interactional effects, 253–254
  protein evolution and, 257
  secondary structure stability, 255
  self-directed mutagenesis, 248–253, 256–257
  structural changes, 254–255
  temperature-sensitive mutants, 247–248, 249, 253, 254–255, 256, 257
  wild-type mutants, 246–247, 248, 249, 254–255, 256–257
Lysozyme–Fab complex, 215–225
  amino acid lysozyme residues, 220–221, 222, 223
  contact sites, 219–220
  crystallography studies, 218–219
  electron density map, 219, 220, 222
  lysozyme inactivator, 220–221, 222

M

M protein, in viral budding, 145–146, 148
Magnetic field effects, in photosynthetic reaction centers, 273–278

Mastigocladus laminosus, photosynthetic reaction centers, 283
Membrane protein
  folding motifs, 319–333
    acetylcholine receptor, 322, 323
    alpha-toxin, 322, 323–324
    bacteriorhodopsin, 322, 324
    colicin E1, 320, 322, 324–325
    crambin, 322, 325
    cytochrome b_5, 322, 326
    fd coat protein, 322, 326–327
    glycophorin, 322, 327
    lactose carrier, 322, 327–328
    porin, 322, 328
    proton-ATPase proteolipid, 322, 328–329
    rhodopsin, 322, 329
  functions, 320
Membrane sorting, enveloped virus budding and, 139–150
  alphaviruses, 141–145
  G protein, 145, 146–147, 148
  M protein, 145–146, 148
  thy-1 protein, 147
  vesicular stomatitis virus, 145–147, 148
Membrane-spanning region, low-density lipoprotein receptor, 340, 344, 345–346
Menaquinone, 271, 272
Mengo virus, 88–89
1-Mercaptopropionate oxytocin, *see* Deaminooxytocin
Merostome, hemacyanin, 163
Microtubule, self-assembly energetics, 240–241
Monoclonal antibody
  antineuraminidase Fab fragment, 125–133
    antigen interaction, 129–132
    binding properties, to monoclonal variants, 129–130
    binding properties, to neuraminidase field strains, 130–131
    electron density map, 126–129
    neuraminidase–Fab fragment, electron microscopic study, 131–132
    structure, 126–129

epitope intramolecular localization, 180–188
Multienzyme complex, 236
Multiple isomorphous replacement, EcoRI endonuclease, 15
Myohemerythrin
  C-helix peptide, 231–232, 233
  hydrated interfaces, 241

N

Neuraminidase, monoclonal antibody Fab fragment, see Monoclonal antibody, antineuraminidase Fab fragment
Nuclear magnetic resonance spectroscopy, of immunogenic peptide structure, 228–234
Nucleocapsid, in viral budding, 140–141, 142–144, 145, 146, 147
Nucleosome, core particle, 13

O

Oligomeric protein, interfaces, 235–236
O-linked sugar, of low density lipoproteins, 340, 343, 345
Ovomucoid inhibitor, 243
Oxyhemoglobin, 239
Oxytocin, see also Deaminooxytocin
  biological response, 301–304
  conformations, 290
  crystal structure, 291
  function, 290
  hydrogen bonding, 290
  1-penicillamine derivatives, 302
  receptor binding, 301–304
  structure in dimethyl sulfoxide, 290, 298–301
  synthesis, 290, 304

P

Panulirus interruptus, hemocyanin, 155–157, 183, 186, 188
1-L-Penicillamine oxytocin, 302
Peptide, immunogenic, nuclear magnetic resonance spectroscopy, 228–234
Photosynthesis, 283

Photosynthetic light-harvesting complex, 283, 284
Photosynthetic reaction centers, 271–286
  components, 283–284
  definition, 271
  energetics, 278–281
  magnetic field effects, 273–278
    data and analysis, 277–278
    physical origins, 273–277
  photo-induced charge separation, 271–273
Phototactic response, of halobacteria, 265–267, 268
Phycobilisome, 284, 285
Phyocyanin, 285
Picomolar dissociation constants, 236
Picornavirus
  assembly, 107–108
  capsid protein cleavage, 86–87
  characteristics, 86
  diseases caused by, 86
  disassembly, 107
  host specificities, 88
  molecular weight, 86
  protomers, 86
  receptor binding site, 88, 104–105
  RNA, 86, 87
  structural homologies, 87
  structural proteins, 86–87
  tissue specificity, 88
  X-ray diffraction studies, 88
Plant virus, capsid proteins, 95, 96, 97, 98, 99
Plasmid, Z-DNA recombination, 7
Poliovirus
  antigenic sites, 88, 102–104
  gene order, 87
  infectivity inhibition, 88
  neutralizing antibodies, 88, 106
  polypeptide chain interweaving, 116
  receptor binding site, 105
  RNA, 87
  serotypes, 86
  X-ray diffraction studies, 88
Porin, secondary structure prediction, 322, 328
  α-helix, 322, 328
  β-sheet, 322, 328
Protease, Staphylococcus griseus, 242–243

Protein(s), *see also* names of specific proteins
  cell localization, 139
  folding, anti-peptide antibodies and, 233
  intracellular transport, 139–140
  membrane sorting, 140–150
Protein interfaces, 235–244
  affinity determinants, 238–239
  categories, 235–236
  complementarity, 236–238
  hydration, 241–242
  hydrophobic effect, 238, 239–240, 241, 242
  mobility, 242–243
  subunit entropy, 240–241
Protein stability, amino acid substitutions and, 245–258
  environmental effects, 253
  global thermal stability and, 253
  interactional effects, 253–254
  protein evolution and, 257
  secondary structure stability and, 255
  self-directed mutagenesis and, 248–253, 256–257
  structural change and, 254–255
  in temperature-sensitive mutants, 247–248, 249, 253, 254–255, 256, 257
  in wild-type mutants, 246–247, 248, 249, 254–255, 256–257
Protein–protein interactions, *see* Protein interfaces
Protein–ribonucleic acid interaction, tobacco mosaic virus, 77–79, 81
Proton-adenosine triphosphatase proteolipid, secondary structure prediction, 322, 328–329

Q

Quinone, of photosynthetic reaction centers, 271, 272, 273, 284, 285

R

Reaction center, *see* Photosynthetic reaction center
Receptor
  acetylcholine, 307–318
    amino acid sequences, 310, 312
    amphipathic membrane crossing, 310–317
    α-helix, 310, 311, 312, 322, 323
    β-sheet, 322, 323
    ion conductance, 307–318
    ionic channel gating, 316–317
    ionic channel location, 309–310, 311
    ionic channel model, 313–316
    ligand-binding site, 317
    quasi-equivalent subunits, 309–310, 311
    secondary structure prediction, 322, 323
    structure, 307–312
  interleukin-2, 343
  low-density lipoprotein, 337–347
  vasopressin, 301–302, 304
Receptor–ligand complex, 236
Retinal proteins, *see* Rhodopsins
Rhinovirus, 113, *see also* Human rhinovirus 2; Human rhinovirus 14
  classification, 86
  gene order, 87
  polypeptide chain interweaving, 116
  RNA, 87
  serotypes, 86
  X-ray diffraction studies, 88–89
Rhodopseudomonas sphaeroides,
  photosynthetic reaction center, 271, 272, 274
Rhodopseudomonas viridis,
  photosynthetic reaction center, 271, 272, 283, 284–285, 286
  electron transfer pathway, 281
Rhodopsins
  of halobacteria, 262–269
  secondary structure prediction, 322, 329
    α-helix, 322, 329
    β-sheet, 322, 329
  sensory, 265–267
Ribonucleic acid
  human rhinovirus, 14, 94, 98, 100
  picornavirus, 86, 87
  poliovirus, 87
  rhinovirus, 87
  tobacco mosaic virus, 69, 70
    protein interfaces, 77–79, 81
    in viral assembly, 80–81

Ribonucleic acid virus
  β-annulus structure, 107–108
  spherical, β-barrel structure, 120, 121

S

Satellite tobacco necrosis virus
  assembly, 107–108
  structure, 96
  X-ray diffraction studies, 88, 113
6-Selenodeaminooxytocin, 291–294, 298
Semliki Forest virus, 141, 142, 143
Sendai virus, 136
Signal peptide, 139
Sindbis virus
  nucleocapsid, 142–143, 147
  structure, 141, 142–143
Southern bean mosaic virus
  assembly, 107–108
  coat protein, 144
  structure, 88, 95, 96, 97, 98, 99, 100, 113
  β-barrel, 121
  polypeptide chain interweaving, 116
Staphylococcus griseus, protease, 242–243
S10/1 antibody, *see* Fab fragment, antineuraminidase monoclonal
Superoxide dismutase dimer, 238
SV40 virus, 6
Systemic lupus erythematosus, 4

T

Thy-1 protein, in viral budding, 147
Tobacco mosaic virus
  assembly, 79–82
    protein aggregates, 79–81
  coat protein, 69–72
  molecular weight, 69
  RNA, 69, 70
    in viral assembly, 80–81
  size, 69
  structure, 69–83
    carboxyl groups, 71–72, 74–77, 81
    disk crystals, 70–71, 75
    fiber diffraction phase problem, 72–73
    α-helix, 69, 70, 71, 76
    protein–RNA interactions, 77–79, 81

    protein structure, 73–77
    salt bridge, 75–76
    structural details, 73–79
    structure determinants, 72–73
    X-ray methods, 72
Tomato bushy stunt virus
  assembly, 107–108
  coat proteins, 144
  structure, 88, 95, 96, 100, 113
    polypeptide chain interweaving, 116
Topoisomerase, 5
Torpedo californica, acetylcholine receptor, 307, 308
Torpedo marmorata, acetylcholine receptor, 307–308
trp repressor
  amino acid sequences, 61
  helix-turn-helix unit, 63
Trypsin–trypsin inhibition complex, 236, 238, 242, 243

U

Ubiquinone, 271, 272
Uca pugnax, hemocyanin, 177, 179
Ustilago, Z-deoxyribonucleic acid-binding protein, 6–7

V

$V_H:V_L$, amino acid interface residue analysis, 194
Vasopressin
  receptor-binding model, 301–302, 304
  synthesis, 304
Vesicular stomatis virus, budding, 145–146, 147
Virus(es), *see also* names of specific viruses
  enveloped, budding, 139–150

W

Water molecule
  in deaminooxytocin, 294, 295–297
  in protein interfaces, 241–242

Z

Zinc superoxide dismutase, 121

DATE DUE

DEMCO 38-297